T0182149

Physical Chemistry Essentials

Andreas Hofmann

Physical Chemistry
Essentials

 Springer

Andreas Hofmann
Griffith University
Eskitis Institute, Nathan, Queensland
Australia

The University of Melbourne
Faculty of Veterinary and Agricultural Sciences, Parkville
Victoria, Australia

ISBN 978-3-030-08928-3   ISBN 978-3-319-74167-3   (eBook)
https://doi.org/10.1007/978-3-319-74167-3

This Springer imprint is published by the registered company Springer International Publishing AG part of Springer Nature.
The registered company address is: Gewerbestrasse 11, 6330 Cham, Switzerland

Preface

This text provides an introduction to physical chemistry, covering the fundamentals of thermodynamics, kinetics and quantum chemistry. The selection of contents was guided by the concept of a two-semester course delivering thermodynamics and kinetics (Chaps. 1–8) and quantum chemistry (Chaps. 8–13) throughout the second year of undergraduate studies in the disciplines of chemistry, biochemistry, engineering and neighbouring disciplines.

The compilation of this text was based strongly by teaching and examination experience of this subject. This includes an appreciation of the fact that students taking this course might have varying background knowledge and need to be guided through physical chemistry concepts in a step-by-step manner. Furthermore, while mathematical analysis is an integral part of physical chemistry, many students taking this subject are not mathematicians. Particular care has therefore been taken in the presentation of the algebraic parts of physico-chemical concepts that should allow the reader to rework the relevant discussion with pen and paper during study time.

Each chapter includes a selection of numerical exercises that serve to revise particular key concepts as well as apply these concepts to physico-chemical problems. Detailed solutions for all exercises are included at the end of this text.

I sincerely thank colleagues for their generous provision of images, in particular Dr Lutz Hammer (Universität Erlangen-Nürnberg, Germany), Prof Chris Rayner (University of Leeds, UK), Prof Anne Simon (Université Claude Bernard Lyon 1, France), Ms Elysia Cave-Freeman and Dr Agatha Garavelas.

I am most grateful to the editorial team at Springer, Heidelberg, for their advice during compilation and their efforts with producing this text. Manuscript and figures for this book have been compiled entirely with open source and academic software under Linux, and I acknowledge the efforts by software developers and programmers who make their products freely available.

Constructive comments from all students who use this book within their studies and from teachers and academics who adopt the book to complement their courses are most welcome.

Brisbane
November 2016

Andreas Hofmann

Contents

Physico-chemical Data and Resources

1

1.1 Periodic Table of the Elements

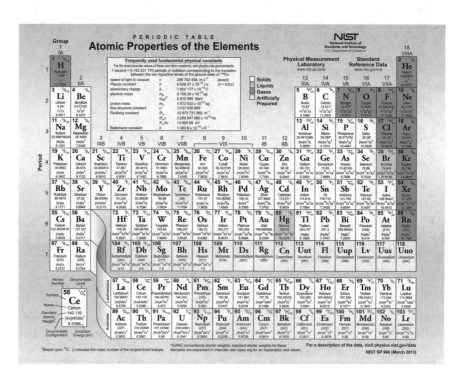

(Obtained from http://www.nist.gov/pml/data/periodic.cfm, 31.01.14)

© Springer International Publishing AG, part of Springer Nature 2018
A. Hofmann, *Physical Chemistry Essentials*,
https://doi.org/10.1007/978-3-319-74167-3_1

1.2   Resources and Databases

There are many widely available resources with various physico-chemical data, in print and online. A selection of resources for general, spectroscopic and structural data is compiled in Table 1.1.

Table 1.1   Select resources of physico-chemical, spectroscopic and structural data

General physico-chemical data		
CRC Handbook of Chemistry and Physics	Haynes, W.M. (edt.) (2012) 93rd edition, CRC Press.	Book
SI Chemical Data	Aylward, G. & Findlay, T. (2013) 6th edition, John Wiley & Sons.	Book
National Institute of Standards and Technology, USA	http://www.nist.gov/srd/onlinelist.cfm	Public
Mass spectrometry		
MassBank (Japan)	http://www.massbank.jp/?lang=en	Public
Mass Spectrometry Data Center (NIST)	http://chemdata.nist.gov/	Public
m/zCloud (HighChem LLC, Slovakia)	http://www.mzcloud.org/	Public
Spectral data		
Spectral Database for Organic Compounds SDBS (National Institute of Advanced Industrial Science and Technology, Japan)	http://sdbs.db.aist.go.jp/sdbs/cgi-bin/cre_index.cgi?lang=eng	Public
Structural data		
Cambridge Structural Database (CSD) Crystal structures of organic compounds	http://www.ccdc.cam.ac.uk/products/csd/	Licence
Inorganic Crystal Structure Database Crystal structures of inorganic compounds	http://www.fiz-karlsruhe.de/icsd_content.html	Licence
CRYSTMET® Crystal structures of metals and alloys	http://www.tothcanada.com/	Licence
Protein Data Bank (PDB) Protein and oligosaccharide structures	http://www.rcsb.org/pdb/	Public
Biological Magnetic Resonance Data Bank (BMRB) NMR structures of biomolecules	http://www.bmrb.wisc.edu/	Public
Nucleic Acids Data Bank Structures of oligonucleotides	http://ndbserver.rutgers.edu/	Public
PDF Databases Powder diffraction data	http://www.icdd.com/	Licence

1.3 Units and Constants

1.3.1 Decimal Factors

Table 1.2 Decimal factors

Factor	Prefix	Symbol	Factor	Prefix	Symbol
10^{-1}	deci	d	10	deka	da
10^{-2}	centi	c	10^2	hekto	h
10^{-3}	milli	m	10^3	kilo	k
10^{-6}	micro	m	10^6	mega	M
10^{-9}	nano	n	10^9	giga	G
10^{-12}	pico	p	10^{12}	tera	T
10^{-15}	femto	f	10^{15}	peta	P
10^{-18}	atto	a	10^{18}	exa	E

1.3.2 The Greek Alphabet

Table 1.3 The Greek alphabet

A, α	alpha	I, ι	iota	Σ, σ	sigma
B, β	beta	K, κ	kappa	T, τ	tau
Γ, γ	gamma	Λ, λ	lambda	Y, υ	upsilon
Δ, δ	delta	M, μ	mu	Φ, ϕ	phi
E, ε	epsilon	N, ν	nu	X, χ	chi
Z, ζ	zeta	Ξ, ξ	xi	Ψ, ψ	psi
H, η	eta	Π, π	pi	Ω, ω	omega
Θ, θ	theta	P, ρ	rho		

1.3.3 SI Base Parameters and Units

Table 1.4 SI base parameters and units

Symbol	Parameter	Unit	Name
I	Electric current	1 A	ampere
I_v	Light intensity	1 cd	candela
l	Length	1 m	metre
m	Mass	1 kg	kilogram
n	Molar amount	1 mol	mole
t	Time	1 s	second
T	Temperature	1 K	kelvin

1.3.4   Important Physico-chemical Parameters and Units

Table 1.5 Important physico-chemical parameters and units

Symbol	Parameter	Unit	Name
b	Molality	1 mol kg^{-1}	
B	Magnetic induction	$1 \text{ T} = 1 \text{ kg s}^{-2} \text{ A}^{-1} = 1 \text{ V s m}^{-2}$	tesla
c	Molar concentration	1 mol l^{-1}	
C	Electric capacitance	$1 \text{ F} = 1 \text{ kg}^{-1} \text{ m}^{-2} \text{ s}^4 \text{ A}^2 = 1 \text{ A s V}^{-1}$	farad
E	Energy	$1 \text{ J} = 1 \text{ kg m}^2 \text{ s}^{-2} = 1 \text{ C V}$	joule
ε	Molar extinction coefficient	$1 \text{ l mol}^{-1} \text{ cm}^{-1}$	
ε	Permittivity	1 F m^{-1}	
F	Force	$1 \text{ N} = 1 \text{ kg m s}^{-2} = 1 \text{ J m}^{-1}$	newton
F	Magnetic flux	$1 \text{ Wb} = 1 \text{ kg m}^2 \text{ s}^{-2} \text{ A}^{-1} = 1 \text{ V s}$	weber
G	Electric conductance	$1 \text{ S} = 1 \text{ kg}^{-1} \text{ m}^{-2} \text{ s}^3 \text{ A}^2 = 1 \text{ }\Omega^{-1}$	siemens
H	Enthalpy	$1 \text{ J} = 1 \text{ kg m}^2 \text{ s}^{-2}$	joule
η	Viscosity	$1 \text{ P} = 0.1 \text{ kg m}^{-1} \text{ s}^{-1} = 0.1 \text{ Pa s}$	poise
i	Current density	1 A m^{-2}	
j	Flux density	$1 \text{ mol m}^{-2} \text{ s}^{-1}$	
κ	Conductivity	1 S m^{-2}	
L	Magnetic inductivity	$1 \text{ H} = 1 \text{ kg m}^2 \text{ s}^{-2} \text{ A}^{-2} = 1 \text{ V A}^{-1} \text{ s}$	henry
Λ_m	Molar conductivity	$1 \text{ S m}^2 \text{ mol}^{-1}$	
M	Molar mass[1]	$1 \text{ g mol}^{-1} = 1 \text{ Da}$	(dalton)
μ	Electric dipole moment	$1 \text{ D} = 3.336 \cdot 10^{-30} \text{ C m}$	debye
ν	Frequency	$1 \text{ Hz} = 1 \text{ s}^{-1}$	hertz
p	Pressure	$1 \text{ Pa} = 1 \text{ kg m}^{-1} \text{ s}^{-2} = 1 \text{ N m}^{-2}$	pascal
P	Power	$1 \text{ W} = 1 \text{ kg m}^2 \text{ s}^{-3} = 1 \text{ J s}^{-1}$	watt
Q	Electric charge	$1 \text{ C} = 1 \text{ A s}$	coulomb
ρ	Density	1 g cm^{-3}	
ρ^*	Mass concentration	1 mg ml^{-1}	
R	Electric resistance	$1 \text{ }\Omega = 1 \text{ kg m}^2 \text{ s}^{-3} \text{ A}^{-2} = 1 \text{ V A}^{-1}$	ohm
S	Entropy	1 J K^{-1}	
θ	Temperature	$1 \text{ }^\circ\text{C}$	degree Celsius
u	Ion mobility	$1 \text{ m}^2 \text{ s}^{-1} \text{ V}^{-1}$	
U, ϕ, E	Electric potential (voltage)	$1 \text{ V} = 1 \text{ kg m}^2 \text{ s}^{-3} \text{ A}^{-1} = 1 \text{ J A}^{-1} \text{ s}^{-1}$	volt
V	Volume	$1 \text{ l} = 1 \text{ dm}^3$	
V_m	Molar volume	1 l mol^{-1}	
v	Partial specific volume	1 ml g^{-1}	
w	Mass fraction Volume fraction	1 (typically given in %w/w, %w/v or %v/v)	
x	Mole fraction	1	
z	Charge number	1	

[1]Note that the molecular mass is the mass of one molecule given in atomic mass units (u, Da). The molar mass is the mass of 1 mol of molecules and thus has the units of g mol^{-1}

1.3.5 Important Physico-chemical Constants

Table 1.6 Important physico-chemical constants

Symbol	Constant	Value
c	Speed of light in vacuo	$2.99792458 \cdot 10^8$ m s^{-1}
e	Elementary charge	$1.6021892 \cdot 10^{-19}$ C
$\varepsilon_0 = (\mu_0 \cdot c^2)^{-1}$	Permittivity in vacuo	$8.85418782 \cdot 10^{-12}$ A^2 s^4 m^{-3} kg^{-1}
$F = e \cdot N_A$	Faraday's constant	$9.648456 \cdot 10^4$ C mol^{-1}
g	Earth's gravity near surface	9.81 m s^{-2}
$g_e = 2\, \mu_e / \mu_B$	Landé factor of free electron	2.0023193134
γ_p	Gyromagnetic ratio of proton	$2.6751987 \cdot 10^8$ s^{-1} T^{-1}
h	Planck's constant	$6.626176 \cdot 10^{-34}$ J s
$k_B = R / N_A$	Boltzmann's constant	$1.380662 \cdot 10^{-23}$ J K^{-1}
m_e	Mass of electron	$9.109534 \cdot 10^{-31}$ kg
m_n	Mass of neutron	$1.6749543 \cdot 10^{-27}$ kg
m_p	Mass of proton	$1.6726485 \cdot 10^{-27}$ kg
μ_0	Magnetic field constant	$4\pi \cdot 10^{-7}$ m kg s^{-2} A^{-2}
$\mu_B = e \cdot h / (4\pi \cdot m_e)$	Bohr magneton	$9.274078 \cdot 10^{-24}$ J T^{-1}
μ_ε	Magnetic moment of electron	$9.284832 \cdot 10^{-24}$ J T^{-1}
$\mu_N = e \cdot h / (4\pi \cdot m_p)$	Nuclear magneton	$5.050824 \cdot 10^{-27}$ J T^{-1}
N_A, L	Avogadro's (Loschmidt's) constant	$6.022045 \cdot 10^{23}$ mol^{-1}
$p^\varnothing$	Standard pressure (IUPAC)	$1.00 \cdot 10^5$ Pa
p_{normal}	Normal pressure (NIST)	1 atm = 1013.25 hPa
R	Gas constant	8.31441 J K^{-1} mol^{-1}
R_∞	Rydberg's constant	$1.097373177 \cdot 10^7$ m^{-1}
$T^\varnothing$, $\theta^\varnothing$	Standard temperature (IUPAC)	273.15 K, 0 °C
T_{normal}, θ_{normal}	Normal temperature	298.15 K, 25 °C
u	Atomic mass unit	$1.6605402 \cdot 10^{-27}$ kg
$V_m^\varnothing = R \cdot T^\varnothing / p^\varnothing$	Molar volume of an ideal gas	22.41383 l mol^{-1}

1.3.6 Conversion Factors

Table 1.7 Conversion factors for energy

	J	cal	eV
1 J	1	0.2390	$6.24150974 \cdot 10^{18}$
1 cal	4.184	1	$2.612 \cdot 10^{19}$
1 eV	$1.60217646 \cdot 10^{-19}$	$3.829 \cdot 10^{-20}$	1

Table 1.8 Conversion factors for pressure

	Pa	bar	atm	mm Hg (Torr)	psi
1 Pa	1	10^{-5}	$9.869\cdot10^{-6}$	$7.501\cdot10^{-3}$	$1.450\cdot10^{-4}$
1 bar	10^5	1	0.9869	750.1	14.50
1 atm	$1.013\cdot10^5$	1.013	1	760.0	14.69
1 mm Hg (Torr)	133.3	$1.333\cdot10^{-3}$	$1.316\cdot10^{-3}$	1	$1.933\cdot10^{-2}$
1 psi	$6.895\cdot10^4$	$6.897\ 10^{-2}$	$6.807\ 10^{-2}$	51.72	1

1.3.7   Thermodynamic Properties of Select Substances

Table 1.9 Enthalpy of formation, Gibbs free energy of formation and molar heat capacity of select substances at normal conditions (T_{normal} = 298 K, p_{normal} = 1.013 bar)

Species	ΔH_f in kJ mol^{-1}	ΔG_f in kJ mol^{-1}	$C_{p,m}$ in J K^{-1} mol^{-1}
$C_{(graphite)}$	0	0	8.53
$HCO_3^-{}_{(aq)}$	−689.9	−586.8	
$H_{2(g)}$	0	0	28.82
$H^+{}_{(aq)}$		0	
$H_2O_{(l)}$	−285.83	−237.14	75.4
$H_2O_{(g)}$	−241.83	−228.61	33.58
$Ni(OH)_{2(s)}$		−444	
$O_{2(g)}$	0	0	29.35
$O_{3(g)}$	142.67	163.19	39.22
$OH^-{}_{(aq)}$	−230.02	−157.22	

1.4   Summary of Important Formulae and Equations

Table 1.10 Important formulae and equations

Thermodynamics	
$p \sim \dfrac{1}{V}$	Boyle's law The pressure exerted by an ideal gas is inversely proportional to the volume it occupies if the temperature and amount of gas remain unchanged within a closed system.
$V \sim T$	Charles' law Gases tend to expand when heated; at constant pressure, the volume is directly proportional to the temperature.
$p \sim T$	Gay-Lussac's law If mass and volume of a gas are held constant, the pressure exerted by the gas increases directly proportional to the temperature.

(continued)

Table 1.10 (continued)

$p \cdot V = n \cdot R \cdot T$	Ideal gas equation The laws by Boyle, Charles and Gay-Lussac combine to the ideal gas equation.
$p_{\text{solute}} = K_{\text{solute}} \cdot x_{\text{solute}}$	Henry's law For solutions at low concentrations, the vapour pressure of the solute is proportional to its mole fraction.
$\left(\dfrac{\delta U}{\delta T}\right)_V = C_V$	Heat capacity at constant volume.
$\left(\dfrac{\delta U}{\delta V}\right)_p = \dfrac{p \cdot C_V}{n \cdot R}$	Variation of the internal energy of an ideal gas with volume at constant pressure.
$\left(\dfrac{\delta H}{\delta T}\right)_p = C_p$	Heat capacity at constant pressure.
Ideal gas: $\left(\dfrac{\delta H}{\delta p}\right)_T = 0$ Otherwise: $\left(\dfrac{\delta H}{\delta p}\right)_T = V - T \cdot \left(\dfrac{\delta V}{\delta T}\right)_p$	Change of enthalpy of a system with respect to a pressure change in an isothermal process. For an ideal gas, there is no change in enthalpy with pressure if the temperature remains the same.
$C_p - C_V = n \cdot R$	Relationship between heat capacities of an ideal gas.
$\left(\dfrac{\delta S}{\delta p}\right)_T = -\left(\dfrac{\delta V}{\delta T}\right)_p$ $\left(\dfrac{\delta S}{\delta V}\right)_T = \left(\dfrac{\delta p}{\delta T}\right)_V$ $\left(\dfrac{\delta T}{\delta V}\right)_S = -\left(\dfrac{\delta p}{\delta S}\right)_V$ $\left(\dfrac{\delta T}{\delta p}\right)_S = \left(\dfrac{\delta V}{\delta S}\right)_p$	Maxwell relations provide a means to exchange thermodynamic functions.
$\left(\dfrac{\delta S}{\delta T}\right)_p = \dfrac{C_p}{T}$	Entropy change of a system with respect to temperature change in an isobaric process.
$\left(\dfrac{\delta G}{\delta p}\right)_T = \Delta V$	The pressure dependence of the Gibbs free energy of an isothermal process defines the volume change of the system during that process.
$\left(\dfrac{\delta G}{\delta T}\right)_p = -\Delta S$	The temperature dependence of the Gibbs free energy of an isobaric process defines the entropy change during that process.
$\dfrac{dp}{dT} = \dfrac{\Delta S_{vap}}{\Delta V_{vap}}$	Clapeyron equation Change of vapour pressure of a one-component system with temperature in terms of entropy.
$\dfrac{d(\ln p)}{dT} = \dfrac{\Delta H_{m,vap}}{R \cdot T^2}$	Clausius-Clapeyron equation Change of vapour pressure of a one-component system with temperature in terms of enthalpy.
$\left(\dfrac{d\ln K}{dT}\right)_p = \dfrac{\Delta H^{\varnothing}}{R \cdot T^2}$ $\left(\dfrac{d\ln K}{dT}\right)_V = \dfrac{\Delta U^{\varnothing}}{R \cdot T^2}$	van't Hoff equations: reaction isobar and isochore Change of the equilibrium constant of a chemical reaction with temperature. The two equations show the case for isothermal and isobaric, or isothermal and isochoric reactions.

(continued)

Table 1.10 (continued)

$\left(\dfrac{\mathrm{d}\ln K}{\mathrm{d}p}\right)_T = -\dfrac{\Delta V^{\varnothing}}{\mathrm{R}\cdot T}$	van Laar-Planck isotherm Change of the equilibrium constant of a chemical equilibrium with pressure in terms of the standard reaction volume.
$\Delta G = \Delta G^{\varnothing} + \mathrm{R}\cdot T\cdot\ln K$	Change of the Gibbs free energy of a chemical reaction with equilibrium constant K and standard Gibbs free energy $\Delta G^{\varnothing}$.
Colligative properties	
$\Pi = i\cdot c\cdot\mathrm{R}\cdot T$	Osmotic pressure
$\Delta T_f = i\cdot K_f\cdot b$	Freezing point depression
$\Delta T_\mathrm{b} = i\cdot K_\mathrm{b}\cdot b$	Boiling point elevation
$p = p^{*}\cdot x$	Raoult's law: vapour pressure depression
Electrochemistry	
$G = \kappa\cdot\dfrac{A}{l} = \dfrac{1}{R}$	Electrical conductance The conductance increases with the cross-sectional area A and decreases with the length l of the conductor; k is the electrical conductivity. The conductance is the inverse of the resistance R.
$\Lambda_\mathrm{m} = \dfrac{\kappa}{c}$	Molar conductivity
$m \sim I\cdot t = Q$	Faraday's first law of electrolysis The mass of a substance altered at an electrode during electrolysis is directly proportional to the quantity of electricity transferred at that electrode.
$m \sim \dfrac{M}{z}$	Faraday's second law of electrolysis For a given quantity of electric charge, the mass of a deposited/generated elementary substance is proportional to the molar mass of that substance divided by the change in oxidation state (i.e. in most cases the charge of the cation in the electrolyte).
$E = E^{\varnothing} - \dfrac{\mathrm{R}\cdot T}{z\cdot F}\cdot\ln K$	Nernst equation Concentration dependence of the Redox potential.
$\Delta G_\mathrm{m}^{\varnothing} = -z\cdot F\cdot E^{\varnothing}$	Change of the standard molar Gibbs free energy of an electrochemical process with the standard electrode potential $E^{\varnothing}$ and the charge state z.
$\mathrm{pH} = \mathrm{p}K_a + \lg\left[\dfrac{c(\mathrm{A}^-)}{c(\mathrm{HA})}\right]$	Henderson-Hasselbalch equation pH of a solution with the buffer system consisting of the weak acid HA and its conjugated base A^-.
$\Lambda_\mathrm{m} = \Lambda_{0\mathrm{m}} - \kappa\cdot\sqrt{c}$	Kohlrausch's law The molar conductivity of strong electrolytes increases with decreasing concentrations (valid for generally low concentrations).
$K_\mathrm{d} = \dfrac{\alpha^2}{1-\alpha}\cdot c_0(\mathrm{AB})$ $K_\mathrm{d} = \dfrac{\Lambda_\mathrm{m}^2}{(\Lambda_{0\mathrm{m}}-\Lambda_\mathrm{m})\cdot\Lambda_{0\mathrm{m}}}\cdot c_0(\mathrm{AB})$	Ostwald's law of dilution The dissociation constant of weak electrolytes is a function of the degree of dissociation α, and with $\alpha = \dfrac{\Lambda_m}{\Lambda_{0m}}$, can be expressed in terms of the molar conductivity.

(continued)

Table 1.10 (continued)

$\Lambda_{0m} = \nu_+ \cdot \lambda_+ + \nu_- \cdot \lambda_-$	Law of the independent migration of ions The limiting molar conductivity is comprised of the two independent limiting molar conductivities of the anions and cations.

Transport

$F = -\dfrac{R \cdot T}{c} \cdot \left(\dfrac{\delta c}{\delta x}\right)_{p,T}$	Thermodynamic force A concentration gradient establishes a thermodynamic force F.
$J = -D \cdot \dfrac{d\,IN}{dx}$	Fick's first law of diffusion Flux of matter is defined by the concentration gradient along x; $IN = \dfrac{N}{V}$ is the molecule density; D is the diffusion coefficient.
$J = -\kappa \cdot \dfrac{dT}{dx}$	Thermal conduction Flux of energy along a temperature gradient; κ is the thermal conductivity.
$J = -\eta \cdot \dfrac{dv_x}{dx}$	Flux of momentum When molecules switch from one flow layer to another, their momentum is also migrating; η is the viscosity.

Kinetics

$k = A \cdot e^{-\frac{E_a}{R \cdot T}}$	Arrhenius equation The rate constants of most reactions depend on the temperature.
$\left(\dfrac{d\ln k}{dT}\right) = \dfrac{E_a}{R \cdot T^2}$	The generalised dependency of rate constants on temperature, which applies to all reactions irrespective of their adhering to the Arrhenius relation or not.
$v = \dfrac{v_{max} \cdot c_0(S)}{c_0(S) + K_m}$	Michaelis-Menten enzyme kinetics

Surface adsorption

$\Gamma = \dfrac{K \cdot p}{K \cdot p + 1}$	Langmuir adsorption isotherm without dissociation The Langmuir constant K is the ratio of the rate constants for adsorption and desorption: $K = \dfrac{k_{ads}}{k_{des}}$.
$\Gamma = \dfrac{\sqrt{K \cdot p}}{\sqrt{K \cdot p} + 1}$	Langmuir adsorption isotherm with dissociation into two species

Table 1.11 Kinetic rate laws in their differential and integrated forms, and the derived half lives

Order	Differential form	Integrated form	Half life
0	$-\dfrac{1}{\nu_A}\cdot\dfrac{dc(A)}{dt}=k$	$c(A)=-\nu_A\cdot k\cdot t+c_0(A)$	$t_{1/2}=\dfrac{c_0(A)}{2\cdot k}$
1	$-\dfrac{1}{\nu_A}\cdot\dfrac{dc(A)}{dt}=k\cdot c(A)$	$-\ln c(A)=\nu_A\cdot k\cdot t-\ln c_0(A)$	$t_{1/2}=\dfrac{\ln 2}{k}$
2	$-\dfrac{1}{\nu_A}\cdot\dfrac{dc(A)}{dt}=k\cdot c(A)^2$	$\frac{1}{c(A)}=\nu_A\cdot k\cdot t+\frac{1}{c_0(A)}$	$t_{1/2}=\dfrac{1}{k\cdot c_0(A)}$
3	$-\dfrac{1}{\nu_A}\cdot\dfrac{dc(A)}{dt}=k\cdot c(A)^3$	$\frac{1}{c(A)^2}=2\cdot\nu_A\cdot k\cdot t+\frac{1}{c_0(A)^2}$	$t_{1/2}=\dfrac{3}{2\cdot k\cdot c_0(A)^2}$

Table 1.12 Interactions between molecules. α: polarisability; ε_0: vacuum permittivity; I: first ionisation potential; μ: dipole moment; r: distance between the two atoms/molecules

Interaction	Potential energy	Explanation	Order of magnitude (kJ mol^{-1})
Covalent bond			$|V| = 200\text{--}800$
Coulomb interaction	$V=-\dfrac{z_A\cdot z_B\cdot e^2}{4\pi\cdot\varepsilon_0\cdot r}$	Interaction between two ions	$|V| = 40\text{--}400$
Ion-dipole interaction	$V=-\dfrac{z\cdot e\cdot\mu}{4\pi\cdot\varepsilon_0\cdot r^2}$	Interaction between ion and permanent dipole	$|V| = 4\text{--}40$
Keesom interaction	$V=-\dfrac{2}{3}\cdot\dfrac{\mu_A^2\cdot\mu_B^2}{(4\pi\cdot\varepsilon_0)^2\cdot r^6}\cdot\dfrac{1}{k\cdot T}$	Interaction between two permanent dipoles	$|V| = 0.4\text{--}4$
Ion-induced dipole interaction	$V=-\dfrac{1}{2}\cdot\dfrac{\alpha\cdot e^2}{4\pi\cdot\varepsilon_0\cdot r^4}$	Interaction between ion and induced dipole	$|V| = 0.4\text{--}4$
Debye force	$V=-\dfrac{\alpha\cdot\mu^2}{(4\pi\cdot\varepsilon_0)^2\cdot r^6}$	Interaction between permanent dipole and induced dipole	$|V| = 0.4\text{--}4$
London dispersion force	$V=-\dfrac{3}{4}\cdot\dfrac{I\cdot\alpha^2}{r^6}$ (pure substance)	Interaction between temporary dipole and induced dipole	$|V| < 0.4$
	$V=-\dfrac{3}{2}\cdot\dfrac{I_A\cdot I_B}{I_A+I_B}\cdot\dfrac{\alpha_A\cdot\alpha_B}{r^6}$ (mixture of A and B)		$|V| < 0.4$
Hydrogen bond			$|V| = 4\text{--}40$

Table 1.13 Interactions of electromagnetic radiation with matter

Model/Transition	Energy	Selection criteria	
Nuclear magnetic resonance	$E = -g_N \cdot \mu_N \cdot m_I \cdot |\vec{B}|$, with $m_I = -I,\ -I+1, \ldots, I-1, I$	$\Delta m_I = \pm 1$	
Electron spin resonance	$E = g_e \cdot \mu_B \cdot m_s \cdot |\vec{B}|$, with $m_s = -s,\ +s = -\frac{1}{2}, +\frac{1}{2}$	$\Delta m_s = \pm 1;\ \Delta m_I = 0$	
Rigid rotor with space-free axis	$E(J) = h \cdot c \cdot B \cdot J \cdot (J+1)$	$\Delta J = \pm 1$	
Harmonic oscillator	$E(v) = h \cdot \nu_0 \cdot \left(v + \frac{1}{2}\right)$	$\Delta v = \pm 1$	
Anharmonic oscillator	$E(v) = h \cdot \nu_0 \cdot \left(v + \frac{1}{2}\right) - h \cdot \nu_0 \cdot x_e \cdot \left(v + \frac{1}{2}\right)^2 + h \cdot \nu_0 \cdot y_e \cdot \left(v + \frac{1}{2}\right)^3 - \ldots$	$\Delta v = \pm 1, \pm 2, \pm 3, \ldots$	
Rota-vibrational absorption (Infrared spectroscopy)	$E = E(J) + E(v)$	$\Delta v = \pm 1\,(\pm 2, \pm 3, \ldots);\ \Delta J = \pm 1$	Singlet molecules
		$\Delta v = \pm 1\,(\pm 2, \pm 3, \ldots);\ \Delta J = 0, \pm 1$	Non-singlet molecules
Rota-vibrational emission (Raman spectroscopy)	$E = E(J) + E(v)$	$\Delta v = \pm 1;\ \Delta J = 0, \pm 2$	
Electronic absorption (UV/Vis spectroscopy)	$E = E(J) + E(v) + E_{electr}$	$M = 2 \cdot S + 1 = \text{const.}$ $\Leftrightarrow \Delta S = 0$	
Electronic emission (Fluorescence)	$E = E(J) + E(v) + E_{electr}$	$M = \text{const.} \Leftrightarrow \Delta S = 0$	
Electronic emission (Phosphorescence)	$E = E(J) + E(v) + E_{electr}$	$M \neq \text{const.} \Leftrightarrow \Delta S \neq 0$	
Optical spectra of atoms	$E = E_{electr}$	$\Delta l = \pm 1, \Delta j = 0, \pm 1$	Alkali metals
		$\Delta J = 0, \pm 1$	Multi-electron atoms
Nuclear resonance	$E^2 < 2 \cdot m \cdot c^2 \cdot k_B \cdot \Theta$		
X-ray spectra of atoms	$E = E_{electr}$	$\Delta l = \pm 1, \Delta j = 0, \pm 1$	
Auger electron spectra	$E = (E_{hole} - E_{second}) - E'_{binding}$	none	

Thermodynamics

2

2.1 Motivation, Revision and Introduction of Basic Concepts

The description of macroscopic phenomena in the area of chemistry, biology, physics and geology is of eminent importance to develop an appreciation and understanding of the molecular processes that give rise to these phenomena. Thermodynamics is a systematic theory that is applicable and valid in a truly general fashion. As such, the knowledge of thermodynamic concepts is a pre-requisite for many neighbouring disciplines, such as materials science, environmental science, biochemistry, forensics, etc.

In particular, thermodynamics is concerned with

- how the macroscopic world behaves
- how energy is transferred
- how and under which conditions equilibrium is achieved
- how and into which direction processes develop.

Despite the fact that the complex looking formulae may be obtained and dealt with at times, the study of thermodynamics is constantly connected to observations made in the real macroscopic world. The laws of thermodynamics are a prime example of how the theory is connected with real world experiences.

The types of question we might want to answer may include:

- If we have a fluid in a sealed container, how will the pressure in the container likely change with temperature?
- Is the solubility of a compound likely to increase or decrease with temperature?

Application of thermodynamic concepts will allow us to answer such questions qualitatively, but, more importantly, we will also be able to derive quantitative answers. It is thus necessary to use some algebra and calculus (see Appendix A).

© Springer International Publishing AG, part of Springer Nature 2018
A. Hofmann, *Physical Chemistry Essentials*,
https://doi.org/10.1007/978-3-319-74167-3_2

2.1.1 Fundamental Terms and Concepts

It seems appropriate to start with the introduction and revision of vocabulary frequently used in thermodynamics. Many fundamental terms and concepts are typically introduced in entry-level chemistry courses, and should thus already be familiar (Table 2.1).

In the following Sects. (2.1.2–2.1.13), we will expand these fundamental concepts so that we can start to explore various thermodynamic concepts in more detail.

All physical measurements are about differences, absolute values of quantities can not be determined. Therefore, when describing quantities such as functions (e.g. energy, entropy) and observables (e.g. temperature, pressure), they are typically characterised by comparing two different states, such as for example the energy difference between two different states that possess different temperatures. When such differences are observed macroscopically (i.e. in a typical bench-top experiment), these differences are denoted with a capital Greek delta (Δ). In other words, the Δ describes the difference between two reasonably spaced discrete points. When an infinitesimally small difference between two very close points is addressed, the lower case Greek δ is used instead. For quantities that can be expressed as continuous mathematical functions, the 'δ' becomes a 'd' to indicate the differential of that quantity. The differential of a quantity with respect to another can be envisaged as the slope of the curve in a plot of the two quantities (see Fig. 2.1).

A basic introduction to differentials is given in Appendix A.2. Throughout this text, we will use 'Δ' to denote a macroscopic difference, and 'δ' or 'd' for infinitesimal small differences. Since differentials ('d') allow application of mathematical formalism and calculus, this will frequently be the notation of choice. It is important to remember, that any differential can be seen as a difference between two states. The

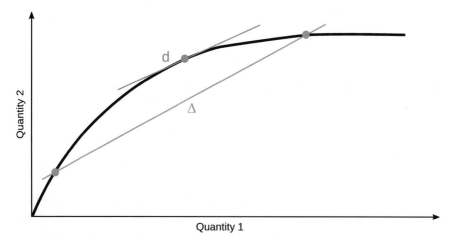

Fig. 2.1 Comparison of the macroscopic difference Δ between two discrete points and the differential as a special case of difference where two neighbouring points merged into one

Table 2.1 Summary of fundamental physico-chemical terms and concepts

Term	Description	Physico-chemical background
System	A part of the universe that can be studied	Isolated, closed, open
Equilibrium	No macroscopic signs of change	
Intensive parameters	Do not depend on the amount of substance present in the system	e.g. p, T
Extensive parameters	Depend on the amount of substance present in the system	e.g. m, V
Temperature	Measure of the average kinetic energy $\bar{\varepsilon}_{kin}$ of particles in a container; state function	$T = \dfrac{\bar{\varepsilon}_{kin}}{k_B}$; $[T] = 1$ K, $[\theta] = 1\ °C$
Pressure	The force of matter exerted onto a surface; state function	$[p] = 1\ \mathrm{N\ m^{-2}}$
Ideal gas	Obeys the ideal gas law: $p \cdot V = n \cdot R \cdot T$	Volume of particles is negligible; interactions between particles is negligible
Partial pressure	Pressure of a gas component if it was alone in a container in the same overall state (p, V, T)	$[p_i] = 1\ \mathrm{N\ m^{-2}}$
Energy	Capacity to do work, state function	$[E] = 1$ J
Work	Object is moved against an opposing force; path function.	$W_{exp} = p \cdot V$; $[W] = 1$ J
Heat	Energy resulting from temperature; path function	$[Q] = 1$ J
Internal energy	The energy possessed by a system; state function	$U = Q + W$; $[U] = 1$ J
Enthalpy	Equal to the heat supplied to a system at constant pressure if there is no work done except for expansion/contraction ; state function	$H = U + p \cdot V$; $[H] = 1$ J
Entropy	Disorderly dispersal of energy; state function	$\Delta S = \dfrac{\Delta Q}{T}$; $[S] = 1\ \mathrm{J\ K^{-1}}$

macroscopic difference of a function X, ΔX, is linked to the many microscopic differences dX within a range from X_{start} to X_{end} by the mathematical process of integration:

$$\Delta X = \int_{X_{start}}^{X_{end}} dX = [X]_{X_{start}}^{X_{end}} = X_{end} - X_{start}$$

System

All scientific study is carried out on a particular system that is the subject of study. A system is the part of the universe that can be conveniently studied with the methods at hand. In many cases, most or all of the properties of the system are under the control of the observer. A system can be as simple as a container filled with water that is placed on a heating plate.

Systems can be subdivided based on their boundaries. It may not always be possible to perfectly reproduce such boundaries in real experiments, but for the development of concepts, idealised versions are required. Typically, three types of systems are considered:

- isolated system: can neither exchange energy nor matter with its environment; it thus cannot elicit any changes in the environment nor can it be changed by its environment.
- closed system: whereas exchange of energy is allowed between the system and its environment, it is not possible to exchange matter.
- open system: the exchange of energy as well as matter is possible, as e.g. between two compartments separated by a semi-permeable membrane.

Equilibrium
A system that has reached equilibrium does not show any macroscopic signs of change, i.e. there is no change of the state of that system.

Intensive and Extensive Parameters
State variables are parameters that describe particular properties or coordinates of a system. They can be classified into two groups:

- extensive parameters—depend on the amount of substance present in the system. The value of extensive parameters can be calculated as the sum of the partial values when the system is subdivided into parts.
- intensive parameters—do not depend on the amount of substance present in the system. The value of intensive parameters can be measured at any point within the system.

Typically, an intensive parameter can be calculated as the quotient of two extensive parameters.

Temperature
Temperature is a measure for the average kinetic energy ($\bar{\varepsilon}_{kin}$) of particles in a container:

$$T = \frac{\bar{\varepsilon}_{kin}}{k_B} \tag{2.1}$$

where k_B is the Boltzmann constant: $k_B = \frac{R}{N_A} = 1.380662 \cdot 10^{-23}$ J K^{-1}, and ε_{kin} is used to indicate the extensive kinetic energy (since it is measured per particle; $[\varepsilon_{kin}] = 1$ J), as opposed to the intensive quantity E_{kin} (measured per mol; $[E_{kin}] = 1$ J mol^{-1}).

Table 2.2 Commonly used temperature scales

Use	Symbol	Scale	Unit
Ambient	θ	degree Celsius	$[\theta] = 1\ °C$
		degree Fahrenheit	$[\theta] = 1\ °F$
Thermodynamic (SI unit)	T	kelvin	$[T] = 1\ K$

Temperature is an intensive property and therefore does not depend on the amount of substance in the container. The temperature indicates the direction of the flow of energy through a thermally conducting wall. If energy flows from object A to B when they are in contact, then A is defined to possess higher temperature. Thermal equilibrium exists, when no energy flow occurs between objects A and B when they are put into contact by a wall that allows heat to flow.

There are various scales (and thus units) of measurement for temperature (Table 2.2):

The fix points of the Celsius scale are defined as $-273.15\ °C = 0\ K$ and the triple point of water being at precisely 273.16 K and 0.01 °C (at standard pressure). Thus, the increments on the degree Celsius and kelvin scale are the same, which led to the recommendation that temperature differences on the degree Celsius scale should be expressed in kelvin; for example:

$$\Delta\theta = \theta_2 - \theta_1 = 20\,°C - 15\,°C = 5.0\,K$$

although this rule has been relaxed by IUPAC (Rossini 1968). According to above definition, temperatures on the kelvin and degree Celsius scales can be converted as per

$$\theta = \left(\frac{T}{1K} - 273.15\right)°C \tag{2.2}$$

The standard temperature as defined by IUPAC is at

$$T^{\varnothing} = 273.15\ K,\ i.e.\ \theta^{\varnothing} = 0\,°C$$

as opposed to the normal temperature of

$$T_{normal} = 298.15\ K,\ i.e.\ \theta_{normal} = 25\,°C$$

which is typically used as a reference temperature for many physico-chemical processes and chemical reactions.

Pressure

Pressure is defined as the force divided by the area to which the force is applied

Table 2.3 Commonly used pressure scales

Use	Scale	Unit
SI unit	pascal	$[p] = 1\,\text{Pa} = 1\,\text{N}\,\text{m}^{-2}$
Historic	mm Hg, torr	$1\,\text{mm Hg} = 1\,\text{torr} = 133.3\,\text{Pa}$
Colloquial	atmosphere	$1\,\text{atm} = 1.013 \cdot 10^5\,\text{Pa}$
Non-SI unit (European)	bar	$1\,\text{bar} = 10^5\,\text{Pa}$
Non-SI unit (Anglo-American)	psi (pound-force per square inch)	$1\,\text{psi} = 6.895 \cdot 10^4\,\text{Pa}$

$$p = \frac{F}{A} \tag{2.3}$$

Pressure is an intensive property and thus does not depend on the amount of matter present.

There are various scales (and thus units) of measurement for pressure (Table 2.3): The IUPAC standard pressure is defined as

$$p^{\varnothing} = 1\,\text{bar} = 10^5\,\text{Pa} \tag{2.4}$$

The normal pressure (which equals the standard pressure used by the National Institute of Standards and Technology, NIST, USA) refers to atmospheric pressure and is defined as

$$p_{\text{normal}} = 1013.25\,\text{hPa} = 1013.25\,\text{mbar} = 1\,\text{atm} = 760\,\text{mm Hg} = 760\,\text{torr} \tag{2.5}$$

Mechanical equilibrium exists when a moveable piston between two compartments filled with gases has no tendency to move.

Boiling points depend on the pressure
Since water boils at a temperature of $\theta_{\text{b,normal}}$ (H_2O) $= 100\,°C$ at normal pressure ($p_{\text{normal}} = 1$ atm $= 1.013$ bar), the boiling temperature is less at IUPAC standard pressure ($p^{\varnothing} = 1$ bar $= 0.9869$ atm).

Ideal Gas
The behaviour of an ideal gas can be explained by the kinetic gas theory which requires the following assumptions:

- the gas is composed of particles whose total volume is negligible compared with the volume of the system
- the interactions between particles are negligible
- the average kinetic energy is proportional to the temperature

Table 2.4 Variables and parameters of the ideal gas equation

Pressure	$[p] = 1$ Pa
Volume	$[V] = 1$ m^3
Molar amount	$[n] = 1$ mol $= 6.022 \cdot 10^{23}$ particles (number of atoms in 12 g of ^{12}C)
Gas constant	$R = 8.3144$ J K^{-1} mol^{-1}
Temperature	$[T] = 1$ K

The ideal gas equation constitutes the equation of state for the ideal gas and informs about the physical state of the system for given conditions:

$$p \cdot V = n \cdot R \cdot T \tag{2.6}$$

The variables and parameters of the ideal gas equation are (Table 2.4):

Partial Pressure

If a system contains more than one gaseous components, the partial pressure describes the pressure of any particular gas if it would reside there alone, under the same total pressure and temperature. In case of ideal gases, the ideal gas equation is valid for each individual gas with its partial pressure:

$$p_i \cdot V = n_i \cdot R \cdot T \tag{2.7}$$

Dalton's law poses that the pressure exerted by a mixture of gases is the sum of the partial pressures of the individual gases:

$$p = p_1 + p_2 + \ldots + p_N = \sum_{i=1}^{N} p_i \tag{2.8}$$

Example of Dalton's law

If we have a mixture of O_2 and N_2 in a flask, where the partial pressure of O_2 is $p(O_2) = 1$ Pa, and the partial pressure of N_2 is $p(N_2) = 2$ Pa, the total pressure in the flask is

$$p = p(N_2) + p(O_2) = 1\,\text{Pa} + 2\,\text{Pa} = 3\,\text{Pa}.$$

Energy

Energy describes the capacity to do work and thus bring about change. Energy is measured in units of joule: $[E] = 1$ J.

Work

Is a form of energy and can be classified either as mechanical or electrical work. Mechanical work is defined as the product of a force and the length along which this force is applied:

$$W = F \cdot x \qquad [W] = 1 \, \text{N m} = 1 \, \text{Pa} \cdot \text{m}^3 = 1 \, \text{J} \tag{2.9}$$

electrical work is done when a current I flows through an electric resistance R over a particular time t.

$$W = R \cdot I^2 \cdot t \qquad [W] = 1 \, \Omega\text{A}^2\text{s} = 1 \, \text{V A s} = 1 \, \text{C V} = 1 \, \text{J} \tag{2.10}$$

Differential expressions

Since energy is all about the potential to change the state of a system, differential expressions are frequently used. The differential expression for mechanical work tells us how much the value of the work changes (dW) when the force F is applied along a particular distance (dx):

$$dW = F \cdot dx$$

Differential expressions need to be integrated in order to obtain a definite value. To integrate this expression, the following integrals need to be resolved:

$$\int_{W_{start}}^{W_{end}} dW = [W]_{W_{start}}^{W_{end}} = W_{end} - W_{start} = \Delta W$$

$$\int_{x_{start}}^{x_{end}} F \cdot dx = F \cdot \int_{x_{start}}^{x_{end}} dx = F \cdot [x]_{x_{start}}^{x_{end}} = F \cdot (x_{end} - x_{start}) = F \cdot \Delta x$$

The value of the work done is thus:

$$W = W_{end} - W_{start} = F \cdot (x_{end} - x_{start}) = \Delta W = F \cdot \Delta x$$

Heat

When energy changes as a result of a temperature difference between the system and its surroundings, we say there is a flow of heat ΔQ. If there is no flow of heat during a process, this is called an adiabatic process. Heat therefore is an energy due to temperature; it is measured in units of joule: $[Q] = 1 \, \text{J}$. Like work, heat is a path function, i.e. the amount of heat flow depends on the way the change occurs.

Internal Energy

Internal energy describes the energy possessed by system, in addition to its kinetic and potential energy:

$$E = E_{kin} + E_{pot} + U \tag{2.11}$$

E_{kin} and E_{pot} are macroscopic parameters of a system and, in general, can be treated as constant. The internal energy U is an extensive state parameter and depends on the internal state variables V, T and n. If the external state variables E_{kin} and E_{pot} are constant, the change in the energy of a system (ΔE) equals the change in the internal energy (ΔU) and comprises of the work done (ΔW) and the exchanged heat (ΔQ):

$$E_{kin} = \text{const.}, E_{pot} = \text{const.} \Rightarrow \Delta E_{kin} = 0, \Delta E_{pot} = 0$$

$$\Delta E = \Delta U = \Delta Q + \Delta W \tag{2.12}$$

If we do not change the amount of substance present in the system (i.e. $n = \text{const.}$), the ideal gas equation relates the three state variables p, V and T (and in order to define the state of a system, it is thus sufficient to experimentally restrain two parameters (e.g. V and T), as the third one (here: p) will follow suit. Since for gases, the volume can be controlled rather conveniently, the internal energy U is generally described as a function of V and T. The internal energy is measured in units of joule: $[U] = 1$ J.

Enthalpy

Whereas the volume of gases can be controlled or changed rather readily in an experimental setting, this is not the case for the condensed phases of liquids and solids. The thermal extension of the volume of solids and liquids can hardly be prevented. From an experimental perspective, it is thus easier to characterise the energy of systems with condensed phases under constant pressure. This is a reasonable criterion, as the ambient pressure can be considered constant for most laboratory experiments. This energy function is the enthalpy H, which is defined as

$$H = U + p \cdot V \tag{2.13}$$

The enthalpy is measured in units of joule: $[H] = 1$ J.

Entropy

Entropy describes the dispersion of energy ('disorder') and is defined as

$$\Delta S = \frac{\Delta Q}{T} \tag{2.14}$$

The entropy is measured in units of joule per kelvin: $[S] = 1$ J K^{-1}.

Units of measurement

Efforts to standardise the units of measurement started around 1800 with the central idea of choosing a natural constant as reference. This resulted in the metric system whereby the length of 1 m was defined as one 10^{-7}-th of the length of the quarter meridian of the earth. The current units of measurements (see Table 1.4) are the result of further efforts into that direction and are defined in the International System of Units (SI) (Bureau international des poids at mesures 2006). In the current system of units, the kilogram is defined by a mass protoype in form of a Pt-Ir cylinder, deposited in the Bureau International des Poids et Mesures (BIPM). By definition, the mass of this protoype is 1 kg, but its real mass has changed over time by contamination; it is estimated to have gained 0.1–0.3 mg as compared to its initial mass. The unit of the molar amount—the mol—is defined via the carbon isotope ^{12}C whereby the molar mass of this isotope is set to be 0.012 kg mol^{-1}. Therefore, the definition of the mol depends on the definition of the kilogram.

In the recent past, it has been debated that a more robust referencing is required to become independent of protoypes which change over time (Bureau international des poids at mesures 2013). One of these suggestions proposes that in the new system, the unit of mol shall be linked to an exact numerical value of Avogadro's constant N_A, which will make it independent of the kilogram (which in turn may be defined via an exact numerical value of Planck's constant h). Consequently, the unit of 1 mol would be the amount of exactly $6.02214129 \cdot 10^{23}$ particles (Stohner and Quack 2015). Conceptually, in this new system, the numerical value of N_A would be fixed and thus no longer carry a statistical uncertainty as in the present system (estimated at $5 \cdot 10^{-7}$). At the same time, the molar mass of ^{12}C, which currently is an exact quantity, will become an experimental quantity in the new system and thus be subject to statistical uncertainty. It is estimated that this uncertainty will be at the order of 10^{-9} which needs to be set into relation with the uncertainty of general molar masses (approx. 10^{-5}). One can thus anticipate that there will be no substantial changes of atomic masses in the periodic system if this new system should be instantiated. Nevertheless, the debate about such changes to the SI are ongoing and, at present, subject to further evaluation by the International Union of Pure and Applied Chemistry (IUPAC) (International Union of Pure and Applied Chemistry 2013).

2.1.2   The 0th Law of Thermodynamics

From the empirical quantity temperature and the equilibrium concept, one can deduce the 0th law of thermodynamics:

▶   If two systems (A and B) each are in thermal equilibrium with a third
    system (C), then they are also in equilibrium among each other (i.e. A
    with B). All three systems share a common property, they have the same
    temperature.

This law implies that the boundaries of a system may possess thermal conductivity. We consider two compartments of a large container that are separated by a styrofoam (i.e. isolating) wall. If we put hot water into one compartment, and cold water into the other compartment, the two systems will have different heat, and this state will not change. If the styrofoam wall is replaced with a metal foil, there will be a change of the states of both compartments. Heat will flow from the hot to the cold water compartment. The heat flow will cease, when both systems are in thermal equilibrium. At that point, both compartments have the same temperature.

2.1.3   Equations of State

The physical state of a system is described by its physical properties. The parameters describing these properties are called state variables or state functions. State variables are defined in terms of the various properties we can attribute to a substance. For example, the state of a pure gas is specified by giving its volume (measured in liter), the amount (measured in moles), pressure (measured in Pa) and temperature (measured in K).

There will be a relationship between the different state variables, i.e. they are not independent from each other. Such relationships are called equations of state. In the case of a gas, this relationship is given by the ideal gas equation.

$$p \cdot V = n \cdot R \cdot T \tag{2.6}$$

It follows from the ideal gas equation that if one specifies three of these four variables, the fourth one will have one discrete value. Each state variable is thus a function of the three others:

$$p = f(T, V, n) = \frac{n \cdot R \cdot T}{V}$$

$$V = f(T, p, n) = \frac{n \cdot R \cdot T}{p}$$

$$T = f(p, V, n) = \frac{p \cdot V}{n \cdot R}$$

$$n = f(T, p, V) = \frac{p \cdot V}{n \cdot R}$$

Important state variables in thermodynamics are (Table 2.5):

Table 2.5 Important thermodynamic state variables	Temperature T	Internal energy U
	Pressure p	Enthalpy H
	Volume V	Entropy S
	Mass m	Density ρ

2.1.4 Energy

Energy describes the capacity of a system to do work. It is a state function and an extensive parameter, since its value depends on the amount of substance. It is measured in units of joule, albeit some reference to the historically used calorie unit may still be found:

$$[E] = 1\,\text{J} = 1\,\text{kg m}^2\text{s}^{-2} = 0.239\,\text{cal}$$

There are various types of energies, including:

- kinetic energy: $E_{\text{kin}} = \frac{1}{2} \cdot m \cdot v^2$ (Newtonian mechanics)
- potential energy: $E_{\text{pot}} = m \cdot g \cdot h$; $g = 9.81$ m s^{-2} (Newtonian mechanics)
- chemical energy: energy present in the bonds between atoms and molecules
- thermal energy: due to the vibration and movement of the atoms and molecules in a substance
- electric energy: potential difference between two half-cells with electron conducting link
- osmotic energy: difference in salt concentration in two half-cells separated by a semi-permeable membrane

Since potential and kinetic energy refer to the position or movement of the system, they are considered external, macroscopic parameters of the system. All other types of energies possessed by the system are combined and constitute its internal energy.

2.1.5 The 1st Law of Thermodynamics

Energy cannot be made or destroyed. Since the internal energy of a system comprises all energy that this system possesses, it follows that any change of the internal energy of a system has to be the result of the work done and the heat transferred:

$$\Delta U = \Delta Q + \Delta W \qquad (2.12)$$

and as such constitutes the 1st law of thermodynamics:

▶   For a closed system with constant external state variables, the internal
    energy U constitutes an extensive state function whose change dU is
    composed of the exchanged heat dQ and work dW done on or by the
    system.

$$dU = dQ + dW \qquad (2.15)$$

Whereas above the 1st law is defined for the differential change dU, the same is
true for the macroscopically measurable change ΔU (see Eq. 2.12), which represents
the integrated form of the differential Eq. 2.15.

The integrated form of the differential equation dU = dQ + dW
Differential changes are infinitesimally small changes in variables. Processes
that are observed macroscopically have definite start and end points. It is thus
necessary to integrate the differential equation.

$$\int dU = \int dQ + \int dW \qquad (2.16)$$

Integration needs to be carried out with respect to each differential variable.
Hence the left side of Eq. 2.16 has one integral (as there is only one differential
variable, dU), but the right side has two integrals (one integrating dQ and one
integrating dW).

In order to resolve $\int dU$, we replace U with x and obtain $\int dx$. This function
has a known integral (see Appendix A.3.1):

$$\int_{x_{start}}^{x_{end}} dx = [x]_{x_{start}}^{x_{end}} = x_{end} - x_{start} = \Delta x, \quad \text{therefore :}$$

$$\int_{U_{start}}^{U_{end}} dU = [U]_{U_{start}}^{U_{end}} = U_{end} - U_{start} = \Delta U.$$

It is immediately obvious, that $\int dQ$ and $\int dW$ can be resolved in the same
fashion:

(continued)

$$\int_{Q_{start}}^{Q_{end}} dQ = [Q]_{Q_{start}}^{Q_{end}} = Q_{end} - Q_{start} = \Delta Q$$

$$\int_{W_{start}}^{W_{end}} dW = [W]_{W_{start}}^{W_{end}} = W_{end} - W_{start} = \Delta W$$

It thus follows:

$$\int dU = \int dQ + \int dW = \Delta U = \Delta Q + \Delta W$$

Note that the 1st law contains one important restraint; it is valid for closed systems which requires that only energy, but no matter can be exchanged. Besides this restriction, the 1st law is generally applicable and in particular applies to reversible as well as irreversible changes of state.

In Sect. 2.1.1, we defined isolated systems by the fact that they cannot exchange any energy with their environment. It follows that the internal energy of an isolated system remains constant, because:

$$\Delta U = \Delta Q + \Delta W = 0 + 0 = 0$$

If there is no change in internal energy ($\Delta U = 0$), the internal energy is constant: $U = \text{const}$.

2.1.6 Work

Work is done when an object is moved against an opposing force. This could for example be electrons moving through an electric conductor (electrical work), molecules moving in space under atmospheric pressure (work of expansion/compression). If a system does work of the value ΔW, its internal energy changes by this amount of energy:

$$\Delta U = \Delta Q + \Delta W \tag{2.12}$$

The amount of work that a system does, depends on how a particular process occurs. The same final state of the system may be reached by different pathways which require different amounts of work. Therefore, work is a path function.

The amount of work to be done for contraction or expansion will depend on the change of the volume of the system ($\Delta V = V_{final} - V_{initial}$) and the surrounding restrictions presented by the external pressure $p = p_{ext}$. The amount of work done by expansion or contraction can macroscopically be calculated as

$$\Delta W = -p \cdot \Delta V \qquad (2.17)$$

If we consider very small changes, we need to move from the macroscopic difference (Δ) to a differential (d):

$$dW = -p \cdot dV \qquad (2.18)$$

The integrated form of the differential equation $dW = -p_{ext} \cdot dV$
To initiate integration of this equation, we add the integral sign $\int$ to both sides of the equation:

$$\int dW = \int (-p_{ext} \cdot dV) \qquad (2.19)$$

The left side of this equation can be resolved easily by

$$\int_{W_{start}}^{W_{end}} dW = [W]_{W_{start}}^{W_{end}} = W_{end} - W_{start} = \Delta W.$$

The right side of Eq. 2.19 takes the form of $\int (a \cdot dx)$ where a is a constant that does not depend on the differential variable dx. The external pressure is the environmental (ambient) pressure and is not affected by the volume change of the system under study, which is considered negligibly small compared to the earth's atmosphere. The integral $\int (a \cdot dx)$ is resolved as:

$$\int_{x_{start}}^{x_{end}} (a \cdot dx) = a \cdot \int_{x_{start}}^{x_{end}} dx = a \cdot [x]_{x_{start}}^{x_{end}} = a \cdot (x_{end} - x_{start}) = a \cdot \Delta x$$

and with $a = -p_{ext}, x = V$ we obtain:

$$\int_{V_{start}}^{V_{end}} (-p_{ext} \cdot dV) = -p_{ext} \cdot \int_{V_{start}}^{V_{end}} dV = -p_{ext} \cdot [V]_{V_{start}}^{V_{end}} = -p_{ext} \cdot (V_{end} - V_{start})$$

$$= -p_{ext} \cdot \Delta V$$

and therefore

$$\int dW = \Delta W = \int (-p_{ext} \cdot dV) = -p_{ext} \cdot \Delta V$$

There are two interesting cases of expansion/compression work, we should look at in more detail. First, if the external pressure is zero (which is the case in a vacuum), Eq. 2.17 yields:

$$\Delta W = -p_{ext} \cdot \Delta V = 0 \cdot \Delta V = 0$$

so there is no work done, despite the fact that the system may expand.

Second, if the external pressure is gradually changed by infinitesimal small amounts so that the internal and external pressures remain equal at all times, the expansion/compression is carried out reversibly. Changes will occur infinitesimally slowly, and the system will be at equilibrium with the surroundings at all times. In that case, the pressure in Eq. 2.18 is no longer constant, but varies with the change in volume: $p = p(V)$. The work done in reversible processes is therefore calculated as per the following integration:

$$\int dW_{rev} = \int -p \cdot dV = \int \frac{-n \cdot R \cdot T}{V} \cdot dV \qquad (2.20)$$

If we consider an ideal gas, we can express the pressure p as a function of the volume V using the ideal gas equation. If we further consider that the reversible volume work is carried out under constant temperature (isothermal process), then the product $n \cdot R \cdot T$ forms a constant that can be taken outside the integral:

$$\int dW_{rev} = -n \cdot R \cdot T \int \frac{1}{V} \cdot dV$$

The integral $\int \frac{1}{V} dV$ is of the form $\int \frac{1}{x} dx = [\ln(x)]$ (see Appendix A.3.1). The reversible expansion/compression work thus resolves to:

$$\int dW_{rev} = \Delta W_{rev} = -n \cdot R \cdot T \cdot [\ln V]_{V_{initial}}^{V_{final}}$$

$$\Delta W_{rev} = -n \cdot R \cdot T \cdot (\ln V_{final} - \ln V_{initial})$$

$$\Delta W_{rev} = -n \cdot R \cdot T \cdot \left(\ln V_{final} + \ln \frac{1}{V_{initial}} \right)$$

$$\Delta W_{rev} = -n \cdot R \cdot T \cdot \left(\ln \frac{V_{final}}{V_{initial}} \right)$$

With the logarithm rule of $-\log_b^a = \log_a^b$ (Appendix A.1.2), this yields:

$$\Delta W_{rev} = n \cdot R \cdot T \cdot \left(\ln \frac{V_{initial}}{V_{final}} \right) \qquad (2.21)$$

Figure 2.2 compares the irreversible (p_{ext} = const.) and reversible (p_{ext} is variable) expansion/compression work under isothermal conditions (T = const.). The work is represented by the indicated area under the p-V-curve. From this

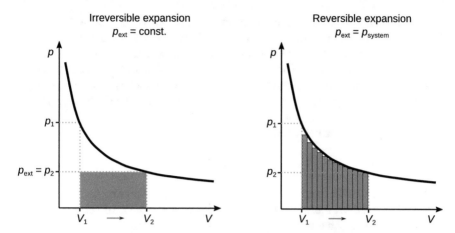

Fig. 2.2 Work done during isothermal irreversible (left) and reversible (right) expansion processes. In irreversible processes, the external pressure is constant and the expansion from V_1 to V_2 happens in a single step, resulting in an amount of work indicated by the shaded area. In reversible processes, the external pressure is at all times equal to the pressure of the system (p), and the amount of work done can be estimated by small incremental areas. The total amount of work for the reversible process is larger than that for the irreversible process

comparison, it is evident that the magnitude of the work is maximised in a reversible process.

2.1.7   Heat Capacity at Constant Volume

The change in the internal energy of a substance when the temperature is raised is described by the heat capacity at constant volume (indicated by the subscript V):

$$C_V = \left(\frac{\delta U}{\delta T} \right)_V \qquad (2.22)$$

It follows straight from Eq. 2.22 that the internal energy of a system varies linearly with temperature, if the volume remains constant:

$$dU = C_V \cdot dT \qquad (2.23)$$

If the heat capacity does not change in the temperature range being considered, then one can use Eq. 2.23 to determine the heat capacity C_V by measuring the heat ΔQ_V provided to the system. Since we still require the volume to be constant, the system will not be able to do any expansion/compression work, i.e. $\Delta W_V = 0$:

$$\Delta U = \Delta Q_V + \Delta W_V = \Delta Q_V + 0 = C_V \cdot \Delta T = \Delta Q_V$$

Experimentally, one can provide a defined amount of heat to a system, for example by electrically heating the system in a bomb and observing the temperature change in the system. The closed bomb case will ensure that the volume remains constant during the process. The bomb will also need to be insulated such that there is no exchange of heat possible with the environment, which is called an adiabatic system. A plot of ΔQ_V versus ΔT will yield a line with the slope C_V. The instrument used for this purpose is an adiabatic bomb calorimeter.

The heat capacity is measured in units of joule per kelvin:

$$[C_V] = 1 \, \mathrm{J\,K^{-1}}$$

Since the internal energy U is an extensive function, the heat capacity at constant volume is extensive, too. Its value increases with the amount of substance in the system. The corresponding intensive properties of matter are:

- molar heat capacity: heat capacity of 1 mol of substance
- specific heat: heat capacity of 1 g of substance.

2.1.8 Enthalpy

If we consider an open container filled with liquid water that is being heated, we know from experience that the water will expand its volume as it gets hotter (Fig. 2.3 left). In addition to the heat ΔQ transferred into this system, we further have to consider the volume work ΔW which the system performs as it pushes back the atmosphere. Thus, the change in internal energy ΔU during the heating process is

$$\Delta U = \Delta Q + \Delta W \qquad (2.12)$$

As the expansion/compression work is a frequently occurring step in many chemical processes, it would help to simplify the description of energy changes of systems, if this work could be accounted for automatically. Therefore, the enthalpy H is defined as

$$H = U + p \cdot V \qquad (2.13)$$

and is measured in units of joule: $[H] = 1 \, \mathrm{J}$. This quantity takes into account the simultaneous loss (or gain) of energy by expansion/compression, when the energy of the system is changed.

In order to calculate an enthalpy change, we need to differentiate Eq. 2.13:

$$dH = dU + d(p \cdot V)$$

With the product rule (Appendix A.2.3) for differentiation this yields:

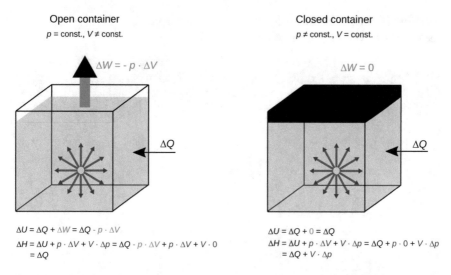

Fig. 2.3 Comparison of the energy changes upon heat transfer into a system residing in either an open or closed container highlights the usefulness of the enthalpy for "everyday experiments", typically conducted under ambient (i.e. constant) pressure

$$dH = dU + p \cdot dV + dp \cdot V$$

If we assume constant external pressure (p = const.), which is the case for processes carried out under ambient pressure, it follows that $dp = 0$:

$$dH = dU + p \cdot dV + 0 \cdot V = dU + p \cdot dV \qquad (2.24)$$

If we again consider the open container with water that is getting heated, we know that this system will expand and thus conduct work against the outside pressure; the internal energy change is thus:

$$dU = dQ + dW = dQ - p \cdot dV$$

from Eqs. 1.15 and 2.18.

Substituting this expression for dU in Eq. 2.24 yields:

$$dH = dU + p \cdot dV = dQ - p \cdot dV + p \cdot dV = dQ \qquad (2.25)$$

Therefore, the enthalpy change of the system during heat transfer under constant pressure is the same as the transferred heat, if no other work (than expansion/compression) is done.

Reactions or processes that involve solids or liquids and gas/vapour are typically characterised by large changes in volume, i.e. dV is not negligible. However, if a reaction or process involves only liquids and solids, the product $p \cdot V$ is rather small, therefore $H \approx U$. With $\Delta H = \Delta Q \approx \Delta U$, the change in internal energy ΔU can be measured by evaluating the heat exchange ΔQ.

Enthalpy change of a closed container with liquid upon heating

We again heat a container filled with liquid, but now require that the container is closed. This means that the volume of the system can not expand during the heating process (Fig. 2.3 right).

Evaluating the change in enthalpy of this system during the heating process, we obtain:

$$\begin{aligned} H &= U + p \cdot V \\ dH &= dU + d(p \cdot V) \end{aligned} \tag{2.13}$$

With the product rule (Appendix A.2.3) for differentiation this yields:

$$dH = dU + p \cdot dV + V \cdot dp$$

Since the container is closed, we know that $V = $ const. and thus $dV = 0$:

$$\begin{aligned} dH &= dU + p \cdot 0 + V \cdot dp \\ dH &= dU + V \cdot dp \end{aligned} \tag{2.26}$$

For the internal energy of the closed container, the following change arises:

$$dU = dQ + dW = dQ - p \cdot dV$$

from Eqs. 2.15 and 2.18

Due to $dV = 0$ it follows that:

$$dU = dQ - p \cdot 0$$

$$dU = dQ \tag{2.27}$$

Since the container is closed, the system cannot expand and thus not do any work; the internal energy equals the transferred heat. This delivers the theoretical basis for the adiabatic bomb calorimeter (Sect. 2.1.7), where the heat capacity C_V of a substance can be determined from the linear relationship between transferred heat and temperature, based on the fact that $dU = dQ$.

However, for the enthalpy, we obtain the following expression by combining Eqs. 2.26 and 2.27:

$$dH = dQ + V \cdot dp \tag{2.28}$$

The enthalpy change of the closed container thus comprises more than just the amount of heat transferred; the pressure change inside the container further adds to the enthalpy.

Enthalpy is a thermodynamic potential. Potentials cannot be measured in absolute terms; rather, one needs to refer to a defined reference point. Practically, one thus measures only changes in enthalpy, ΔH.

Furthermore, enthalpy is an extensive property, i.e. it depends on the amount of substance in a system. The corresponding intensive property is the molar enthalpy H_m, which is normalised with respect to 1 mol: $[H_m] = 1 \text{ J mol}^{-1}$.

2.1.9   Heat Capacity at Constant Pressure

Like the internal energy, the enthalpy of a system can vary with temperature, and this variation is described by the heat capacity. Whereas the heat capacity derived from the internal energy refers to processes at constant volume (C_V), the heat capacity derived from the enthalpy refers to processes at constant pressure:

$$C_p = \left(\frac{\delta H}{\delta T}\right)_p \qquad (2.29)$$

Since enthalpy characterises processes at constant pressure, it can be evaluated by monitoring the change in temperature of a system in cells that are not closed ($p = p_{ext} = $ const.). This type of analysis is called calorimetry. Whereas differential scanning calorimetry (DSC) evaluates enthalpy changes as a function of temperature (e.g. to determine the heat capacity C_p), isothermal titration calorimetry (ITC) is used to monitor enthalpy changes due to chemical processes (binding, reactions) at constant temperature.

Relationship between heat capacities for an ideal gas
From the definition of the heat capacity at constant pressure (Eq. 2.29), we obtain by resolving for dH:

$$dH = C_p \cdot dT$$

From Eq. 2.24 we know that

$$dH = dU + p \cdot dV$$

Therefore, we obtain

$$C_p \cdot dT = dU + p \cdot dV$$

Using Eq. 2.22 we can substitute for dU:

(continued)

$$C_p \cdot dT = dU + p \cdot dV = C_V \cdot dT + p \cdot dV$$
$$C_p = C_V + \frac{p \cdot dV}{dT}$$

By resolving the ideal gas equation (with the differentials dV and dT)

$$p \cdot dV = n \cdot R \cdot dT$$

for $\frac{p \cdot dV}{dT}$, it becomes clear that

$$\frac{p \cdot dV}{dT} = n \cdot R$$

Therefore:

$$C_p = C_V + \frac{p \cdot dV}{dT} = C_p = C_V + n \cdot R$$

or

$$C_p - C_V = n \cdot R$$

2.1.10 Entropy

Entropy describes the disorderly dispersal of energy. The change in entropy is related to how much energy is reversibly transferred as heat (see Eq. 2.14). A system that disperses heat into its surroundings without doing any other expansion/contraction work ($dW = 0$) therefore changes the entropy of the surroundings by:

$$dS_{surr} = \frac{dQ_{surr}}{T_{surr}} = \frac{-dQ_{sys}}{T_{surr}} = -dS_{sys} \tag{2.30}$$

For a process at constant pressure, the enthalpy change in the system dH_{sys} can be used to determine the entropy change of the surroundings, since we know from Eq. 2.25 that $dH = dQ$; therefore:

$$dS_{surr} = \frac{-dH_{sys}}{T_{surr}} \tag{2.31}$$

For reversible adiabatic processes, which do not allow the exchange of heat with the environment, it follows from Eq. 2.30 that there is no change in entropy:

$$dQ = 0 \quad \Rightarrow \quad dS = 0 \quad \text{for reversible adiabatic processes} \qquad (2.32)$$

Reversible processes refer to processes that occur while the system remains at equilibrium at all times. They are generally hypothetical, unless there is no overall change occurring in the system. An equilibrium can be considered as a state of the system where there is cross-over from the forward to the reverse reaction, both of which occur spontaneously. Obviously, for such processes there will be no change in the entropy:

$$dS = 0 \quad \text{for processes in equilibrium} \qquad (2.33)$$

Important equilibrium process are those of phase transitions, such as a substance boiling at its boiling point. The entropy change during this process is defined by the enthalpy of vapourisation ΔH_{vap}:

$$\Delta S_{sys} = \frac{\Delta H_{vap}}{T_b}; p = \text{const.} \qquad (2.34)$$

Similarly, the entropy change of a substance freezing at its freezing point is given by:

$$\Delta S_{sys} = \frac{\Delta H_{fusion}}{T_m}; p = \text{const.} \qquad (2.35)$$

In both Eqs. 2.34 and 2.35, we are using 'Δ' instead of 'd' to indicate that these are macroscopically observable changes. As given, both equations are expressions for the entropy change of the system (ΔS_{sys}) that undergoes the phase transition. One should keep in mind that the entropy change of the system is related to that of the surroundings by:

$$\Delta S_{sys} = -\Delta S_{surr} \qquad (2.30)$$

2.1.11 The 2nd Law of Thermodynamics

Observation of general phenomena tells us that there is a direction associated with all processes, for example a compound dissolving in a solvent or hot coffee getting cold. It then becomes obvious that no process is possible in which the sole result is the absorption of heat from a reservoir and its complete conversion into work; energy is always dissipated. In the example of the hot coffee getting cold, the heat disappearing from the cup does not result in any useful work. It is rather dissipated into the environment and cannot be re-captured.

These concepts are summarised in the 2nd law of thermodynamics:

▶ The entropy of the universe increases in the course of every natural
change:

$$\Delta S_{universe} > 0 \tag{2.36}$$

As a consequence, this law predicts what direction a process or reaction will
follow. It further produces interesting implications, such as the phenomenon that the
direction of a macroscopic process is always the same. The hot coffee gets colder if it
is let to stand—it never gets hotter.

Another intriguing aspect of social importance and the way science is
communicated is the problem commonly referred to as 'energy crisis' which refers
to the dwindling energy resources on earth. A critical appraisal of the 1st law of
thermodynamics shows that such concerns are unwarranted. Energy cannot be
destroyed, so there is no worry that it would be depleted. Rather, it is the way in
which energy is being dispersed that constitutes the problem. By consuming energy,
we increase the entropy (i.e. disperse the energy) and thus destroy its availability. In
this sense, there is an 'entropy crisis' rather than an 'energy crisis'.

2.1.12 The 3rd Law of Thermodynamics

A relationship of practical importance for the determination of entropy changes of
substances at constant pressure can be derived from

$$dH = dU + p \cdot dV = dQ - p \cdot dV + p \cdot dV = dQ \tag{2.25}$$

and

$$dS = \frac{dQ}{T} \quad \Rightarrow \quad dQ = T \cdot dS \tag{2.30}$$

The enthalpy change during a process can then be expressed by

$$dH = T \cdot dS \tag{2.37}$$

whereby the temperature dependence of this process is given by the heat capacity at
constant pressure, C_p:

$$C_p = \left(\frac{\delta H}{\delta T} \right)_p \tag{2.29}$$

This yields with Eq. 2.37:

$$\left(\frac{\delta H}{\delta T}\right)_p = C_p = T \cdot \left(\frac{\delta S}{\delta T}\right)_p \tag{2.38}$$

and upon integration in the temperature interval from 0 to T_{end}:

$$\int_{S_0}^{S_{end}} dS = \int_0^{T_{end}} \frac{C_p}{T} dT$$

$$\Delta S = \int_0^{T_{end}} \frac{C_p}{T} dT \tag{2.39}$$

If one knew the entropy of a substance at $T = 0$ K, it would be possible to determine absolute entropy values. For this reason, Planck postulated that the entropy of a pure, homogeneous phases in internal equilibrium at zero temperature would be zero.

In crystals, which represent pure homogeneous phases, the state of internal equilibrium is attained when there are no defects in the lattice. This postulate is an extension of the Nernst heat theorem:

$$\lim_{T \to 0} S = 0 \tag{2.40}$$

which states that the entropy of pure substances in their perfect crystalline state at $T = 0$ tends to zero. This forms the basis of the 3rd law of thermodynamics:

▶   If the entropy of every element in any crystalline state at $T = 0$ is set to zero ($S_0 = 0$), then all substances have a positive entropy. At $T = 0$, the entropy of substances in their ideal crystalline states is zero.

Since real crystals of substances are virtually impossible to attain, there is a remaining non-zero entropy of such real systems. A prominent example in this context includes glasses. In substances, where there is a possibility of varying molecular orientation within the crystal lattice (e.g. CO: C—O vs. O—C; N_2O: N—N—O vs. O—N—N), a non-zero entropy at $T = 0$ is observed. The same is true for crystalline mixtures.

2.1.13 Entropy and the Gibbs Free Energy Change

Whether or not a particular process or reaction has a tendency to occur under specified conditions depends on the value of the total entropy change for this process. The total entropy change is composed of two parts, the entropy change in the reaction vessel ($dS_{process}$) and the extent of dispersal into the surroundings (dS_{surr}):

$$dS_{\text{total}} = dS_{\text{process}} + dS_{\text{surr}} \tag{2.41}$$

We already discussed the extent of energy dispersal into the surroundings (Sect. 2.1.10); for processes at constant pressure it is the enthalpy difference dH_{sys} at a given temperature T:

$$dS_{\text{surr}} = \frac{-dH_{\text{sys}}}{T} \tag{2.31}$$

Therefore:

$$dS_{\text{total}} = dS_{\text{process}} - \frac{dH_{\text{sys}}}{T}$$

After multiplying with T on both sides of the equation, this yields:

$$T \cdot dS_{\text{total}} = T \cdot dS_{\text{process}} - dH_{\text{sys}}$$

or

$$-T \cdot dS_{\text{total}} = dH_{\text{sys}} - T \cdot dS_{\text{process}}$$

The product $-T \cdot dS_{\text{total}}$ describes a new over-arching energy change for the underlying process and is called the Gibbs free energy change, which is defined as:

$$dG = -T \cdot dS_{\text{total}} = dH - T \cdot dS \tag{2.42}$$

Since spontaneously occurring processes have a positive change in total entropy, the value of the Gibbs free energy change needs to be negative:

$$dS_{\text{total}} > 0 \quad \Rightarrow \quad dG < 0 \tag{2.43}$$

2.1.14 Inter-relatedness of Thermodynamic Quantities

From discussions in the previous sections we come to appreciate that the thermodynamic functions U, H, S, G and A of a given system vary with changes in the environmental functions T, p and V (assuming that the composition of the system remains constant). It also became clear that all thermodynamics properties are interrelated, and one can therefore describe any property as a function of two others.

A rigorous treatment of this issue reveals that there exist relationships between properties that are not apparently related. These relationships have been derived from basic equations of state by James Clerk Maxwell in 1870 and are thus termed the Maxwell relations. They are of tremendous practical importance when determining experimental values of quantities that are difficult to measure, such as entropy.

2.2 Free Energy

2.2.1 The Gibbs Function

In the previous Sect. (2.1.13), we have introduced the Gibbs free energy, by means of its change during a process. Named after Josiah Willard Gibbs (1839–1903), who introduced this function in 1875, the free energy acknowledges that not all energy turned around during a process is intricately linked to molecular and atomic interactions (and thus 'free' to be harvested); there is a component of the entire energy change that is required to be dissipated into the surroundings.

When we characterised the energy of a system, we found it feasible to distinguish between two types of processes, those that occur at constant volume, and those occur under constant pressure. The former situation is best described by the internal energy U, whereas the latter, more frequently occurring situation is best described by the enthalpy H. In both cases, we also had to consider that the temperature during the ongoing process remained constant.

Concomitantly, when establishing the free energy, the same pre-requisites will need to be applied. The Gibbs free energy is useful for characterising process that occur under constant pressure and temperature, as it is intimately tied to the enthalpy change dH during the process.

For the change of the Gibbs free energy during a process we had obtained:

$$dG = dH - T \cdot dS \text{ or, for macroscopic processes}: \quad \Delta G = \Delta H - T \cdot \Delta S. \quad (2.42)$$

The definition of the state function G in terms of the state functions enthalpy and entropy is thus:

$$G = H - T \cdot S \qquad (2.44)$$

Differentiation of the Gibbs state function
Differentiation of Eq. 2.36 yields:

$$dG = dH - d(T \cdot S) \qquad (2.45)$$

For the second term on the right hand side of the equation, the product rule (see Appendix A.2.3) needs to be applied:

$$dG = dH - T \cdot dS - S \cdot dT$$

If we consider processes at constant pressure and temperature:

(continued)

$$p = \text{const.} \quad \Rightarrow \quad dp = 0$$
$$T = \text{const.} \quad \Rightarrow \quad dT = 0$$
$$dG = dH - T \cdot dS - S \cdot 0$$

which yields the relationship we had established earlier:

$$dG = dH - T \cdot dS \tag{2.42}$$

and we emphasise that this is true only for $T, p = \text{const.}$, since we made this assumption during the above calculation.

2.2.2 Gibbs Free Energy and the Entropy of the Universe

Any process that is accompanied by a change in entropy ultimately also changes the entropy in the universe. The entropy change in the universe with respect to a process comprises two components, the entropy change in the system and the entropy change in the surroundings:

$$dS_{\text{universe}} = dS_{\text{sys}} + dS_{\text{surr}} \tag{2.46}$$

We derived in Sect. 2.1.10 that

$$dH_{\text{sys}} = -T \cdot dS_{\text{surr}} \tag{2.31}$$

and therefore obtain from the above equation:

$$dS_{\text{universe}} = dS_{\text{sys}} - \frac{dH_{\text{sys}}}{T} \tag{2.47}$$

Using the relationship that links the change in the Gibbs free energy with the change in entropy (Eq. 2.42) one obtains

$$dG_{\text{sys}} = dH_{\text{sys}} - T \cdot dS_{\text{sys}} \quad \Rightarrow \quad \frac{dG_{\text{sys}}}{T} = \frac{dH_{\text{sys}}}{T} - dS_{\text{sys}} \tag{2.48}$$

This can be substituted into Eq. 2.47 which yields:

$$dS_{\text{universe}} = -\frac{dG_{\text{sys}}}{T} \tag{2.49}$$

The above equation delivers a fundamental paradigm. From the 2nd law of thermodynamics, we know that any process that happens spontaneously needs to increase the entropy in the universe, i.e.

$$dS_{universe} > 0 \quad \Rightarrow \quad dG_{sys} < 0 \text{ for spontaneous processes.} \tag{2.50}$$

Therefore, the Gibbs free energy change of a spontaneous process *has* to be negative.

For a reversible process, we have discussed earlier that an equilibrium can be considered as a state of the system where there is cross-over from the forward to the reverse reaction, both of which occur spontaneously. Reversible processes therefore occur while the system remains at equilibrium at all times. For such processes, we know from Eq. 2.33 that $dS = 0$; therefore:

$$dS_{universe} = 0 \quad \Rightarrow \quad dG_{sys} = 0 \text{ for processes at equilibrium.} \tag{2.51}$$

2.2.3 The Helmholtz Free Energy

For processes that occur at constant volume, the internal energy U is used to describe the energy and energy changes of a system. One can then define a free energy that linked to internal energy changes, and thus useful for characterising process that occur under constant volume and temperature. This free energy is named after Hermann von Helmholtz (1821–1894), who introduced this function in 1882 independently of Gibbs. This function is thus called the Helmholtz free energy A. The state function A is defined as

$$A = U - T \cdot S \tag{2.52}$$

Differentiation of the Helmholtz free energy function
The differential of the state function A is obtained as:

$$dA = dU - d(T \cdot S)$$

For the second term on the right hand side of the equation, the product rule (see Appendix A.2.3) needs to be applied:

$$dA = dU - T \cdot dS - S \cdot dT$$

If we consider processes at constant volume and temperature:

$$\begin{aligned} V &= \text{const.} \quad \Rightarrow \quad dV = 0 \\ T &= \text{const.} \quad \Rightarrow \quad dT = 0 \\ dA &= dU - T \cdot dS - S \cdot 0 \\ dA &= dU - T \cdot dS; T, V = \text{const.} \end{aligned} \tag{2.53}$$

The requirement for $V = \text{const.}$ is a condition of using the internal energy U which describes the energy of systems at conditions of constant volume.

2.2.4 Free Energy Available to Do Useful Work

For practical applications, it is useful to obtain an expression that describes the free energy under conditions of thermodynamic equilibrium, i.e. reversible processes.

Gibbs Free Energy
From the definition of entropy we know that the heat exchange at constant temperature is

$$dQ = T \cdot dS. \tag{2.54}$$

Using the definition of the Gibbs free energy (Eq. 2.42), we therefore obtain:

$$dG = dH - T \cdot dS = dH - dQ$$

The value obtained for dG in above equation describes the amount of energy that originates from the process and could be used to perform 'useful' work, i.e. work that is not compression or expansion. We have already established that the maximum work is done when a process is carried out reversibly (Sect. 2.1.6); this can easily be understood by recalling that in reversible processes, there is no change of entropy:

$$dS_{rev} = 0 \quad \Rightarrow \quad dQ_{rev} = 0 \quad \Rightarrow \quad dG_{rev} = dH_{rev}$$

Since $dS = 0$, there is no further term subtracted from the enthalpy change dH. This shows that the change of the Gibbs free energy of a reversible process equals the enthalpy change, which represents the maximum additional non-expansion work done by a system during this process:

$$dG_{rev} = dH_{rev} = dW_{max} \tag{2.55}$$

Helmholtz Free Energy
In the previous section, we derived the differential of the Helmholtz free energy as

$$dA = dU - T \cdot dS \tag{2.53}$$

in which we can substitute dQ for $T \cdot dS$, and therefore obtain

$$dA = dU - T \cdot dS = dU - dQ$$

As we discussed above, in reversible processes, there is no change in entropy ($dS = 0$), and therefore there is no heat exchange ($dQ = 0$). Therefore, the Helmholtz free energy is maximised in reversible processes and equals the change in the internal energy for this process:

$$dA_{rev} = dU_{rev} = dW_{max} \tag{2.56}$$

2.2.5 Gibbs Free Energy Change for a Reaction (Part 1)

So far, we have mainly been considering unspecified 'processes' which comprise physical as well as chemical transformations. We now want to consider a chemical reaction with the stoichiometry coefficients ν_A, ν_B, ν_C and ν_D, and the physical states indicated by the subscripts (l) for liquid and (g) for gaseous:

$$\nu_a A_{(l)} + \nu_b B_{(g)} \rightarrow \nu_c C_{(l)} + \nu_d D_{(g)}$$

for which we can obtain the macroscopically measurable change in the Gibbs free energy in a generic expression as:

$$\Delta G = G_{\text{products}} - G_{\text{reactants}} \tag{2.57}$$

This expression does not take into account that the reaction may be conducted with varying concentrations of the reactants A and B. Once the concentrations are considered, the following expression is obtained:

$$\Delta G = \Delta G^{\varnothing} + R \cdot T \cdot \ln Q$$

$$\text{with } Q = \frac{\prod\limits_{i=1}^{N_{\text{products}}} \left(\frac{c_i}{c^{\varnothing}}\right)^{\nu_i}}{\prod\limits_{j=1}^{N_{\text{reactants}}} \left(\frac{c_j}{c^{\varnothing}}\right)^{\nu_j}} = \frac{\left(\frac{c(C)}{c^{\varnothing}}\right)^{\nu_c} \cdot \left(\frac{c(D)}{c^{\varnothing}}\right)^{\nu_D}}{\left(\frac{c(A)}{c^{\varnothing}}\right)^{\nu_A} \left(\frac{c(B)}{c^{\varnothing}}\right)^{\nu_B}} = \frac{[C]^{\nu_c} \cdot [D]^{\nu_D}}{[A]^{\nu_A} \cdot [B]^{\nu_B}} \tag{2.58}$$

Q is called the reaction coefficient. We will derive this equation later using the chemical potential (Sect. 3.2.4). For the moment, we just consider the fact that the change of the Gibbs free energy during a reaction is dependent on the concentrations of the individual reactants. Equation 2.58 also introduces a reference value for the Gibbs free energy of the reaction, $\Delta G^{\varnothing}$, and the standard molar concentration $c^{\varnothing}$. The symbol 'Ø' (stroked letter O) is used to denote a property or quantity under standard conditions; $c^{\varnothing} = 1$ mol l^{-1}.

If all reactants are present at a concentration of 1 mol l^{-1} each, then the reaction coefficient Q equals 1. The term $R \cdot T \cdot \ln Q = R \cdot T \cdot \ln 1$ then becomes zero, as all logarithmic functions are zero when their argument is 1 (see Fig. 2.4). In this case, the Gibbs free energy change for the reaction, ΔG, equals the reference value $\Delta G^{\varnothing}$; it is therefore called the standard Gibbs free energy change for this reaction.

If the reaction has reached equilibrium, all reactants exist with particular concentrations that characterise that state of equilibrium, and there is no net change:

$$\nu_A A_{(l)} + \nu_B B_{(g)} \rightleftharpoons \nu_C C_{(l)} + \nu_D D_{(g)}$$

We derived earlier in Sect. 2.2.2 that there is no change in the Gibbs free energy at equilibrium, $\Delta G = 0$ (Eq. 2.51). The relationship that introduced the reaction coefficient Q (Eq. 2.58) therefore yields for the equilibrium state:

Fig. 2.4 All logarithmic
functions assume the value of
zero when their argument is 1

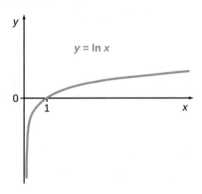

$$\Delta G_{eq} = \Delta G^{\emptyset} + R \cdot T \cdot \ln Q_{eq} = 0$$

and thus

$$\ln Q_{eq} = \ln K = -\frac{\Delta G^{\emptyset}}{R \cdot T}, \text{ as well as } K = e^{-\frac{\Delta G^{\emptyset}}{R \cdot T}} \tag{2.59}$$

In the state of equilibrium, the reaction quotient $Q = Q_{eq}$ is termed the equilibrium constant K. From the above relationship it is obvious that knowledge of the value of the equilibrium constant allows calculation of the standard free energy $\Delta G^{\emptyset}$ of a reaction. Vice versa, if the value of the standard free energy of a reaction is known, the equilibrium constant of the reaction at any temperature can be calculated.

2.2.6   Temperature Dependence of the Equilibrium Constant

If we are interested in the temperature dependence of the equilibrium constant, we can use Eq. 2.59 to evaluate the variation of K with temperature. This variation is expressed as the differential of $\ln K$ with T:

$$\left(\frac{\delta \ln K}{\delta T}\right)_p$$

We denote the differential with 'δ' to indicate that K is dependent on several parameters (T, p and V). Using Eq. 2.59 one then obtains:

$$\left(\frac{\delta \ln K}{\delta T}\right)_p = -\frac{1}{R} \cdot \left(\frac{\delta \frac{\Delta G^{\emptyset}}{T}}{\delta T}\right)_p = -\frac{\Delta G^{\emptyset}}{R} \cdot \left(\frac{\delta \frac{1}{T}}{\delta T}\right)_p$$

where the gas constant R can be excluded from the differential as it is a constant. The same is true for $\Delta G^{\emptyset}$ which is the standard free energy of the reaction and thus a characteristic constant for the process. The temperature differential on the right hand side of the above equation can easily be resolved by setting $x = T$ and $f(x) = \frac{1}{x}$. The derivative of $f(x)$ is $f'(x) = -\frac{1}{x^2}$. Therefore, $\left(\frac{\delta\frac{1}{T}}{\delta T}\right)$ resolves to $-\frac{1}{T^2}$:

$$\left(\frac{\delta \ln K}{\delta T}\right)_p = -\frac{\Delta G^\emptyset}{R \cdot T^2}$$

With knowledge of the following relationship between enthalpy and free energy

$$H = G - T \cdot \left(\frac{\delta G}{\delta T}\right)_p$$

we obtain the van't Hoff equation for an isobaric process:

$$\left(\frac{\delta \ln K}{\delta T}\right)_p = \frac{\Delta H^\emptyset}{R \cdot T^2} \tag{2.60}$$

which describes the reaction isobar, i.e. the variation of the equilibrium constant with temperature under these conditions.

For a reaction that occurs under constant volume and temperature—an isochoric process—the van't Hoff equation is analogous to Eq. 2.60, with the enthalpy being replaced by the internal energy:

$$\left(\frac{\delta \ln K}{\delta T}\right)_V = \frac{\Delta U^\emptyset}{R \cdot T^2} \tag{2.61}$$

Finally, a reaction that occurs under constant temperature, the van Laar-Planck reaction isotherm delivers the variation of the equilibrium constant with pressure:

$$\left(\frac{\delta \ln K}{\delta p}\right)_T = -\frac{\Delta V^\emptyset}{R \cdot T} \tag{2.62}$$

The three Eqs. 2.60–2.62 describe numerically what is known as the principle of Le Châtelier (see also Sect. 6.1.2):

▶   A system at equilibrium, when subject to a disturbance, responds in a way that tends to minimise the effect of the disturbance.

2.2.7   Pressure Dependence of the Gibbs Free Energy

We are now interested in the pressure dependence of the Gibbs free energy. To evaluate the variation of the Gibbs function with pressure, we attempt to find a direct relationship between G and p, using the set of thermodynamic functions and relationships we have encountered so far.

Starting with the definition of the Gibbs free energy, we recall that

$$G = H - T \cdot S \tag{2.44}$$

We also remember that

$$H = U + p \cdot V \tag{2.13}$$

and therefore

$$G = U + p \cdot V - T \cdot S$$

Since we are interested in a variation of G with p, we differentiate the above equation, and obtain

$$dG = dU + d(p \cdot V) - d(T \cdot S)$$

The differentials of the products, $d(p \cdot V)$ and $d(T \cdot S)$, need to be resolved with the product rule (Appendix A.2.3), and yield $p \cdot dV + V \cdot dp$ and $T \cdot dS + S \cdot dT$, respectively. Therefore:

$$dG = dU + p \cdot dV + V \cdot dp - T \cdot dS - S \cdot dT \tag{2.63}$$

From the 1st law of thermodynamics we know that

$$dU = dQ + dW \tag{2.15}$$

The heat exchange dQ can be expressed in terms of the entropy change $dQ = T \cdot dS$, and the work done by the system at constant pressure is the volume work $dW = p \cdot dV$. Equation 2.15 then becomes

$$dU = T \cdot dS - p \cdot dV$$

which we can use to substitute in Eq. 2.63:

$$dG = T \cdot dS - p \cdot dV + p \cdot dV + V \cdot dp - T \cdot dS - S \cdot dT$$

This simplifies to:

$$dG = V \cdot dp - S \cdot dT \tag{2.64}$$

At constant temperature, $dT = 0$ which renders the expression for the Gibbs free energy change as

$$dG = V \cdot dp \tag{2.65}$$

For two distinct pressure values, $p_{initial}$ and p_{final}, the Gibbs free energy change then yields

$$\Delta G = G(p_{\text{final}}) - G(p_{\text{initial}}) = \int\limits_{p_{\text{initial}}}^{p_{\text{final}}} dG = \int\limits_{p_{\text{initial}}}^{p_{\text{final}}} V \cdot dp$$

Importantly, the volume is itself dependent on the pressure and thus cannot be isolated from the integral! Instead, we will attempt to find an expression for the volume of the system in terms of the pressure. For an ideal gas this can conveniently be achieved by using the ideal gas equation

$$p \cdot V = n \cdot R \cdot T \quad \Rightarrow \quad V = \frac{n \cdot R \cdot T}{p} \tag{2.6}$$

which yields for the Gibbs free energy change:

$$\Delta G = \int\limits_{p_{\text{initial}}}^{p_{\text{final}}} \frac{n \cdot R \cdot T}{p} \cdot dp$$

Since n, R and T are not dependent on the pressure, they can be isolated from the integral, yielding:

$$\Delta G = n \cdot R \cdot T \cdot \int\limits_{p_{\text{initial}}}^{p_{\text{final}}} \frac{dp}{p}$$

The integral is of the type $\int \frac{1}{x} dx$ and resolves to $\ln x$ (see Table A.2). One thus obtains:

$$\Delta G = n \cdot R \cdot T \cdot \left[\ln p\right]_{p_{\text{initial}}}^{p_{\text{final}}}$$
$$\Delta G = n \cdot R \cdot T \cdot \left(\ln p_{\text{final}} - \ln p_{\text{initial}}\right)$$
$$\Delta G = n \cdot R \cdot T \cdot \ln \frac{p_{\text{final}}}{p_{\text{initial}}}$$

The standard state of a gas is defined as the gas at a pressure of $p^{\varnothing} = 1$ bar. If we define any change of the system with respect to standard conditions, the initial pressure becomes the standard pressure: $p_{\text{initial}} = p^{\varnothing}$. Above equation then reads:

$$\Delta G = G(p_{\text{final}}) - G(p^{\varnothing}) = n \cdot R \cdot T \cdot \ln \frac{p_{\text{final}}}{p^{\varnothing}}$$

and resolves to the general form of

$$G(p) = G^{\emptyset} + n \cdot R \cdot T \cdot \ln \frac{p}{p^{\emptyset}} \qquad (2.66)$$

with the standard Gibbs free energy $G^{\emptyset} = G(1 \text{ bar})$.

Equation 2.66 allows us to calculate the Gibbs free energy of an ideal gas at any given pressure, provided that we have a value for the standard Gibbs free energy of that gas.

2.3   Properties of Real Systems

In the previous sections, we have focused on single component systems whose composition does not change. This is clearly a limitation, since many real systems (as opposed to ideal systems) consist of multiple components and we thus need to consider varying compositions.

When considering the free energy of real systems it proves useful to introduce three new properties:

- chemical potential
- fugacity
- activity

2.3.1   Chemical Potential

In the previous section, we derived a relationship that enables calculation of the free energy of an ideal gas at any given pressure:

$$G(p) = G^{\emptyset} + n \cdot R \cdot T \cdot \ln \frac{p}{p^{\emptyset}} \qquad (2.66)$$

If we consider 1 mol of the ideal gas, this becomes:

$$G(p) = G^{\emptyset} + 1\text{mol} \cdot R \cdot T \cdot \ln \frac{p}{p^{\emptyset}}$$

$$\frac{G(p)}{1\text{mol}} = \frac{G^{\emptyset}}{1\text{mol}} + R \cdot T \cdot \ln \frac{p}{p^{\emptyset}}$$

The reference to the molar amount of 1 mol is indicated by the subscript 'm':

$$G_{\text{m}}(p) = G_{\text{m}}^{\emptyset} + R \cdot T \cdot \ln \frac{p}{p^{\emptyset}} \qquad (2.67)$$

thereby introducing G_{m} as the molar Gibbs free energy and thus the chemical potential μ:

$$G_m(p) = \mu(p) = \mu^{\varnothing} + R \cdot T \cdot \ln\frac{p}{p^{\varnothing}} \tag{2.67}$$

The term 'potential' is chosen based on the idea that for mechanical systems, the lowest potential energy is the most stable state. Similarly, the state of a system with the lowest chemical potential constitutes the most stable state, since for all spontaneous processes

$$\Delta G < 0 \implies \Delta G_m < 0 \implies \mu < 0$$

From the point of view of varying compositions, the change of the chemical potential μ equals the change of the free energy of the system, if a molar amount of $n = 1$ mol more substance is added.

The chemical potential μ is thus the molar free energy of a substance, and therefore an intensive property; its value is independent of the amount, since it has been normalised against 1 mol substance.

2.3.2 Open Systems and Changes of Composition

If we are dealing with chemical and biological systems comprising of multiple components, the composition will likely change. Sometimes, substances may even leave the system under observation; such systems are called open systems. We thus need to consider changes in the molar amount of substance in the system.

So far, we have assumed that the state of a system can be expressed as a function of two variables, say T and p. However, if the composition changes, the state also depends on the molar amount of each substance:

$$G = G(p, T, n_1, n_2, n_i, \ldots, n_N)$$

To evaluate the change of the free energy of the system, we therefore need to consider the change in all properties, including the amount of each component. This can be expressed as the differential of the Gibbs free energy with respect to each of the variable parameters (p, T, n_i):

$$dG = \left(\frac{\delta G}{\delta p}\right)_{T,n_i} dp + \left(\frac{\delta G}{\delta T}\right)_{p,n_i} dT + \sum_{i=1}^{N}\left(\frac{\delta G}{\delta n_i}\right)_{p,T,n_{j\neq i}} dn_i \tag{2.68}$$

Form earlier considerations in Sect. 2.2.7, we know how the Gibbs free energy varies with pressure and temperature:

$$dG(p, T) = V \cdot dp - S \cdot dT \tag{2.64}$$

And we also know that the chemical potential m is the change of the free energy of a substance when changing its molar amount ('adding 1 mol'):

$$G = n \cdot G_{\mathrm{m}} = n \cdot \left(\frac{\delta G}{\delta n} \right) = n \cdot \mu \qquad (2.67)$$

If we substitute Eqs. 2.64 and 2.67 into Eq. 2.68, we obtain the fundamental equation of thermodynamics:

$$dG = V \cdot dp - S \cdot dT + \sum_{i=1}^{N} \mu_i \cdot dn_i \qquad (2.69)$$

In chemical and biological systems (e.g. electrochemical or biological cells), T and p often remain constant, but the composition changes. For example, the intra- and extracellular metal ion concentration in cells is markedly different (Table 2.6):

Muscle cells expend energy to transport calcium ions to the outside of the cells. The calcium ions that flow into the muscle cells promote the cross-bridging between two fibre proteins (actin and myosin) which ultimately causes contraction. It is thus the change of chemical composition that drives this biological reaction.

2.3.3   Fugacity: A Pressure Substitute for Non-ideal Gases

When we derived the chemical potential in Sect. 2.3.1, we have been considering a system that contained an ideal gas:

$$\mu(p) = \mu^{\varnothing} + R \cdot T \cdot \ln \frac{p}{p^{\varnothing}} \qquad (2.67)$$

Real gases, i.e. such that behave in a non-ideal fashion, don't fulfil some or all of the criteria we have requested for an ideal gas. The ideal gas equation poses that the product $(p \cdot V)$ remains constant at constant temperature; the value of $(p \cdot V)$ should thus be independent of the pressure. However, experimental observation for real gases shows that this is not the case. To account for these deviations, The ideal gas equation may be extended by additional terms containing virial coefficients (B, C, D, ...):

$$p \cdot V = n \cdot R \cdot T + n \cdot B \cdot p + n \cdot C \cdot p^2 + n \cdot D \cdot p^3 + \ldots \qquad (2.70)$$

These deviations include the observation that the volume of a gas may change under certain circumstances, even if pressure and temperature remain constant, e.g. when the gas condenses to a liquid. Since the existence of a condensed phase

Table 2.6 Extra- and intra-cellular concentrations of biologically important metals

Extra-cellular	$c(Na^+) = 140$ mM	$c(K^+) = 4$ mM	$c(Ca^{2+}) = 9$ mM
Intra-cellular	$c(Na^+) = 12$ mM	$c(K^+) = 139$ mM	$c(Ca^{2+}) < 0.2$ mM

Table 2.7 Fugacities and fugacity coefficients for N_2 at $T^{\varnothing} = 273.15$ K

p (bar)	ϕ	f (bar)
50	0.98	49
100	0.97	97
200	0.97	194
400	1.06	424
600	1.22	732
800	1.47	1176
1000	1.81	1810

is not possible without intermolecular interactions, it becomes clear that a central requirement for an ideal gas is being violated.

In order to adjust the chemical potential such that one can also accommodate non-ideal behaviour, a new property called fugacity f is defined. The fugacity constitutes an adjusted pressure and is related to the pressure by a scalar factor called the fugacity coefficient ϕ:

$$p = \phi \cdot f \tag{2.71}$$

Fugacity is thus measured in units of pressure: $[f] = 1$ Pa.

The fugacity of a gas expresses its tendency to escape (being fugitive). This is intimately connected with how compressible the real gas is, compared to the ideal gas.

It is common to use an implicit definition of the fugacity: it is the property that replaces the pressure p in the definition of the chemical potential, thus ensuring that the following equation remains valid even in the case of non-ideal behaviour:

$$\mu(p) = \mu^{\varnothing} + R \cdot T \cdot \ln \frac{f}{p^{\varnothing}} \tag{2.72}$$

The difference between pressure and fugacity, and thus the extend of deviation from ideal gas behaviour, is illustrated in Table 2.7 for nitrogen at various pressures.

2.3.4   Solutions

When we consider solutions or mixtures, characteristic properties for the system under observation are the concentrations of individual components in the system. In addition to the well-known molar concentration that defines the molar amount of a substance in a particular volume of solution

$$c_{\text{solute}} = \frac{n_{\text{solute}}}{V_{\text{solution}}}, \text{ with } [c] = 1 \text{ mol } l^{-1} \tag{2.73}$$

it may be more convenient to use further measures of concentration, depending on the system under study (for a summary see Sect. 4.3.2). Here, we want to focus in particular on

- the mole fraction: x_i
- the partial molar volume: v_i

Mole Fraction
The mole fraction is defined as the molar amount of an individual component divided by the total molar amount in a system:

$$x_i = \frac{n_i}{n_1 + n_2 + n_3 + \ldots + n_N} = \frac{n_i}{\sum\limits_{j=1}^{N} n_j} \tag{2.74}$$

In the above notation, the index i denotes an individual compound from the set $(1, 2, 3, \ldots, N)$. The running index on the sum in the denominator of the quotient thus needs to be a different character (here j) to indicate that it is independent of i.

For gas mixtures, the mole fraction can be calculated based on the partial pressure of the individual components:

$$x_i = \frac{p_i}{p_1 + p_2 + p_3 + \ldots + p_N} = \frac{p_i}{\sum\limits_{j=1}^{N} p_j} = \frac{p_i}{p} \tag{2.75}$$

The mole fraction is a frequently used measure of concentration in gas, liquid and solid phase systems.

Partial Molar Volume
The partial molar volume of a particular substance is defined as the change in volume of a mixture, when 1 mol of the particular substance is added. One can thus express the partial molar volume as the change of the volume of the entire system when changing the molar amount of the particular substance in the system:

$$v_i = \left(\frac{\delta V}{\delta n_i} \right)_{p,T,n_{j \neq i}} \tag{2.76}$$

In the above expression, the index i denotes an individual substance, and the subscripts to the right of the differential inform us that not only the pressure and temperature need to remain constant during the addition of further amounts of the i-th substance, but also the amounts of all other substances in the system (denoted by the index j which has to be different from i) need to remain constant.

The partial molar volume is a function of the composition of the mixture to which it is added, i.e. the partial molar volume is not a constant for a given substance!

Notably, gases and liquids behave differently, due to the existence of inter-molecular interactions: If two different ideal gases are combined, the total volume is the sum of the individual volumes of each gas. However, if two different liquids are combined, the total volume is not the sum of each, due to the interactions between the liquids.

Example

If 1 mol of water is added to a very large volume of water, the change in volume is 18 ml.

However, if 1 mol of water is added to a very large volume of ethanol (so large that each water molecule is surrounded by ethanol molecules) then the increase is just 14 ml.

The partial molar volume of water in pure water is $v(H_2O, H_2O) = 18$ ml mol^{-1}, the partial molar volume of water in pure ethanol is $v(H_2O, EtOH) = 14$ ml mol^{-1}.

Once determined, the partial molar volume can be used to determine the change in volume of system composed of two components (A and B), when a known amount of A is added to a solution:

$$dV_{system} = v_A(x_A, x_B) \cdot dn_A + v_B(x_A, x_B) \cdot dn_B$$

Integration of this equation yields the total volume of a system with known composition:

$$\int_0^V dV_{system} = \int_0^{n_A} v_A(x_A, x_B) \cdot dn + \int_0^{n_B} v_B(x_A, x_B) \cdot dn$$

All integrals in above equation are of the type $\int_{x_{initial}}^{x_{final}} dx = [x]_{x_{initial}}^{x_{final}} = x_{final} - x_{initial}$, so the equation resolves to:

$$V = n_A \cdot v_A(x_A, x_B) + n_B \cdot v_B(x_A, x_B)$$

2.3.5   The Gibbs-Duhem Equation

In the previous section, by means of describing concentrations, we have expressed the volume of a multi-component system as function of the molar amounts of the individual components.

In the same way, other properties such as e.g. the Gibbs free energy of a mixture can be expressed in terms of the molar composition:

$$G = n_A \cdot G_{mA} + n_B \cdot G_{mB} = n_A \cdot \mu_A + n_B \cdot \mu_B, \text{ or generally } G = \sum_{i=1}^{N} n_i \cdot \mu_i$$

$$(2.77)$$

The above expression allows calculation of the Gibbs free energy (extensive!) based on knowledge of the molar amounts of the individual components as well as their chemical potentials (intensive!).

The molar functions we have introduced so far include:

- molar Gibbs free energy = chemical potential (a partial molar function when it refers to an individual component)
- partial molar volume
- mole fraction (a partial molar function as it relates to an individual component)

▶   If the partial molar function (e.g. the chemical potential or the partial molar volume) of one component of a mixture changes, it must be balanced by the opposing change in the partial molar functions of the other.

For example, when adding an additional volume of liquid to a solution, the partial molar volume of that substance increases, while the partial molar volumina of all other components decrease:

$$\sum_{i=1}^{N} n_i \cdot dv_i = 0$$

Applying the same considerations to the chemical potential, one obtains the Gibbs-Duhem equation:

$$\sum_{i=1}^{N} n_i \cdot d\mu_i = 0. \tag{2.78}$$

2.4   Exercises

1. Calculate the work done per mole when an ideal gas is expanded reversibly by a factor of three at 120 °C.

2. A friend thinking of investing in a company that makes engines shows you the company's prospectus. Analysing the machine, you realise that it works by harnessing the expansion of a gas. The operating temperature is approx. constant at 120 °C, and the gas doubles in volume during the power extraction phase of operation. It is claimed that the machine produces 5.5 kJ mol^{-1} of work during expansion. What advice would you give your friend? Briefly explain the thermodynamic rationale.

3. Which of the following equations embodies the first law of thermodynamics:

   (a) $dS = \dfrac{dQ_{rev}}{T}$

   (b) $U = Q + W$

   (c) $\Delta S_{universe} > 0$

   (d) $S_{system} > 0$

   (e) $C_p = \left(\dfrac{\delta H}{\delta T}\right)_p$

4. If the pressure is constant, the system is in mechanical equilibrium with its surroundings and no work is done other than work due to expansion and compression, which of the following are true:

   (a) $\Delta H = \Delta Q$ and $\Delta W = -p \cdot \Delta V$

   (b) $\Delta U = \Delta Q$ and $\Delta W = -p \cdot \Delta V$

   (c) $\Delta U = \Delta Q - p \cdot \Delta V$ and $\Delta W = p \cdot \Delta V$

   (d) $\Delta H = \Delta Q - p \cdot \Delta V$ and $\Delta W = -p \cdot \Delta V$

   (e) $\Delta U = \Delta Q$ and $\Delta W = p \cdot \Delta V$

5. Which of the following equations define fugacity?

   (a) $\mu(p) = \mu^{\varnothing}(p) + R \cdot T \cdot \ln\dfrac{f}{p^{\varnothing}}$

   (b) $f = \dfrac{p}{p^{\varnothing}}$

   (c) $\mu_A = \mu_A^* + R \cdot T \cdot \ln a_A$

6. The chemical potential is defined as:

$$\text{(a)} \quad \mu = G$$

$$\text{(b)} \quad \mu = G_m$$

$$\text{(c)} \quad \mu = G_m^{\varnothing}$$

$$\text{(d)} \quad \mu = n \cdot G_m$$

$$\text{(e)} \quad \mu = G^{\varnothing} + n \cdot R \cdot T \cdot \ln \frac{p}{p^{\varnothing}}$$

$$\text{(f)} \quad \mu = G_m^{\varnothing} + R \cdot T \cdot \ln \frac{p}{p^{\varnothing}}$$

Mixtures and Phases

3

3.1 Why Do or Don't Things Mix

In every-day life, we frequently encounter systems that consist of more than one component and thus represent mixtures. An important fundamental question in this context is whether two different components will spontaneously mix.

3.1.1 Gases

Experience tells us that two gases contained in the same physical container will inter-mix. Since all spontaneous processes (at constant pressure) must be accompanied by a negative change of the Gibbs free energy (Eq. 2.50), we can assess the every-day experience by thermodynamic means.

The Gibbs free energy change for the mixing process can be calculated as per:

$$\Delta G_{\text{mix}} = G_{\text{final}} - G_{\text{initial}} \tag{3.1}$$

The Gibbs free energies for the initial state of the system is given by the sum of the chemical potentials of each of the components that are about to mix, A and B, multiplied by their respective molar amounts n_A and n_B. The chemical potential of a gas at a pressure p was introduced with Eq. 2.67, and $\mu^{\varnothing}$ indicates represents the chemical potential under standard conditions:

$$G_{\text{initial}} = n_A \cdot \mu_A\left(p_{\text{initial}}\right) + n_B \cdot \mu_B\left(p_{\text{initial}}\right)$$

$$= n_A \cdot \left(\mu_A^{\varnothing} + R \cdot T \cdot \ln\frac{p}{p^{\varnothing}}\right) + n_B \cdot \left(\mu_B^{\varnothing} + R \cdot T \cdot \ln\frac{p}{p^{\varnothing}}\right)$$

When the mixing process starts (i.e. in the initial state), we are releasing two gases A and B into a container with an atmosphere of pressure p. In the final state, when the

© Springer International Publishing AG, part of Springer Nature 2018
A. Hofmann, *Physical Chemistry Essentials*,
https://doi.org/10.1007/978-3-319-74167-3_3

inter-mixing of the gases is complete, we know that each component can be described by its partial pressure, p_A and p_B (see page 28). The chemical potential for each component can thus be calculated by using the partial pressures, and one thus obtains for the Gibbs free energy in the final state:

$$G_{final} = n_A \cdot \mu_A(p_A) + n_B \cdot \mu_B(p_B)$$

$$= n_A \cdot \left(\mu_A^\emptyset + R \cdot T \cdot \ln\frac{p_A}{p^\emptyset} \right) + n_B \cdot \left(\mu_B^\emptyset + R \cdot T \cdot \ln\frac{p_B}{p^\emptyset} \right)$$

Both expressions can now be used to substitute in Eq. 3.1:

$$\Delta G_{mix} = n_A \cdot \left(\mu_A^\emptyset + R \cdot T \cdot \ln\frac{p_A}{p^\emptyset} \right) - n_A \cdot \left(\mu_A^\emptyset + R \cdot T \cdot \ln\frac{p}{p^\emptyset} \right)$$

$$+ n_B \cdot \left(\mu_B^\emptyset + R \cdot T \cdot \ln\frac{p_B}{p^\emptyset} \right) - n_B \cdot \left(\mu_B^\emptyset + R \cdot T \cdot \ln\frac{p}{p^\emptyset} \right)$$

The molar amounts n_A and n_B, the chemical potentials in the standard state, as well as R and T are all constants and therefore separated from the unknown parameters p, p_A and p_B:

$$\Delta G_{mix} = n_A \cdot \mu_A^\emptyset - n_A \cdot \mu_A^\emptyset + n_A \cdot R \cdot T \cdot \left(\ln\frac{p_A}{p^\emptyset} - \ln\frac{p}{p^\emptyset} \right)$$

$$+ n_B \cdot \mu_B^\emptyset - n_B \cdot \mu_B^\emptyset + n_B \cdot R \cdot T \cdot \left(\ln\frac{p_B}{p^\emptyset} - \ln\frac{p}{p^\emptyset} \right)$$

We see that the above expression can be simplified since some terms cancel:

$$\Delta G_{mix} = n_A \cdot R \cdot T \cdot \left(\ln\frac{p_A}{p^\emptyset} - \ln\frac{p}{p^\emptyset} \right) + n_B \cdot R \cdot T \cdot \left(\ln\frac{p_B}{p^\emptyset} - \ln\frac{p}{p^\emptyset} \right)$$

With the logarithm rule of $\log a - \log b = \log\frac{a}{b}$ (see Appendix A.1.2) one obtains:

$$\Delta G_{mix} = n_A \cdot R \cdot T \cdot \ln\frac{p_A \cdot p^\emptyset}{p^\emptyset \cdot p} + n_B \cdot R \cdot T \cdot \ln\frac{p_B \cdot p^\emptyset}{p^\emptyset \cdot p}$$

$$\Delta G_{mix} = n_A \cdot R \cdot T \cdot \ln\frac{p_A}{p} + n_B \cdot R \cdot T \cdot \ln\frac{p_B}{p}$$

As we have introduced earlier (Eq. 2.75), the quotient of partial pressure and total pressure yields the mole fraction x, when dealing with gases. The above equation can thus be re-written in terms of mole fractions x_A and x_B:

$$\Delta G_{mix} = n_A \cdot R \cdot T \cdot \ln x_A + n_B \cdot R \cdot T \cdot \ln x_B$$

The molar amounts of A and B, n_A and n_B, are related to the mole fractions via the total molar amount of compounds in the system:

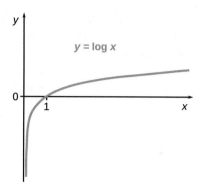

Fig. 3.1 Any logarithmic function results in negative values between 0 and 1

$$\frac{n_A}{n} = x_A \ \Rightarrow \ n_A = n \cdot x_A \ \text{and} \ \frac{n_B}{n} = x_B \ \Rightarrow \ n_B = n \cdot x_B$$

Therefore we arrive at:

$$\Delta G_{mix} = (n \cdot R \cdot T) \cdot x_A \cdot \ln x_A + (n \cdot R \cdot T) \cdot x_B \cdot \ln x_B < 0 \qquad (3.2)$$

Since x_A and x_B are positive numbers between 0 and 1, it becomes clear that $\ln x_A$ and $\ln x_B$ are negative numbers, since any logarithmic function is negative between 0 and 1 (Fig. 3.1). With $(n \cdot R \cdot T)$ yielding a positive value, we can then conclude that ΔG_{mix} from Eq. 3.2 will always deliver a negative value. Since all processes with $\Delta G < 0$ are spontaneous processes, the mixing of two gases is a spontaneously occurring process!

Considering that the Gibbs free energy change at constant pressure and constant temperature is composed of enthalpic and entropic changes as per:

$$\Delta G = \Delta H - T \cdot \Delta S \qquad (2.42)$$

we can conclude that the mixing two ideal gases is a purely entropic effect, since there is no enthalpy change ($\Delta H = 0$) when two ideal gases intermix. This follows from the requirement of the gas being ideal; in the ideal gas there are no intermolecular forces between the molecules.

In contrast, in real gases inter-molecular interactions do occur and there is therefore a change of enthalpy upon mixing ($\Delta H \neq 0$), due to intermolecular forces between molecules. It also needs to be considered that ΔH and ΔS depend on p and T. However, the Gibbs free energy change for the inter-mixing of real gases is still generally negative.

3.1.2 Liquids

A similar approach can be taken, when considering the mixing of two liquids. As in the previous section, we establish the Gibbs free energy of the system before mixing occurs. This can be expressed as the sum of the chemical potentials of the pure liquids (indicated by the asterisk '*'), multiplied with the respective molar amounts of each liquid in the system (since the chemical potential is the Gibbs free energy per 1 mol of substance):

$$G_{\text{initial}} = n_A \cdot \mu^*_{A(l)} + n_B \cdot \mu^*_{B(l)}$$

When applying Eq. 2.67 to liquid systems, we appreciate that the chemical potential of a liquid at any concentration can be expressed with reference to the chemical potential in its pure state ('*') by adjusting for the concentration of the liquid with the term $R \cdot T \cdot \ln x$, where x is the mole fraction of the liquid:

$$\mu = \mu^* + R \cdot T \cdot \ln x \tag{3.3}$$

The final state of the liquid mixture is attained when both liquids are present in their respective concentrations, expressed as mole fractions x_A and x_B. With Eq. 3.3 we thus obtain:

$$G_{\text{final}} = n_A \cdot \left(\mu^*_{A(l)} + R \cdot T \cdot \ln x_A\right) + n_B \cdot \left(\mu^*_{B(l)} + R \cdot T \cdot \ln x_B\right)$$

and can now proceed to calculate the change in the Gibbs free energy during the mixing process:

$$\Delta G_{\text{mix}} = G_{\text{final}} - G_{\text{initial}}$$
$$\Delta G_{\text{mix}} = n_A \cdot \mu^*_{A(l)} + n_A \cdot R \cdot T \cdot \ln x_A + n_B \cdot \mu^*_{B(l)} + n_B \cdot R \cdot T \cdot \ln x_B$$
$$- n_A \cdot \mu^*_{A(l)} - n_B \cdot \mu^*_{B(l)}$$

Since several terms cancel, this simplifies to:

$$\Delta G_{\text{mix}} = n_A \cdot R \cdot T \cdot \ln x_A + n_B \cdot R \cdot T \cdot \ln x_B$$

We then express the individual molar amounts n_A and n_B as mole fractions x_A and x_B (and thus with respect to the total molar amount n in the system):

$$\frac{n_A}{n} = x_A \Rightarrow n_A = n \cdot x_A \text{ and } \frac{n_B}{n} = x_B \Rightarrow n_B = n \cdot x_B$$

which yields for the Gibbs free energy change of the mixing process:

$$\Delta G_{\text{mix}} = n \cdot R \cdot T \cdot x_A \cdot \ln x_A + n \cdot R \cdot T \cdot x_B \cdot \ln x_B$$
$$\Delta G_{\text{mix}} = n \cdot R \cdot T \cdot (x_A \cdot \ln x_A + x_B \cdot \ln x_B) \tag{3.4}$$

As we have discussed in the previous section, due to the mole fractions x_A and x_B taking values between 0 and 1, Eq. 3.4 we always deliver a negative value for the Gibbs free energy change, and the mixing of two liquids should thus always occur spontaneously. However, as in the case of gases, this is only true for ideal liquids, i.e. such that do not possess any intermolecular forces between molecules. With

$$\Delta G = \Delta H - T \cdot \Delta S \qquad (2.42)$$

we can then conclude that in ideal solutions, at constant temperature and constant pressure, there is no enthalpy change ($\Delta H = 0$) and the mixing of ideally behaving liquids is thus a purely entropic effect: $\Delta G_{mix} = - T \cdot \Delta S$.

In contrast to gases, non-ideal effects are a lot more common in liquids, and therefore we frequently need to consider changes in ΔH due to intermolecular forces, which differ when comparing interactions between A–A, B–B and A–B. A positive change of the enthalpy during the mixing ($\Delta H > 0$) of two liquids is therefore possible and may eventually exceed the entropic effect. The Gibbs free energy of mixing of non-ideal liquids may thus be positive or negative; this will determine whether or not the liquids mix.

3.2 Liquids

3.2.1 Chemical Potential of Liquid Solutions

A closer look at liquids and liquid solutions shows that there is not only the liquid phase that needs to be considered, but also a vapour phase above the liquid. If we envisage a container in which there is a liquid in equilibrium with its vapour, we can describe the entire contents of the container as one system. Since we assume equilibrium conditions, the chemical potential must be uniform throughout the system, i.e. the chemical potential in the liquid phase is the same as in the vapour phase:

$$\mu_{(g)} = \mu_{(l)}$$

For an ideal gas A, we know that chemical potential at any pressure (p_A^*) can be described as per Eq. 2.67:

$$\mu_{A(g)}^* = \mu_A^{\varnothing} + R \cdot T \cdot \ln \frac{p_A^*}{p^{\varnothing}}$$

$$\mu_{A(g)}^* = \mu_A^{\varnothing} + R \cdot T \cdot \ln \frac{p_A^*}{p^{\varnothing}} = \mu_{A(l)}^* \qquad (3.5)$$

where the asterisks '*' indicate the pure component (i.e. here pure component A in either liquid or vapour form).

When two components A and B are mixed, they no longer exist in their pure forms, and we indicate this by omitting the asterisk '*'. However, for each component A and B in the liquid mixture, the chemical potential in the different phases can still be calculated as in Eq. 3.5; we thus obtain:

$$\mu_{A(l)} = \mu_{A(g)} = \mu_A^{\varnothing} + R \cdot T \cdot \ln\frac{p_A}{p^{\varnothing}} \tag{3.6}$$

$$\mu_{B(l)} = \mu_{B(g)} = \mu_B^{\varnothing} + R \cdot T \cdot \ln\frac{p_B}{p^{\varnothing}} \tag{3.7}$$

The pressures p_A and p_B now indicate the partial pressure of component A and B in the vapour above the liquid.

Inspection of Eqs. 3.5 and 3.6 shows that the chemical potential of component A is different when we compare the pure state with a mixture where A is not the exclusive component of the system. The difference in the chemical potential between both states is available by subtracting Eq. 3.5 from 3.6:

$$\mu_{A(l)} - \mu_{A(l)}^* = \mu_{A(g)} - \mu_{A(g)}^*$$

$$\mu_{A(l)} - \mu_{A(l)}^* = \mu_A^{\varnothing} + R \cdot T \cdot \ln\frac{p_A}{p^{\varnothing}} - \left(\mu_A^{\varnothing} + R \cdot T \cdot \ln\frac{p_A^*}{p^{\varnothing}}\right)$$

$$\mu_{A(l)} - \mu_{A(l)}^* = R \cdot T \cdot \ln\frac{p_A}{p^{\varnothing}} - R \cdot T \cdot \ln\frac{p_A^*}{p^{\varnothing}}$$

$$\mu_{A(l)} - \mu_{A(l)}^* = R \cdot T \cdot \left(\ln\frac{p_A}{p^{\varnothing}} - \ln\frac{p_A^*}{p^{\varnothing}}\right)$$

Using the log rule of $\log_b(uv) = \log_b u + \log_b v$ and $\log_b\left(\frac{u}{v}\right) = \log_b u - \log_b v$ (Appendix A.1.2), this yields:

$$\mu_{A(l)} - \mu_{A(l)}^* = R \cdot T \cdot \ln\frac{p_A \cdot p^{\varnothing}}{p^{\varnothing} \cdot p_A^*}$$

which simplifies to

$$\mu_{A(l)} = \mu_{A(l)}^* + R \cdot T \cdot \ln\frac{p_A}{p_A^*} \tag{3.8}$$

This equation delivers a relationship between the chemical potential of a particular component in a liquid mixture (μ_A) and its partial vapour pressure (p_A). It is found that p_A always takes smaller values than p_A^* (see next section), which results in $\frac{p_A}{p_A^*} < 1$ and thus $\ln\frac{p_A}{p_A^*} < 0$; the chemical potential μ of a solvent is therefore lowered as a result of the presence of the solute.

3.2.2 Raoult's Law

Eq. 3.8 can be transformed with some basic algebra to yield:

$$\frac{p_A}{p_A^*} = e^{\frac{\mu_{A(l)} - \mu_{A(l)}^*}{R \cdot T}} \tag{3.9}$$

The French chemist Francois-Marie Raoult (1830–1901) discovered experimentally that there is a relationship between the partial vapour pressure of component and its mole fraction in the liquid mixture, and summarised his results in 1886 in a rule known as Raoult's law:

▶   The partial vapour pressure of a component in a liquid mixture is equal to the vapour pressure of the pure component multiplied by its mole fraction in the liquid phase

$$p_A = x_A \cdot p_A^* \tag{3.10}$$

Equation 3.9 can thus be extended to read:

$$\frac{p_A}{p_A^*} = e^{\frac{\mu_{A(l)} - \mu_{A(l)}^*}{R \cdot T}} = x_A \tag{3.9}$$

Since the mole fraction x_A takes values less or up to 1 ($0 \leq x_A \leq 1$), an important effect of Raoult's law is that the vapour pressure of a solution is always lower than that of the pure solvent at any particular temperature. This effect is also known as vapour pressure depression.

Considering that the boiling point of a solution is the temperature at which the vapour pressure of the liquid is equal to the surrounding environmental pressure, another consequence of the above effects is the boiling point elevation observed with liquid solutions. A solution therefore always has a higher boiling point than the pure solvent. More generally, several properties of liquids (T_m, T_b, p, etc) depend on the presence of solute molecules, in particular the number of particles in solution. Such properties are called colligative properties and will be further discussed in Sect. 4.3.2.

3.2.3   Activity: Mole Fraction for a Non-ideal Solution

In Sect. 2.3.3, we have defined the fugacity f for a non-ideal gas to ensure that the chemical potential of the gas at any pressure can be calculated according to the equation

$$\mu(p) = \mu^{\varnothing} + R \cdot T \cdot \ln \frac{f}{p^{\varnothing}}. \tag{2.72}$$

In Sect. 3.2.1 of this chapter, we established a relationship between the chemical potential of a liquid component ($\mu_{A(l)}$) in a mixture and its partial vapour pressure (p_A) above the liquid mixture:

$$\mu_{A(l)} = \mu_{A(l)}^* + R \cdot T \cdot \ln\frac{p_A}{p_A^*} \tag{3.8}$$

For an ideal ideal solution, it then follows with Raoult's law that $\ln\frac{p_A}{p_A^*} = \ln x_A$, and therefore:

$$\mu_{A(l)} = \mu_{A(l)}^* + R \cdot T \cdot \ln x_A \tag{3.11}$$

Analogous to the fugacity in the case of gases, we can now define a property that will account for non-ideal behaviour in the liquid mixtures. This property is called activity (a) and replaces the mole fraction x_A in the above case, such that the chemical potential of compound A at any concentration in a non-ideal solution varies as per the relationship:

$$\mu_{A(l)} = \mu_{A(l)}^* + R \cdot T \cdot \ln a_A \tag{3.12}$$

The activity coefficient γ indicates the deviation from the ideal behaviour:

$$a_A = \gamma_A \cdot x_A. \tag{3.13}$$

3.2.4   Gibbs Free Energy Change for a Reaction (Part 2)

In Sect. 2.2.5, we discussed the Gibbs free energy change of the following reaction

$$\nu_A A_{(l)} + \nu_B B_{(g)} \rightarrow \nu_C C_{(l)} + \nu_D D_{(g)}$$

by means of the reaction quotient Q, before introducing the chemical potential.

After discussing mixtures of components by means of their chemical potentials in the preceding sections, we can now apply this knowledge and revise our discussion of the Gibbs free energy change for a reaction.

As before, the change in free energy is given by

$$\Delta G = G_{products} - G_{reactants} = (G_C + G_D) - (G_A + G_B)$$

This free energy change is an extensive property, as it depends on the actual amounts of compounds. We can introduced the corresponding intensive state functions which are the molar Gibbs free energies (G_m):

$$\Delta G = (\nu_C \cdot G_{mC} + \nu_D \cdot G_{mD}) - (\nu_A \cdot G_{mA} + \nu_B \cdot G_{mB})$$

The molar Gibbs free energies G_m are indeed the chemical potentials, so we can substitute and obtain:

$$\Delta G = \left(\nu_C \cdot \mu_C^* + \nu_C \cdot R \cdot T \cdot \ln x_C + \nu_D \cdot \mu_D^{\varnothing} + \nu_D \cdot R \cdot T \cdot \ln p_D\right)$$
$$- \left(\nu_A \cdot \mu_A^* + \nu_A \cdot R \cdot T \cdot \ln x_A + \nu_B \cdot \mu_B^{\varnothing} + \nu_B \cdot R \cdot T \cdot \ln p_B\right)$$

By combining the constant terms ($\nu_i \cdot \mu_i^{*,\varnothing}$ represent the chemical potentials of the components in their pure or standard states), we obtain:

$$\Delta G = \nu_C \cdot \mu_C^* + \nu_D \cdot \mu_D^{\varnothing} - \nu_A \cdot \mu_A^* - \nu_B \cdot \mu_B^{\varnothing}$$
$$+ R \cdot T \cdot (\nu_C \cdot \ln x_C + \nu_D \cdot \ln p_D - \nu_A \cdot \ln x_A - \nu_B \cdot \ln p_B)$$

The constant terms $\nu_i \cdot \mu_i^{*,\varnothing}$ represent a standard Gibbs free energy change for this reaction and are thus substituted by $\Delta G^{\varnothing}$. Using the logarithm rules of $a \cdot \log x = \log x^a$, $\log a + \log b = \log(a \cdot b)$ and $\log a - \log b = \log\frac{a}{b}$ (see A.1.2) one obtains:

$$\Delta G = \Delta G^{\varnothing} + R \cdot T \cdot \ln \frac{x_C^{\nu_C} \cdot p_D^{\nu_D}}{x_A^{\nu_A} \cdot p_B^{\nu_B}}$$

which is a relationship between the Gibbs free energy change for a reaction at varying concentrations of components, here expressed as mole fractions for the liquids A and C, and as partial pressures for the gases B and D. The argument of the logarithm is the reaction coefficient Q introduced earlier. Based on the above rigorous assessment of the chemical potentials, we can thus confirm the relationship introduced in Sect. 2.2.5:

$$\Delta G = \Delta G^{\varnothing} + R \cdot T \cdot \ln Q \tag{2.58}$$

We recall that when the reaction has reached equilibrium, there is no change the Gibbs free energy observed, therefore:

$$\Delta G = 0$$

Also, at equilibrium, the reaction coefficient Q becomes the equilibrium constant

$$Q_{eq} = K$$

One thus obtains from Eq. 2.58 for equilibrium conditions:

$$\Delta G = \Delta G^{\varnothing} + R \cdot T \cdot \ln K = 0$$

$$\ln K = -\frac{\Delta G^{\varnothing}}{R \cdot T} \quad \text{and} \quad K = e^{-\frac{\Delta G^{\varnothing}}{R \cdot T}} \tag{2.59}$$

This means that if we can determine the change in the Gibbs free energy of a process, we can calculate the equilibrium constant.

3.3 Phase Equilibria

3.3.1 Phase Diagrams and Physical Properties of Matter

Phase diagrams tell us the state of a system under various conditions. The states of a system at various pressures and temperatures can be depicted in a p-T diagram (see for example Fig. 3.2). In phase diagrams, lines separate regions where various phases are thermodynamically stable. These lines are called phase boundaries and show values of p and T where two phases coexist in equilibrium. When changing one of the two parameters (p or T), in order to maintain the state of equilibrium, the other parameter needs to follow suit. Phase boundaries are therefore univariant (one parameter can be freely chosen).

From the ideal gas equation, we know that

$$p \cdot V = n \cdot R \cdot T \tag{2.6}$$

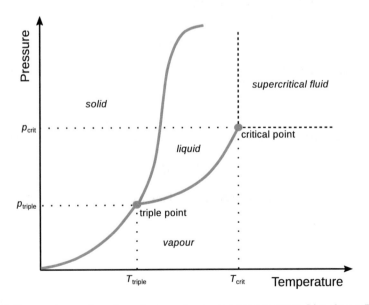

Fig. 3.2 Pressure-temperature phase diagram of a one-component system. Lines in a p-T phase diagrams have the same volume all along the way (isochores)

If we assume a system with constant molar amount n, then the above equation relates the three variables p, V and T. For a given state, the third variable is determined by choosing the two other—a characteristic of equations of state that we have introduced earlier. As a consequence, the lines in a p-T phase diagram need to describe conditions of constant volumes; these lines are thus called isochores.

There are several points of interest in a phase diagram, illustrated in the p-T diagram in Fig. 3.2.

Boiling Point

The boiling point is the temperature at which the vapour pressure of the liquid is equal to the external pressure. Note that the frequently used term 'boiling point' is strictly speaking not correct, as it is not a single point in the phase diagram. It is a temperature extrapolated from the isochore branch that separates liquid and vapour phases. If the external pressure is the normal pressure

$p_{normal} = 1$ atm $= 101.3$ kPa   then this is called the normal boiling point.

If the external pressure is the standard pressure

$p^{\varnothing} = 1$ bar $= 100.0$ kPa   then this is called the standard boiling point.

Melting Point

The melting point is the temperature at which the liquid and solid phases coexist; it equals the freezing temperature. As above, please note that the frequently terms 'boiling/freezing points' are strictly speaking not correct, as it is they are not single points in the phase diagram. It is a temperature extrapolated from the isochore branch that separates solid and liquid phases. If the external pressure is the normal pressure

$p_{normal} = 1$ atm
$\qquad = 101.3$ kPa   then this is called the normal freezing or melting point.

If the external pressure is the standard pressure

$p^{\varnothing} = 1$ bar
$\qquad = 100.0$ kPa   then this is called the standard freezing or melting point.

Critical Point

At the critical point, there is no physical interface between the liquid and the vapour; both phases coalesce and there is no liquid phase. At pressures and/or temperatures beyond the critical point, the system is said to be in the supercritical state, which is neither vapour nor liquid. Note that the isochore branch stops at the critical point, i.e. this isochore branch does not continue beyond the critical point. Since the critical point is defined by two discrete values on the p- and T-axes, it is indeed a point in the p-T diagram. The critical point of water occurs at:

$$T_{crit} = 647.10 \text{ K}, p_{crit} = 22.1 \text{ MPa}$$

Triple Point

At a single definite pressure and temperature, three phases can exist in equilibrium; this is called a triple point. There may be more than one triple points in a phase diagram. Since triple points are defined by two discrete values on the p- and T-axes, they are indeed points in the p-T diagram. One important triple point of water is observed at

$$T_{triple} = 273.16 \text{ K}, p_{triple} = 611 \text{ Pa}$$

Here, solid, liquid and gaseous water exist at the same time; this triple point has a general importance as it is used to define the thermodynamic temperature scale.

3.3.2   Phase Diagrams with Isotherms

Phase diagrams may also be constructed by plotting the pressure versus the volume of a system; a useful method especially when analysing gases. As discussed in the previous section, the third parameter, now the temperature, needs to be constant when characterising a system under varying pressures and volumes. The lines in the p-V diagram are thus called isotherms; the temperature is the same all the way along (see Fig. 3.3).

In Fig. 3.3 (left), the behaviour of a real gas is compared to that of an ideal gas. The real gas undergoes condensation. At a certain pressure, the volume collapses (B–C–D) and the gas turns into a liquid, but the system maintains the same pressure (Fig. 3.4). The pressure at B–C–D is called the vapour pressure of the liquid. For condensation to occur, the molecules must be close enough and slow enough to aggregate.

High temperatures imply high molecular velocities. Therefore, at sufficiently high temperatures (above T_{crit}), the individual gas molecules possess too high velocity in order to engage in intermolecular interactions and thus no condensation will occur, no matter how small the volume is made.

At the critical point, when the boundaries between liquid and gas phase vanish, the isotherm in the p-V diagram has zero slope (Fig. 3.3, right; red isotherm). Since the critical point is indeed a point in the phase behaviour of compounds, it possesses three discrete values for pressure, temperature and volume: p_{crit}, T_{crit} and V_{crit}. The critical point is a characteristic of a particular substance (Table 3.1).

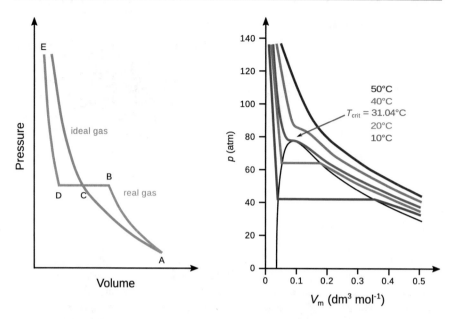

Fig. 3.3 Lines in phase diagrams that plot pressure versus volume have the same temperature all along the way (isotherms). Left: The condensation observed with real gases gives rise to a horizontal section in the p-V diagram. Right: Isotherms of CO_2 at different temperatures. At sufficiently high temperature, the condensation behaviour of real gases disappears

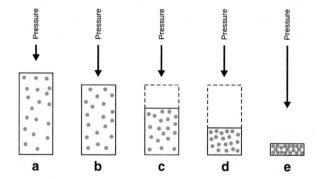

Fig. 3.4 In real gases, condensation occurs due to inter-molecular interactions, which gives rise to a volume decrease at constant external pressure (points B–C–D in Fig. 3.3). The lengths of the vectors indicate the magnitude of the pressure

Table 3.1 Critical constants for selected gases

Gas	p_{crit} (kPa)	V_{crit} (cm^3 mol^{-1})	θ_{crit} (°C)
Ar	4862	75.3	−123
CO_2	7385	94.0	31.0
He	228.9	57.8	−268
O_2	5075	78.0	−118

Fig. 3.5 Phase transitions

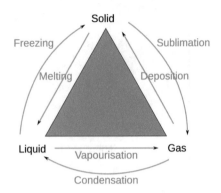

3.3.3 Phase Transitions

A form of matter that is uniform throughout in chemical composition and physical state of matter is called a phase. The fundamentally different phases are solid, liquid and gas; with water, for example, ice, liquid water and water steam. However, there may also be various solid phases, conducting and superconducting phases, super-fluid phases. A phase transition is the conversion of one phase into another. The possible phase transitions between the three states of matter—solid, liquid and gas—are illustrated in Fig. 3.5. Some compounds possess more than one liquid or solid phase; transitions are then also possible between those.

When two or more phases are in equilibrium, the chemical potential of a sub-stance is the same in each of the two phases and at all points in each of the phases. This is a consequence of the 2nd law of thermodynamics. For a given pressure, the temperature at which two phases are in equilibrium (and thus matter spontaneously transitions from one to the other) is called transition temperature.

3.3.4 The Gibbs Phase Rule

In Sect. 3.3.1, we introduced phase boundaries is univariant regions in the phase diagram: it is possible to freely choose one parameter in order to maintain this particular state of the system. The Gibbs phase rule allows calculation of the number of intensive parameters (i.e. independent of the amount of substance) that can be varied independently (F) while the number of phases (P) remains constant, given a system with a particular number of components (C) of a system:

$$F = C - P + 2 \tag{3.14}$$

For example, if two phases ($P = 2$) in a system consisting of one component ($C = 1$) are in equilibrium, then

$$F = 1 - 2 + 2 = 1$$

parameter (e.g. T) can be changed, but the other parameters (e.g. p) will follow suit. Systems with $F = 1$ are called univariant.

Gibbs phase rule examples
Water steam
  Water steam describes one phase ($P = 1$) in the one-component system water ($C = 1$). Therefore:

$$F = 1 - 1 + 2 = 2$$

  With two degrees of freedom, water steam constitutes a bivariant system. It corresponds to an area in an x–y plot (for example, in a p-T diagram). We can vary the temperature of the steam without having to change the pressure at the same time, but still maintain the gas phase.

  Liquid water in equilibrium with its vapour
  Here, we again deal with the one-component system water ($C = 1$), but now have to consider two phases, liquid and vapour, i.e. $P = 2$. Therefore:

$$F = 1 - 2 + 2 = 1$$

  With one degree of freedom, this constitutes a univariant system. It corresponds to a line in an x–y plot. The temperature can be varied, but the pressure needs to be varied accordingly to in order to preserve the equilibrium between liquid and vapour phase.

3.3.5 One Component Systems: Carbon Dioxide

The phase diagram of carbon dioxide is shown in Fig. 3.6, annotated with the three states of matter. as well as the triple and critical points. The degrees of freedom for the different areas in the phase diagram are also shown.

  A closer look at the numerical values of the p- and T-axes shows that solid CO_2 sublimes (i.e. transitions to the vapour phase). Since the solid and liquid phases of CO_2 do not co-exist at normal pressure, solid CO_2 is also called dry ice. The liquid phase does not exist below a pressure of 518 kPa. In other words, if liquid CO_2 is required, a pressure of at least 518 kPa must be applied.

  At a pressure of 101.3 kPa, which constitutes the normal atmospheric pressure, the transition temperature between solid and gas phase is 195 K ($-78\,°C$), so solid CO_2 evaporates under normal conditions.

  Inspection of the phase boundary between solid and liquid CO_2 shows that an increase in pressure results in an increase in the melting temperature; the slope of the solid–liquid isochore is positive. This is due to solid CO_2 being denser than the

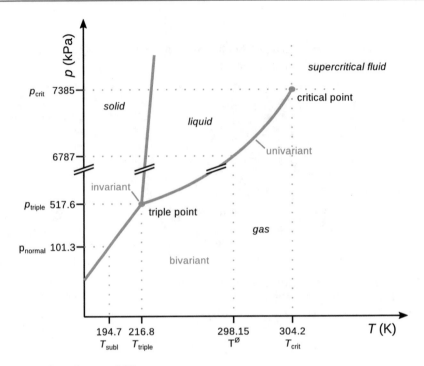

Fig. 3.6 Phase diagram of CO_2

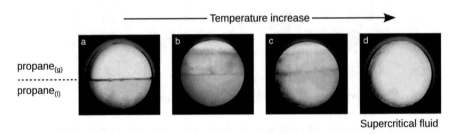

Fig. 3.7 Formation of supercritical propane. (**a**) Liquid and gas phases of propane are visible and the meniscus is easily observed. (**b**) With an increase in temperature the meniscus begins to diminish. (**c**) Increasing the temperature further causes the gas and liquid densities to become more similar. The meniscus is less easily observed but still evident. (**d**) Once the critical temperature and pressure have been reached, the two distinct phases of liquid and gas are no longer visible and the meniscus can no longer be observed. Images obtained from CM Rayner, AA Clifford and KD Bartle, University of Leeds, UK and reproduced with permission

liquid. An increase in pressure promotes the denser phase, so higher pressures stabilise the solid state. This behaviour is typical of most substances.

The phase changes when approaching the critical point and formation of super-critical propane is illustrated in Fig. 3.7. In the supercritical fluid, one homogenous phase is formed which shows properties of both liquids and gases.

3.3.6 One Component Systems: Water

The p-T phase diagram for water is shown in Fig. 3.8. The thick lines indicate the main phase boundaries between solid, liquid and vapour states. Especially at high pressures, water can form several different solid phases, the boundaries of which are shown by the thin lines. Therefore, the phase diagram contains several triple points, where three phases are in co-existence. The main triple point, where solid, liquid and vapour water are in equilibrium, is observed at $T = 273.16$ K and $p = 611$ Pa. This triple point has fundamental importance, as it is used to define the zero point on the Celsius temperature scale.

Of particular importance is the rather exceptional property of water arising from the steep slope of the solid–liquid phase boundary (the melting temperature curve): it not only has a very steep, but a negative slope. The negative slope indicates that the liquid phase has a higher density than the solid phase. Therefore, solid ice floats on liquid water (due to the lower density of the former; see also Sect. 3.4.6), and the possibility of ice skating also arises form this unusual behaviour. A skater on ice of about 70 kg exerts a pressure of about 7 MPa (assuming a contact area of 1 cm^2). At

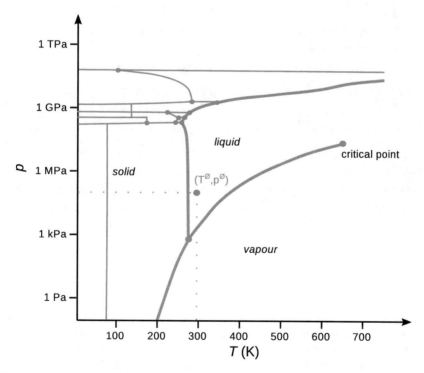

Fig. 3.8 Phase diagram of H$_2$O. Thick lines show the main phase boundaries; thin lines indicate phase boundaries of the various solid phases. The critical point is labelled; all other indicated points are triple points. The phase at normal conditions ($T_{normal} = 298.15$ K, $p_{normal} = 101.3$ kPa) is marked

that high pressure, the melting temperature of ice is no longer $0\,°C$ but $-1\,°C$. The high pressure leads to disruption of hydrogen bonds that hold the water molecules in the solid structure. The generated film of liquid water enables the smooth skating process.

Water: not one, but two liquids?
The list of anomalies of water is constantly increasing, and currently includes some 72 properties that distinguish water from conventional liquids (Chaplin 2014). Among the best known anomalies are the density maximum of water in its liquid state at $3.98\,°C$, as well as the so-called Mpemba effect which describes the phenomenon that hot water freezes faster than cold water (Mpemba and Osborne 1969).

The reason for such extra-ordinary behaviour lies in discontinuities in the heat capacity C_p (see also Sect. 3.4.9). Such a discontinuity gives rise to the critical point, where fluctuations happen at all length scales and thus light cannot penetrate the system—it becomes opaque. The critical point indicates the end of the boiling curve, and liquid and gas state are no longer distinguishable (supercritical fluid). In supercritical water, salts can no longer be dissolved, but mixture with apolar solvents is observed. However, the C_p discontinuity at the critical point cannot explain anomalies that occur at lower pressures and temperatures.

Based on simulations (Poole et al. 1992), it has thus been proposed that a so far non-identified equilibrium curve exists in water that separates the liquid phase into two liquids, a high-density and a low-density liquid. Similar to the boiling curve, this postulated equilibrium curve may contain a critical point at approx. $-50\,°C$ and atmospheric pressure.

This hypothesis is a matter of ongoing debate, but may have notable implications. If this second critical point exists, then water under standard conditions would constitute a supercritical fluid that fluctuates between a low- and a high-density liquid state. At lower temperatures, the mixture of the two phases should spontaneously separate into a low-density liquid that flows on top of the high-density liquid.

3.3.7   One Component Systems: Helium

Compared to the two substances we have looked at so far, the phase diagram for helium shows further unusual behaviour. The two isotopes of helium, ^3He and ^4He, have different phase diagrams; the p-T diagram of ^4He is illustrated in Fig. 3.9. In contrast to its lighter isotope, ^4He has two liquid phases, called He-I and He-II, with a transition between them. This transition line is called the λ-line. Whereas the liquid phase He-I forms a normal liquid, He-II has properties of a superfluid: it flows with

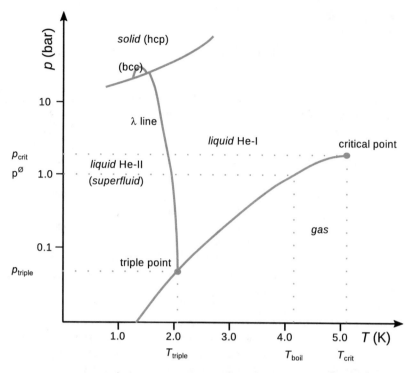

Fig. 3.9 Phase diagram of ^{4}He. hcp: hexagonal closed packing, bcc: body-centred cubic packing

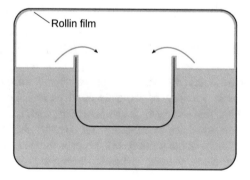

Fig. 3.10 Superfluid He-II flows with zero viscosity. It will "creep" along surfaces in order to equally level the two compartments in the container. The Rollin film also covers all interior surfaces of the container. If the container was not closed, He-II would creep out and escape

zero viscosity (Figure 3.10). At the low temperature triple point of helium, the phases He-II$_{(l)}$, He-I$_{(l)}$ and He$_{(g)}$ coexist. Another unusual observation is that solid and gas are never in equilibrium, due to the light He atoms having large amplitude vibrations. Therefore, helium needs very high pressures to form a solid.

3.3.8 Other Phases

In addition to the three fundamental phases (solid, liquid, gas), there a few unusual phases:

- Plasma is a phase formed by an ionised gas when electrons are stripped from atoms at high temperatures. Plasmas are an important phase in high-temperature processes such as nuclear fusion and stellar atmospheres. They have a practical importance in spectroscopic instruments where they are used as ionisation devices (e.g. inductively-coupled plasma atomic emission spectroscopy, ICP-AES).
- Supercooled liquids constitute a phase that is established by cooling a liquid below the freezing temperature but without crystallisation. The most prominent example for a supercooled liquid is glass. Albeit these materials appear solid, the particles form an amorphous (formless) structure.
- Liquid crystals flow like ordinary liquids, but the molecules form swarms with some low-order structure. As such, the liquid crystalline phase is an intermediate between the crystalline and liquid phase. The liquid crystalline phase is of eminent importance for the physics of membranes. It also has important practical applications since the low-order structure of liquid crystalline materials can be changed by orienting the molecules by applying a small electric potential; the change of orientation results in different optical properties of the liquid crystal. These phenomena are used in liquid crystalline displays (LCDs), which makes this phase a highly important phenomenon for microelectronic devices (displays, computer and TV screens).

3.4   Thermodynamic Aspects of Phase Transitions

The locations of phase boundaries (lines) in phase diagrams are determined by the relative thermodynamic stability of the individual phases, i.e. by the chemical potential μ (molar Gibbs free energy G_m). The chemical potential can thus be thought of as the potential of a substance to bring about a physical change. Under different conditions, different forms of the same system become more stable. A system that is in equilibrium has the same chemical potential throughout in all actually existing phases at that time.

In this section, we will discuss how the stability of phases are affected by particular environmental conditions. Knowledge about these relationships will allow us to predict phase transitions based on thermodynamic parameters.

3.4.1 Temperature Dependence of Phase Stability

We have previously (2.2.1) learned that the Gibbs free energy of a system is given by:

$$G_{sys} = H_{sys} - T \cdot S_{sys} \tag{2.44}$$

and thus its derivative is calculated as per

$$dG_{sys} = dH_{sys} - d(T \cdot S_{sys}) \tag{2.45}$$

which can be resolved by considering the product rule (A.2.3):

$$dG_{sys} = dH_{sys} - S_{sys} \cdot dT - T \cdot dS_{sys}$$

After dividing the equation by dT, one obtains:

$$\frac{dG_{sys}}{dT} = \frac{dH_{sys}}{dT} - \frac{S_{sys} \cdot dT}{dT} - \frac{T \cdot dS_{sys}}{dT}$$

which simplifies to:

$$\frac{dG_{sys}}{dT} = \frac{dH_{sys}}{dT} - S_{sys} - \frac{T \cdot dS_{sys}}{dT} \tag{3.15}$$

In Sect. 2.2.2, it was established that for processes at equilibrium, there is no net change in the entropy of the universe:

$$dS_{universe} = 0 \tag{2.51}$$

The entropy of the universe was given by Eq. 2.47:

$$dS_{universe} = dS_{sys} - \frac{dH_{sys}}{T}$$

which combines to:

$$dS_{universe} = dS_{sys} - \frac{dH_{sys}}{T} = 0$$

From this equation, it follows that

$$dS_{sys} = \frac{dH_{sys}}{T} \quad \text{and} \quad \text{thus}$$

$$dH_{sys} = T \cdot dS_{sys}$$

We can now use this expression to substitute dH_{sys} in Eq. 3.15 and obtain:

$$\frac{dG_{sys}}{dT} = \frac{T \cdot dS_{sys}}{dT} - S_{sys} - \frac{T \cdot dS_{sys}}{dT}$$

which simplifies to

$$\frac{dG_{sys}}{dT} = -S_{sys} \tag{3.16}$$

When considering a system consisting of 1 mol of substance (so we can use $G_m = \mu$), and remembering that the Gibbs free energy is not only dependent on the temperature, but also on the pressure, we obtain the following expression from Eq. 3.16 and also introduce the molar entropy S_m:

$$\left(\frac{\delta G_m}{\delta T}\right)_p = \left(\frac{\delta \mu}{\delta T}\right)_p = -S_m \tag{3.17}$$

Since G (as well as G_m, μ) is also dependent on the pressure, we need to request constant pressure when calculating the temperature differential; this is done by calculating a partial differential (indicated by 'δ'), and denoting the pressure as a subscript.

The molar entropy S_m is a characteristic parameter of a substance and always positive. Therefore, the change of the chemical potential μ with increasing temperature at constant pressure is always negative.

The phase with the lowest chemical potential μ at a particular temperature is the most stable one for that temperature. The transition temperatures (T_m/T_f and T_b) are the temperatures, at which the chemical potentials of the two interfacing phases are equal and the phases are thus at equilibrium.

3.4.2 Entropies of Substances

In the previous section, we have introduced the molar entropy S_m in the context of different phases of a one-component system. We also noted that molar entropies are characteristic parameters of a substance.

In gases, molecules can freely diffuse through the volume they are contained in and the individual molecules (as well as their energies) are dispersed across the entire volume. Therefore, the degree of disorder is rather large, compared to that of liquids or solids. The molar entropy S_m of gases is thus larger than that of liquids or solids (see Table 3.2).

In contrast, the molecules in a solid are confined to a small volume and their degrees of freedom are restricted to vibrational motion. The molar entropies S_m of solids are thus fairly low. Solids comprising of large molecules (e.g. sucrose) or complexes ($CuSO_4 \cdot 5\ H_2O$) have a large number of atoms and may thus possess comparatively high molar entropies, since the energy may be shared among the many atoms.

We remember that we have defined the explicit entropy change dS in Sect. 2.1.10 as:

Table 3.2 Molar entropies S_m in J K^{-1} mol^{-1} of selected solids, liquids and gases at 25 °C

Solids		Liquids		Gases	
C (diamond)	2.4	Hg	76.0	H_2	130.6
C (graphite)	5.7	H_2O	69.9	N_2	192.1
Fe	27.3	$H_3C-COOH$	159.8	O_2	205.0
Cu	33.1	C_2H_5OH	160.7	CO_2	213.6
AgCl	96.2	C_6H_6	173.3	NO_2	239.9
Fe_2O_3	87.4			NH_3	192.3
$CuSO_4 \cdot 5\ H_2O$	300.4			CH_4	186.2
sucrose	360.2			N_2O_4	304.0

$$dS = \frac{dQ}{T} \qquad (2.30)$$

where dQ is the reversibly transferred heat at a particular temperature T. From this equation, it becomes clear that there is a smaller change of entropy when a given quantity of energy (dQ) is transferred to an object at high temperature than at low temperature. More disorder is induced when the object is cool rather than hot.

The ability of a substance to distribute energy over its molecules is related to the heat capacity. As we have discussed in Sect. 2.1.12, this link can be used to establish a relationship between the entropy change and the heat capacity C_p:

$$\Delta S = \int_{T_{start}}^{T_{end}} \frac{C_p}{T} dT \qquad (2.39)$$

which forms the basis for entropy determination of substances by heat capacity measurements (i.e. calorimetrically). Alternatively, electrochemical cells can be used to determine entropies of substances.

3.4.3 Pressure Dependence of Melting

In Sect. 4.1 above, we looked at the change of the Gibbs free energy with temperature (at constant pressure) and derived the definition of the molar entropy S_m. From the Maxwell equations describing the relations between different state variables (Sect. 2.1.14), the variation of the Gibbs free energy with pressure (at constant temperature) can be derived, leading to the definition of the molar volume V_m:

$$\left(\frac{\delta G_m}{\delta p}\right)_T = \left(\frac{\delta \mu}{\delta p}\right)_T = V_m \qquad (2.65)$$

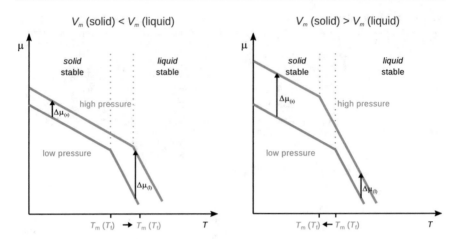

Fig. 3.11 The blue and green lines show the chemical potential in dependence of the temperature. The branch at lower temperature describes the chemical potential for the solid and the branch at higher temperature describes the chemical potential for the liquid phase. The transition temperature is the temperature where the two branches intersect. An increase of the chemical potential (when the pressure is increased) results in a vertical shift of the branches. The phase with the smaller volume is less impacted by a change of pressure and thus has a lesser change of the chemical potential

Like the molar entropy, the molar volume V_m is a characteristic parameter for a substance and always positive. Therefore, the change of the chemical potential μ ($= G_m$) with increasing pressure at constant temperature is always positive. Generally, the molar volume V_m is larger for liquids than for solids (exception: water).

We remember that the transition temperatures (T_m/T_f and T_b) are the temperatures, at which the chemical potentials of the two interfacing phases are equal and the phases are thus at equilibrium. The transition temperature between the solid and liquid phases (T_m/T_f) is generally larger at higher pressures (exception: water).

Conceptually, this is illustrated in Fig. 3.11. If the molar volume V_m of the solid is smaller than that of the liquid (Fig. 3.11 left)—an observation made for most substances—the chemical potential of the solid phase $\mu_{(s)}$ increases less than the chemical potential of the liquid phase, $\mu_{(l)}$, when the pressure is increased. This situation leads to a shift of the intersect between the two branches to higher temperatures; therefore, the melting temperature increases when the pressure is increased.

In contrast, water, for example, shows the behaviour illustrated in the right panel of Fig. 3.11: the molar volume V_m of the solid is larger than that of the liquid. Therefore, when the pressure is increased, the liquid phase experiences a lesser change of the chemical potential than the solid phase. The intersect between the two branches thus migrates to lower temperatures. This means that at higher pressure, substances like water freeze at lower temperatures.

3.4.4   Pressure Dependence of Vapour Pressure

In Sect. 3.2.2, we arrived at an expression of Raoult's law that relates the vapour pressure of a solution with that of the pure solvent as per:

$$p_A = p_A^* \cdot e^{\frac{\mu_{A(l)} - \mu_{A(l)}^*}{R \cdot T}} \qquad (3.11)$$

$\mu_{A(l)}$ and $\mu_{A(l)}^*$ describe the chemical potentials of the solution and the pure solvent, respectively. In the previous section, we have seen that the differential of the chemical potential with respect to pressure changes is

$$\left(\frac{\delta \mu}{\delta p} \right)_T = V_m. \qquad (2.65)$$

We then realise that the chemical potential difference in Eq. 3.11 can be expressed in terms of the molar volume and the pressure change:

$$\mu_{A(l)} - \mu_{A(l)}^* = \Delta \mu = V_m \cdot \Delta p$$

which we can substitute in Eq. 3.11 and obtain:

$$p = p^* \cdot e^{\frac{V_m \cdot \Delta p}{R \cdot T}} \qquad (3.18)$$

The pressure difference Δp in the exponential term describes the pressure of two different states. Let the initial state be at $p_{initial} = p^\varnothing$, and the final state at a much high pressure. In that case, $\Delta p > 0$ and thus the exponent is positive and the entire exponential term a factor greater than 1 (see Fig. 3.12). This means that the vapour pressure of a pressurised liquid is higher than that of the system under standard pressure.

Similarly, if we consider the case where p_{final} describes the system at lower pressure (e.g. the evacuated system), then we appreciate that $\Delta p < 0$, the exponent

Fig. 3.12 For positive arguments x, the function $y = e^x$ yields values larger than 1. For negative arguments, the function yields values less than 1

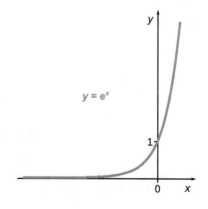

$y = e^x$

becomes negative and thus the exponential factor has a value less than 1. Therefore, the vapour pressure will be less in an evacuated system than in one that contains ambient atmosphere.

3.4.5   Phase Boundaries

As we have established earlier, the lines in a phase diagram separating two neighbouring phases are called phase boundaries. Along those boundaries, two phases (we will here call them 'a' and 'b') co-exist. Being lines in a two-dimensional plot, the phase boundaries are best discussed in terms of their slopes. So for a p-T diagram, phase boundaries are characterised by the differential

$$\left(\frac{\mathrm{d}p}{\mathrm{d}T}\right).$$

Since the two phases co-exist, there must be equilibrium and the changes in the chemical potentials must be equal:

$$\mathrm{d}\mu_a(p, T) = \mathrm{d}\mu_b(p, T)$$

From earlier discussions we know that

$$\mathrm{d}\mu = \mathrm{d}G_m = V_m \cdot \mathrm{d}p - S_m \cdot \mathrm{d}T \qquad (2.45)$$

and can therefore substitute this expression on both sides of the equilibrium equation above:

$$V_{m,a} \cdot \mathrm{d}p - S_{m,a} \cdot \mathrm{d}T = V_{m,b} \cdot \mathrm{d}p - S_{m,b} \cdot \mathrm{d}T$$

We group together the volumes and entropies on opposing sides:

$$(S_{m,b} - S_{m,a}) \cdot \mathrm{d}T = (V_{m,b} - V_{m,a}) \cdot \mathrm{d}p$$

and consider that the differences of the volumes and entropies of phases 'a' and 'b' describe the transition from one phase to the other:

$$\Delta S_{m,\text{trans}} \cdot \mathrm{d}T = \Delta V_{m,\text{trans}} \cdot \mathrm{d}p$$

This equation can be re-arranged to yield the slope of the phase boundary through the differential of pressure and temperature:

$$\left(\frac{\mathrm{d}p}{\mathrm{d}T}\right) = \frac{\Delta S_{m,\text{trans}}}{\Delta V_{m,\text{trans}}} \qquad (3.19)$$

This equation is of fundamental importance as it describes the phase transitions (phase boundaries) in p-T phase diagrams; it is also known as the Clapeyron equation, developed by the French physicist an engineer Benoît Clapeyron in 1834 (Clapeyron 1834). Equipped with this equation, we can now discuss the three phase boundaries solid–liquid, liquid–vapour and solid–vapour.

3.4.6   Phase Boundaries: Solid–Liquid

We can apply the general form of the Clapeyron equation above to particular phase transitions, such as the melting or fusion process where solid and liquid phases are in equilibrium. The slope of the solid–liquid phase boundary in a p-T diagram is described as the differential of pressure (y-axis) with respect to temperature (x-axis), $\frac{dp}{dT}$. The differences in the molar entropies and volumes between solid and liquid states are macroscopically measurable; we thus use 'Δ' instead of 'd', and obtain the Clapeyron equation for the melting (fusion) process:

$$\frac{dp}{dT} = \frac{\Delta S_{m,\text{melt}}}{\Delta V_{m,\text{melt}}} \tag{3.19}$$

We remember from earlier discussions that

$$\Delta S = \frac{\Delta H}{T} \tag{2.35}$$

and therefore can express the phase transition in terms of the molar enthalpy change:

$$\frac{dp}{dT} = \frac{\Delta S_{m,\text{melt}}}{\Delta V_{m,\text{melt}}} = \frac{\Delta H_{m,\text{melt}}}{T_{\text{melt}} \cdot \Delta V_{m,\text{melt}}} \tag{3.20}$$

For the melting process, we can evaluate the Clapeyron equation in an approximative fashion. The change in molar enthalpies

$$\Delta H_{m,\text{melt}} = H_{m,\text{liquid}} - H_{m,\text{solid}}$$

is generally positive, and the change in molar volumes

$$\Delta V_{m,\text{melt}} = V_{m,\text{liquid}} - V_{m,\text{solid}}$$

is generally positive and rather small (the liquid and solid states of most substances have a similar volume, with the liquid phase typically having a slightly larger volume; exception: water). Also, $\Delta V_{m,\text{melt}}$ can be considered independent of the temperature. Therefore, the differential

$$\frac{dp}{dT}$$

Fig. 3.13 Using the Clapeyron equation, the slope of the phase boundary between solid and liquid phases can be calculated. Most substances have a lesser density in their liquid than in their solid states, hence the molar volume difference upon melting is positive, and the phase boundary therefore has a positive slope

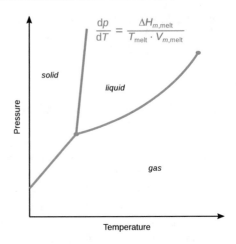

is generally positive and large. This means we will obtain steep boundaries at the solid–liquid interfaces in a p-T diagram (Fig. 3.13), with this phase boundary having a positive slope.

3.4.7   Phase Boundaries: Liquid–Vapour

In analogy to the discussion in the previous section, we can formulate, based on the Clapeyron equation, an expression for the vapourisation process, i.e. the transition from the liquid to the gas phase:

$$\frac{dp}{dT} = \frac{\Delta S_{m,\text{vap}}}{\Delta V_{m,\text{vap}}} = \frac{\Delta H_{m,\text{vap}}}{T \cdot \Delta V_{m,\text{vap}}} \tag{3.19}$$

For the vapourisation process, we can not assume that $\Delta V_{m,\text{vap}}$ is independent of the temperature. We remember that for an ideal gas, there is a relationship between the gas volume and the temperature and pressure, given by the ideal gas equation:

$$V = \frac{n \cdot R \cdot T}{p} \tag{2.6}$$

When we consider the molar volume V_m, we normalise the volume with respect to the molar amount n ($V_m = \frac{V}{n}$), so we obtain:

$$V_m = \frac{R \cdot T}{p}$$

Since the gas phase of a substance occupies a much larger volume than the liquid state, we can assume that the volume of the liquid is negligible compared to the volume of the gas:

$$V_{m,gas} \gg V_{m,liquid} \;\Rightarrow\; \Delta V_{m,vap} = V_{m,gas} - V_{m,liquid} \approx V_{m,gas}$$
$$\Delta V_{m,vap} \approx V_{m,gas} = \frac{R \cdot T}{p}$$

Note that we now assume conditions of an ideal gas and negligible volume of the solid compared with the gas. This expression for $\Delta V_{m,vap}$ can be used to substitute in Eq. 3.19 which then yields:

$$\frac{dp}{dT} = \frac{\Delta H_{m,vap}}{T \cdot \Delta V_{m,vap}} = \frac{\Delta H_{m,vap}}{T \cdot \left(\frac{R \cdot T}{p}\right)}$$

This simplifies to

$$\frac{dp}{dT} = \frac{p \cdot \Delta H_{m,vap}}{R \cdot T^2}.$$

We isolate the two independent variables, p and T, on opposite sides of the equation, and use the formality of $\int \frac{1}{x} dx = \ln x \;\Leftrightarrow\; \frac{1}{x} dx = d(\ln x)$ to achieve a more convenient notation (see Appendix A.3.1). This results in a relationship known as the Clausius-Clapeyron equation:

$$\frac{dp}{p \cdot dT} = \frac{d(\ln p)}{dT} = \frac{\Delta H_{m,vap}}{R \cdot T^2}. \tag{3.21}$$

This equation was first derived by the German physicist and mathematician Rudolf Clausius in 1850 (Clausius 1850).

3.4.8 Phase Boundaries: Solid–Vapour

The phase boundary between the solid and vapour phases describes the sublimation (or deposition) process. For this process, the Clapeyron equation yields:

$$\frac{dp}{dT} = \frac{\Delta S_{m,subl}}{\Delta V_{m,subl}} = \frac{\Delta H_{m,subl}}{T \cdot \Delta V_{m,subl}} \tag{3.19}$$

As in the vapourisation process (previous section), we can not assume that ΔV_m is independent of the temperature, because a gas phase is involved. Therefore, we again replace V_m with the expression from the ideal gas equation. We thus assume conditions of an ideal gas and negligible volume of the solid compared with the gas. The substitution yields:

$$\frac{dp}{dT} = \frac{\Delta H_{m,subl}}{T \cdot \Delta V_{m,subl}} = \frac{\Delta H_{m,subl}}{T \cdot \frac{R \cdot T}{p}}$$

Which simplifies to the following equation, also called the Clausius-Clapeyron equation for the sublimation process:

$$\frac{dp}{p \cdot dT} = \frac{d(\ln p)}{dT} = \frac{\Delta H_{m,\,subl}}{R \cdot T^2}. \tag{3.22}$$

3.4.9   The Ehrenfest Classifications

In Fig. 3.11, we visualised the change of the chemical potential μ with temperature in the region where the solid–liquid phase transition occurs. At the transition, the chemical potential function shows a kink (the function is continuous, but the first derivative of the function is not), so the change in the chemical potential μ is not smooth with a change in the temperature T. This is observed at all major phase transitions

- solid↔liquid (melting/freezing)
- liquid↔gas (vapourisation/condensation)
- solid↔gas (sublimation/deposition)

which are therefore called first order phase transitions. At a molecular level, first-order transitions involve the relocation of atoms, molecules or ions, accompanied by a change of the interaction energies.

What is the implication of this for material properties? As mentioned above, a mathematical function with a kink is characterised by an abrupt change of direction of the plotted function. Whereas the function is continuous and possesses discrete values all the way along, the first derivative is discontinuous. At the kink, the first derivative is infinite.

We remember that the Clapeyron equation (3.19, 3.20) is a function of enthalpy and temperature, in fact $f = \frac{H}{T}$, and the first derivative of this function, $f' = \frac{dH}{dT}$, leads us to the heat capacity for constant pressure, C_p:

$$C_p = \left(\frac{\delta H}{\delta T}\right)_p \tag{2.29}$$

At a first-order transition, H changes by a finite amount whereas T changes by an infinitesimal small amount. The heat capacity C_p (the first derivative) thus becomes infinite.

Following this classification concept, a transition for which the first derivative of the chemical potential μ with respect to the temperature T is continuous, but the second derivative is discontinuous, is called second order phase transition. This implies that volume, entropy and enthalpy do not change at the transition. The heat capacity C_p is discontinuous but not infinite. At a molecular level, second-order transitions are often associated with changes of symmetry in a crystal structure. Rather than the molecular interaction energies, it is the long-range order that varies. Examples for second order phase transitions include the conducting-superconducting transition in metals at low temperatures.

Phase transitions that are not first order, but where nevertheless the heat capacity C_p becomes infinite, are called λ-transitions. In such instances, the heat capacity typically increases well before the actual transition occurs. At a molecular level, λ-transitions are associated with order-disorder transitions. Examples include:

- order-disorder transitions in alloys
- onset of ferromagnetism
- fluid-superfluid transition in liquid helium (hence the name λ-line in the He phase diagram, Fig. 3.9)

3.5 Mixtures of Volatile Liquids

3.5.1 Phase Diagrams of Mixtures of Volatile Liquids

We assume a mixture of two volatile liquids, A and B, where the liquid and vapour phases are in equilibrium. Even though, the compositions in the two phases are not necessarily the same; the vapour phase will contain more of the more volatile component.

Raoult's law enables calculation of the vapour pressure of a particular liquid (A) in a mixture, for different concentrations of that liquid in the mixture (expressed as molar fraction x):

$$p_A = x_A \cdot p_A^*$$
(3.11)

p_A^* is the vapour pressure of the pure liquid A (i.e. at $x_A = 1$). Since we will have to consider mole fractions for several different phases in the following discussion, in this current chapter, we will denote the mole fraction of substances in the liquid phase as 'z' instead of 'x'.

The total vapour pressure in an ideal mixture of A and B is then given by Dalton's law (Eq. 2.8), which poses that the total vapour pressure of a vapour mixture is the sum of the partial vapour pressures of all components (here, the partial vapour pressures are given by Raoult's law for A and B):

$$p = z_A \cdot p_A^* + z_B \cdot p_B^* = p_B^* + \left(p_A^* - p_B^*\right) \cdot z_A$$
(3.23)

where z_A and z_B are the mole fractions of A and B in the liquid, and p_A^* and p_B^* the vapour pressures of the pure liquids, respectively. This relationship may be plotted in a diagram (see Fig. 3.4, upper left) where the total pressure p provides the ordinate (y-axis) and the mole fraction z_A the abscissa (x-axis). It is obvious, that at $z_A = 0$, there is pure liquid B and hence the total vapour pressure is given by the vapour pressure of pure liquid B, p_B^*.

With Dalton's law informing us that the total pressure of a gas mixture equals the sum of the partial pressures of the individual components, we have access to the mole fraction of the individual components in the vapour phase, which, for clarity, we call 'y' instead 'x' in this section:

$$\frac{p_A}{p} = \frac{\frac{n_A \cdot R \cdot T}{V}}{\frac{(n_A + n_B) \cdot R \cdot T}{V}} = \frac{n_A}{n_A + n_B} = y_A \quad \text{and} \quad \frac{p_B}{p} = \frac{\frac{n_B \cdot R \cdot T}{V}}{\frac{(n_A + n_B) \cdot R \cdot T}{V}} = \frac{n_B}{n_A + n_B} = y_B \quad (3.24)$$

y_A and y_B are the mole fractions of A and B in the vapour. This relationship may also be plotted in a phase diagram (Fig. 3.14, upper right) with pressure as the ordinate (y-axis), but this time the mole fraction y_A as abscissa (x-axis). It is no surprise that the resulting phase boundary is now a different line than before; after all we are no longer plotting against the mole fraction of A in the liquid, but in the vapour. From Eq. 3.24 it is obvious that at $y_A = 1$, the vapour contains pure substance A, hence the total vapour pressure p is the vapour pressure of A, p_A^*.

Equipped with these relationships, we can now calculate the mole fraction of each component in the vapour phase (y_i) of the liquid mixture, knowing their mole fractions in the liquid phase (z_i).

From Raoult's law we know:

$$p_A = z_A \cdot p_A^*$$

and from Dalton's law:

$$y_A = \frac{p_A}{p}$$

Therefore:

$$y_A = z_A \cdot \frac{p_A^*}{p}$$

For a mixture consisting of two components only, the mole fraction of the second component is then available as

$$y_B = 1 - y_A.$$

From a practical perspective, it will be very inconvenient to plot two phase diagrams, one each for liquid and vapour mole fractions. Hence the two diagrams are combined into one (Fig. 3.14, bottom).

The combined phase diagram is plotted with the pressure as ordinate and the mole fraction of the total composition as abscissa. The mole fraction of the total composition (x_i) can easily be obtained from the mole fractions in the liquid (z_i) and vapour phases (y_i); an example is shown in Table 3.3.

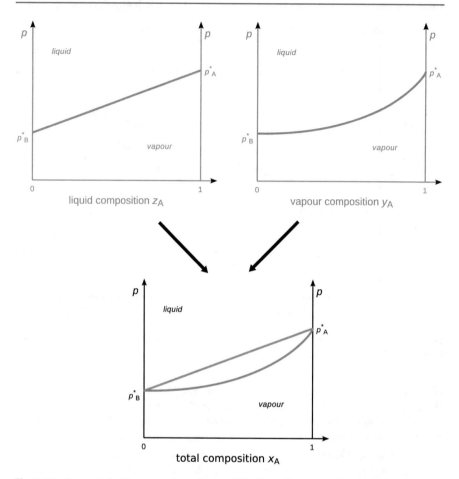

Fig. 3.14 Concept of a binary *p-x* phase diagram. The phase diagram plotting total composition as the abscissa can be thought of as a superposition of phase diagrams for liquid and vapour composition

Table 3.3 Example calculation to obtain the total composition of a binary system, when the compositions of liquid and vapour phases are known

	$n(A)$	$n(B)$	Mole fraction of A	Mole fraction of B
Liquid	2 mol	3 mol	$^2/_5 = z_A$	$^3/_5 = z_B$
Vapour	1 mol	2 mol	$^1/_3 = y_A$	$^2/_3 = y_B$
Total	3 mol	5 mol	$^{(2+1)}/_{(5+3)} = {}^3/_8 = x_A$	$^{(3+2)}/_{(5+3)} = {}^5/_8 = x_B$

From binary phase diagrams such as the one shown in Fig. 3.15, composition data of the system at various conditions can be easily obtained.

For example, a mixture of two liquids A and B, each present with a mole fraction of 0.5 and a phase diagram as shown in Fig. 3.15 is held at a pressure $p = p_0$. In the

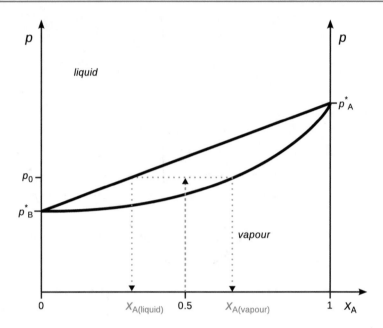

Fig. 3.15 Illustration of how to obtain molar compositions of liquid and vapour phases of a binary liquid mixture

diagram this situation is depicted as a point at $(x_A; p) = (0.5; p_0)$. This point lies in the region between the two lines marking the liquid–vapour phase boundary; the upper line which delivers the molar composition of the liquid phase and the lower line which represents the composition of the vapour phase. This area is called the two-phase area, where the liquid and vapour phases co-exist. Therefore, at the identified point, a line parallel to the x-axis is used to extrapolate to the phase boundary lines (such a line is called a conode or tie line). Where the tie line intersects, drop lines to the x-axis are used to determine the molar fraction of A in the vapour and liquid phases. The molar fractions for B are available as per:

$$x_{B(\text{liquid})} = 1 - x_{A(\text{liquid})} \text{ and } x_{B(\text{vapour})} = 1 - x_{A(\text{vapour})}$$

As a result of the mixing with substance B, the vapour pressure of A in the mixture is lowered as compared to the vapour pressure of pure liquid A. The mole fraction of A in the vapour phase of a 1:1 mixture is therefore larger than 50%, despite it possessing the higher vapour pressure when comparing the pure liquids.

Any point located in the two-phase indicates a mixture composition where separation into the two co-existing phases occurs. The tie line further yields information about the relative molar amounts of substance A (since x_A is plotted as abscissa) in the liquid and vapour phases. The ratio of molar amounts of substance A in the liquid and vapour phases is given by the lever rule, graphically illustrated in Fig. 3.16:

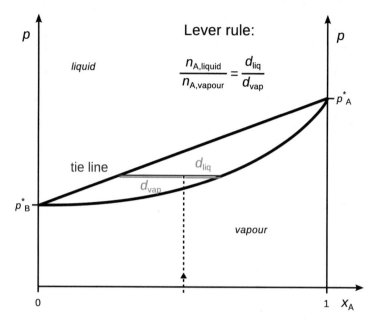

Fig. 3.16 Illustration of the lever rule to determine the molar amounts of substance A in the liquid and vapour phases

$$\frac{n_{A,\text{liquid}}}{n_{A,\text{vapour}}} = \frac{x_A - x_{A,\text{vapour}}}{x_{A,\text{liquid}} - x_A} = \frac{d_{\text{liq}}}{d_{\text{vap}}}. \tag{3.25}$$

The examples and illustrations above were all concerned with pressure-composition (p-x) phase diagrams, where phase boundaries are characterised in dependence of pressure at a particular constant temperature.

Of practical importance are also temperature-composition (T-x) phase diagrams, which show phases at a single pressure. A typical T-x diagram found with many real mixtures is shown in Fig. 3.17. The way of determining the mole fractions of a substance in the liquid and vapour phases in T-x diagrams is the same as discussed above for p-x diagrams. The lever rule can also be applied in an analogous fashion.

Temperature-composition phase diagrams are particularly useful when analysing distillation processes.

3.5.2 Simple Distillation

Distillation procedures are based on vapour and liquid having different compositions. In a simple distillation apparatus, mixtures comprising two components of low (component A) and high (component B) volatility can be separated to some degree. Since the distillation experiments are carried out at constant pressure (ambient pressure

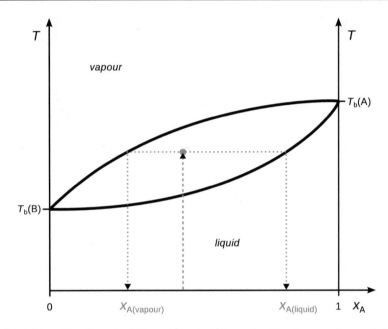

Fig. 3.17 Illustration of how to obtain molar compositions of liquid and vapour phases of a binary liquid mixture from a T-x phase diagram. The process is the same as described above for p-x diagrams. However, note that the location of liquid and vapour phases in T-x diagrams is different to those in p-x diagrams

or under vacuum), temperature-composition phase diagrams such as the one shown in Fig. 3.18, can be used to track the process of simple distillation.

We are starting at ambient temperature with a liquid mixture that has relatively low concentration of the high volatility component B ($x_A \approx 0.75$, so $x_B \approx 0.25$). The mixture is heated (arrow 1), and at the intersection of arrow 1 with the boiling curve a vapour phase appears. The composition of the vapour can be obtained where the tie line (arrow 2) intersects with the condensation curve. In a basic distillation apparatus, vapour with this composition is condensed on the fractionation side of the apparatus; the condensate consists of a liquid with enriched component B (Fig. 3.19).

3.5.3   Fractional Distillation

In the fractional distillation, vapour is continually removed from the boiling equilibrium system, thus allowing enrichment of the more volatile component to very high purity. This is illustrated in Fig. 3.20.

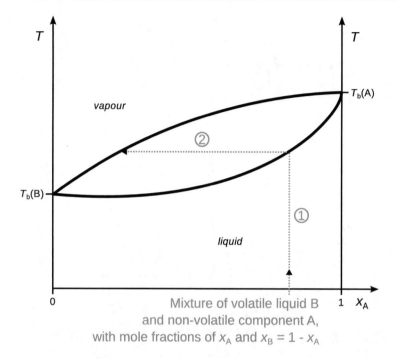

Fig. 3.18 Illustration of a simple distillation process

- Step 1: A mixture of less volatile component (A; higher boiling temperature) and more volatile component (B; lower boiling temperature) is heated. The mole fraction of component B in the initial mixture is $x_B \approx 0.15$.
- Step 2: The boiling point of the mixture at this molar composition is reached, and a vapour phase with a much higher mole fraction x_B is obtained.
- Step 3: The vapour from (2) condenses as it cools at a fractionation plate and reaches the boiling temperature of the liquid mixture at the new molar composition.
- Step 4: A new vapour phase is formed, that is further enriched in component B.
- Steps 5 onwards repeat this process, until an endpoint in the phase diagram is reached.

The efficiency of a fractionating column is expressed in terms of the number of theoretical plates. A theoretical plate is a hypothetical zone in which two phases establish an equilibrium with each other. This is the number of effective condensation and vapourisation steps that are required to achieve a condensate with desired composition from a given distillate. In the example in Fig. 3.20, five theoretical plates are required to obtain pure component B from the initial liquid mixture with $x_B \approx 0.15$.

Fig. 3.19 Photograph of a distillation apparatus. For a simple distillation process as shown in Fig. 3.18, the small fractionation column is omitted

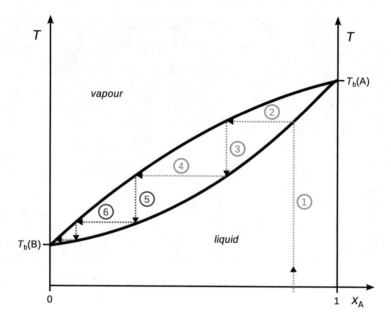

Fig. 3.20 Illustration of the fractional distillation process. Five theoretical plates (steps 2, 4, 6, 8, 10) are required to obtain pure component B from the initial mixture. For clarity, steps 7–10 are not individually labelled

In distillation experiments, this process can be achieved by using Vigreux fractionation columns (Fig. 3.21) that contain spikes which form the contact devices (physical plates) between liquid and vapour and thus provide for a number of separation steps. In industrial applications, so-called bubble-cap or valve-cap trays are used. The trays are perforated, thus allowing efficient flow of vapour upwards through the column.

The efficiency of physical plates is non-ideal and therefore the number of physical plates needed for a desired separation step is more than the calculated number of theoretical plates:

$$N_{\mathrm{a}} = \frac{N_{\mathrm{t}}}{E}$$

N_{a} is the number of actual plates, N_{t} the number of theoretical plates, and E is the plate efficiency. Obviously, in order to be able to calculate the number of theoretical plates for a distillation process, substantial liquid–vapour equilibrium data (i.e. phase diagrams) need to be available.

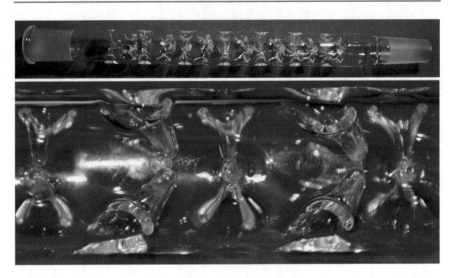

Fig. 3.21 Photograph of a Vigreux fractionation column (top) and close-up of a spike section (bottom)

3.5.4 Mixtures of Volatile Liquids with Azeotropes

The phase diagrams of liquid mixtures we have encountered so far featured monotonous boiling and condensation curves. Some mixtures, though, show additional features in their phase diagrams. The phase diagram of the mixture illustrated in Fig. 3.22 possesses a point where boiling and condensation curves touch in one point. At this point, the liquid and vapour phases have the same composition; the point is called an azeotrope. For example, an ethanol–water mixture with 4% water forms an azeotrope that boils at 78 °C at ambient pressure.

When boiling a liquid mixture that forms an azeotrope, evaporation will proceed without a changing composition of liquid and vapour phases. The mixture behaves as if it were a pure substance. If mixture with azeotropes are subjected to fractionated distillation, the distillation process stops being useful when the azeotrope is reached.

Two types of azeotropes can be distinguished. In low boiling azeotropes, the interactions between the two mixture components are unfavourable compared to ideal mixing. The azeotrope thus boils at the lowest temperature of all possible mixtures of the two components. An example for mixtures with low-boiling azeotropes is the ethanol–water system. In contrast, high boiling azeotropes are mixtures where the interactions between both components are more favourable when compared to the ideal case. Such azeotropes boil at the highest temperature of all possible mixtures of the two components. An example of such behaviour is a system comprising of chloroform and acetone.

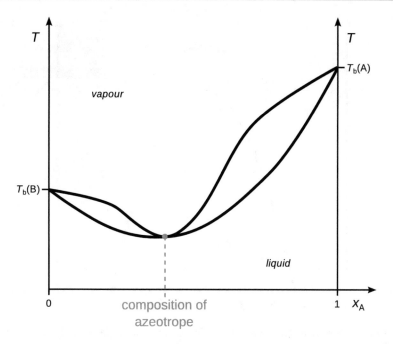

Fig. 3.22 Phase diagram of a mixture that forms a low boiling azeotrope

3.5.5 Mixtures of Immiscible Liquids

Not all liquids can be mixed. If liquids are immiscible, then we can treat the solutions separately, but the vapour pressure of a system that comprises both components is the sum of the two vapour pressures (p_A^*, p_B^*). Boiling occurs when the vapour pressure of the liquid phase equals the atmospheric pressure:

$$p = p_A^* + p_B^* = p_{normal} = 1 \, atm$$

This results in an interesting consequence: When two immiscible liquids are put together, the pair of them possesses a lower boiling point than either pure liquid alone.

This behaviour is useful when heat-sensitive compounds need to be distilled. When put together with an immiscible liquid, the distillation can proceed at lower temperature than the boiling point of the pure compound. This process is typically carried out as steam distillation.

3.5.6 Phase Diagrams of Two-component Liquid/Liquid Systems

In the previous sections, we have discussed liquid mixtures that were either fully or not miscible. The mutual solubility or miscibility of two liquids is a function of temperature and composition. Of course, there are systems where the two liquid

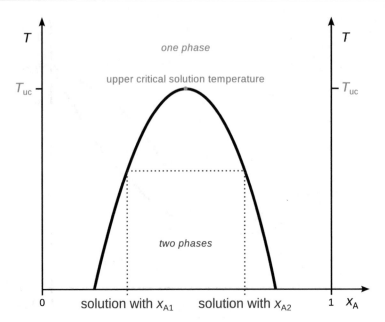

Fig. 3.23 Phase diagram for two partially miscible liquids with an upper critical solution temperature (T_{uc})

components mix under some but not all conditions. When two liquids are partially soluble in each other, two liquid phases can be observed. These are partially miscible liquids (e.g. methanol–cyclohexane, nicotine–water, phenol–water, triethylamine–water, and others). A typical phase diagram for the most common types of partially miscible liquids is shown in Fig. 3.23. The phase diagram indicates that the two liquids are fully miscible and form a one-phase liquid) at high temperatures (above T_{uc}), but separate into two liquid phases at lower temperatures (below T_{uc}). T_{uc} is called the upper critical temperature. The tie line is used to determine the composition of the two phases.

There are two other cases of liquid–liquid mixtures, which are less common. The left panel in Fig. 3.24 shows the phase diagram of a mixture that possesses a lower critical solution temperature T_{lc}. Below this temperature, the mixture forms one liquid phase, i.e. the two components are fully miscible. Above T_{lc}, two liquid phases exist. This type of behaviour is observed when there are weak interactions between both components (below the critical solution temperature), such as for example in water-triethylamine.

There also exist some systems that possess both a lower and an upper critical solution temperature, the most prominent example being water and nicotine (Fig. 3.24 right panel). These mixtures are characterised by weak interactions between both components that ensure full miscibility below the lower critical solution temperature. Above T_{lc}, these interactions are disrupted and there is only partial miscibility, and accordingly, two phases. Above the upper critical solution temperature, the mixture is homogenised and exists as a single liquid phase.

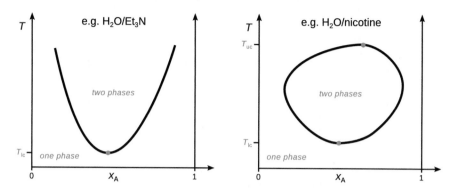

Fig. 3.24 Left: Phase diagram for mixture with a lower critical solution temperature (T_{lc}). Right: Phase diagram for system with both lower (T_{lc}) and upper (T_{uc}) critical solution temperatures

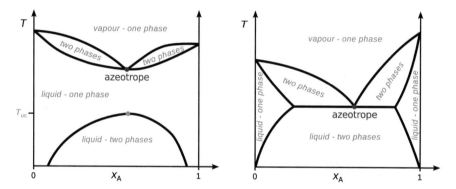

Fig. 3.25 Phase diagrams of two-component liquid–vapour systems of partially miscible liquids with a low-boiling azeotrope. Left: The liquids fully mix before the mixture starts to boil. Right: The mixture starts to boil before the two components become fully miscible

3.5.7 Phase Diagrams of Two Component Liquid–Vapour Systems

If we now consider a mixture of two partially miscible liquids and also form a low-boiling azeotrope, we arrive at a fairly common behaviour of real substances. Both properties, partial miscibility and azeotrope formation, emphasise the fact that the molecules of the two components tend to avoid each other.

This behaviour is possible with two different options:

- the liquids may become fully miscible before they boil, i.e. the azeotrope is well separated from the upper critical solution temperature (Fig. 3.25 left), or
- boiling occurs before the two liquids are fully mixed, i.e. the azeotrope and the upper critical solution temperature merge (Fig. 3.25 right).

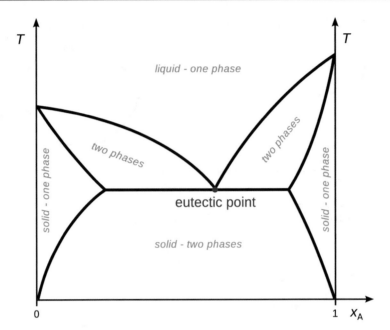

Fig. 3.26 Phase diagram of two partially miscible solids. The mixture with the lowest melting point has eutectic composition

3.5.8 Phase Diagrams of Two-component Solid–Liquid Systems

In the previous sections, we considered systems consisting of two components and discussed their liquid and vapour phase behaviour. In the same fashion, solid and liquid phases can be characterised. Conceptually, there is no difference in the way such phase diagrams are interpreted. As an example, Fig. 3.26 illustrates the phase diagram of two partially miscible solids whose melting point occurs before the two solids are fully mixed.

It is immediately obvious that this phase diagram is highly similar to the one we have seen before (Fig. 3.25 right) in the case of a two-component liquid–vapour system where the two liquids were partially miscible and boiling occurred before full mixing of the liquids. For solids, the mixture with the lowest melting point is called the eutectic point (as opposed to the azeotrope in the liquid–vapour systems).

3.5.9 Phase Diagrams of Three-component Systems

Of course, mixtures can be made of more than just two components, but visualisation of phase diagrams for higher order systems becomes challenging. For three-component systems, it is still possible to plot two-dimensional phase diagrams. Instead of a Cartesian x–y diagram, a triangular coordinate system can be used,

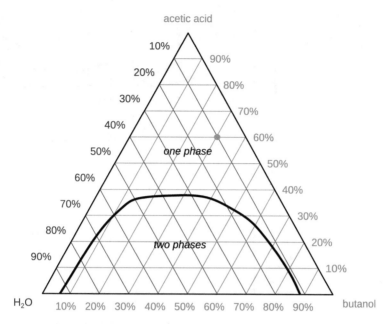

Fig. 3.27 Qualitative ternary phase diagram for the three liquids water, acetic acid and butanol. The one and two phase regions of this system are separated by the black line. The arbitrary point marked in the diagram represents 10% water, 60% acetic acid and 30% butanol

that allows to plot the composition of each of the three components (A, B, C) in the system. However, such phase diagrams are restricted to a particular temperature and pressure.

Like for binary systems before, we can state that the sum of the individual mole fractions is 1:

$$x_A + x_B + x_C = 1$$

Figure 3.27 illustrates such a triangular phase diagram, showing the mole fractions for a ternary system, at a certain pressure and temperature. The phase separation of the water-butanol-acetic acid system is shown with the grid of the coloured coordinate system in the background.

3.5.10 Cooling Curves

Cooling curves show how the temperature in a system changes during the time course of the cooling process. Pure liquid, solid or gas phases have smooth, monotonous changes in temperature, until the process approaches a phase transition.

The phase transitions of pure substances proceed along the pathway of gas → liquid → solid, occur at single temperatures and are exothermic (they are endothermic along the opposite pathway). During the phase transition, the temperature does not

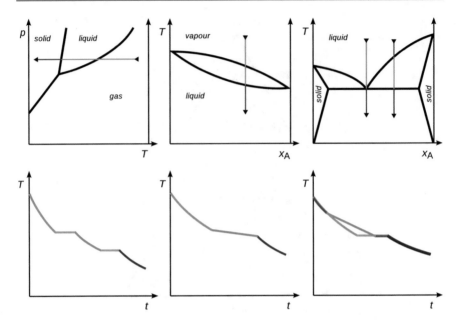

Fig. 3.28 Cooling processes of a pure substance with three phases (left), a pure substance with two phases (middle), and a binary solid–liquid mixture with eutectic point. The upper row shows the phase diagrams, and the lower row the cooling curves as time-temperature diagrams

change until the phase of the substance has been fully converted. This is illustrated in Fig. 3.28 for a phase diagram with three (left panel) and two phases (middle panel).

Phase transitions during cooling of mixtures also proceed along the pathway of gas → liquid → solid, but occur over a temperature range (see Fig. 3.28 right panel). As with pure substances, the phase transitions during cooling of mixtures are exothermic processes. Therefore, the temperature during these transitions is not constant, but abrupt changes in the overall cooling curve are still visible when the cooling process begins and ends (i.e. the phase transition still leads to occurrence of kinks in the graph). Notably, azeotropes and eutectics behave like a pure substance and show a constant temperature during the phase transition.

3.6   Exercises

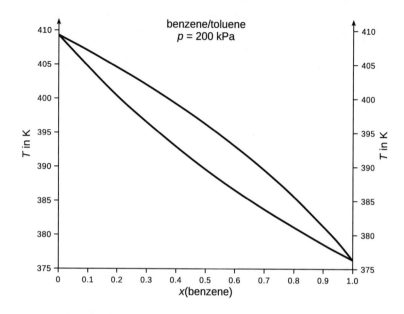

1. Is it possible for a one-component system to exhibit a quadruple point?

2. Henry's law is valid for dilute solutions. Using the Henry's law constant for oxygen (solute) and water (solvent) of $K(O_2) = 781 \cdot 10^5$ Pa M^{-1}, calculate the molar concentration of oxygen in water at sea level with an atmospheric pressure of $p_{atm} = p^{\varnothing}$.

3. Below is the T-x phase diagram of the benzene/toluene system acquired at a constant pressure of two bar. A mixture that contains 40% benzene is heated steadily to 122 °C. How many phases are present at this point and what are their compositions? If the total amount of 1 mol of substances was in the initial mixture with 40% benzene, how many moles of substances are in the phase(s) at 122 °C?

4. A mixture of benzene and toluene with x(benzene) = 0.4 is subjected to fractional distillation at 2 bar (see Exercise 3.3 above). What is the boiling temperature of this mixture? How many theoretical plates are required as a minimum to obtain pure benzene in the distillate?

Solutions of Electrolytes

4

4.1 Fundamental Concepts

4.1.1 Ions in Solution

In the previous sections, we have mostly assumed that the systems under study consisted of non-dissociating solutes, i.e. non-electrolytes. We now want to expand considerations to substances that dissociate in solution. Such substances are called electrolytes and form ions when being dissolved in solvents. Electrolytes can be classified into strong and weak electrolytes.

Strong electrolytes completely dissociate in solution:

$$
\begin{aligned}
NaCl_{(s)} + H_2O &\rightarrow Na^+{}_{(aq)} + Cl^-{}_{(aq)} + H_2O \\
HCl_{(g)} + H_2O &\rightarrow H_3O^+{}_{(aq)} + Cl^-{}_{(aq)} \\
AgCl_{(s)} + H_2O &\rightarrow Ag^+{}_{(aq)} + Cl^-{}_{(aq)} + H_2O
\end{aligned}
$$

In contrast, weak electrolytes do not fully dissociate in solution, due to inter-ionic interactions:

$$
H_3C - COOH_{(l)} + H_2O \rightarrow H_3O^+{}_{(aq)} + H_3C - COO^-{}_{(aq)}
$$

The occurrence of ions in solutions of electrolytes is not dependent on the flow of current; electrolytes dissociate readily upon dilution in solvent. The degree of dissociation α (see Sect. 4.3.3) describes the fraction of solute present as ions.

In aqueous solution, ions have water molecules associated with them; this is called the hydration shell. Ions change the structure of the water hydrogen bond network. In the presence of an ion, a water molecule reorients such that its polarised charge faces the opposite charge of the ion. During this reorientation process, the hydrogen bonds of the water molecule to its nearest neighbours is broken. The orientation of the water molecules that form the hydration shell directly around the ion is called the inner solvation shell (see Fig. 4.1, orange molecules) and results in a

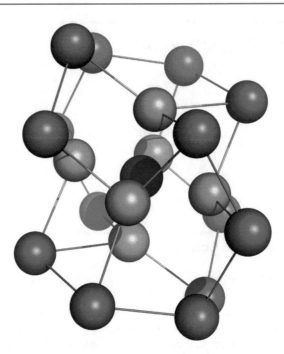

Fig. 4.1 Quantum-chemical calculations with the semi-empirical AM1 method show that mono-atomic ions (such as e.g. Na^+, blue) exist in a hydration shell resulting in the complex $[Na(OH_2)_{20}]^+$ (Peslherbe et al. 2000). The shell consists of an inner solvation shell where six water molecules take the vertex positions of an octahedron (orange). The remaining 14 water molecules (red) form the second solvation shell. The water structure around the cation takes the form of a puckered dodecahedron. The formation of this structure can be thought of as a 'pulling in' of the inner solvation shell molecules from their initial positions on the vertices of a regular dodecahedron

net charge on the outside of this shell. This charge is of the same sign as that of the ion in the centre. The charge on the outside of this inner hydration shell causes further water molecules in the vicinity to reorient, leading to a second solvation shell (see Fig. 4.1, red molecules).

The occurrence of such hydration shells explains the freezing point depression of solutions. The hydration shells of dissolved ions disrupt the hydrogen bonding network of water that would otherwise form the hexagonal structure of ice.

4.1.2 Charge and Electroneutrality

The charge Q describes the quantity of electricity; it can be positive (cations, protons) or negative (anions, electrons). The charge is measured in units of Coulomb: $[Q] = 1$ C.

The elementary charge is the charge of the electron: $|Q(e^-)| = e = 1.602 \cdot 10^{-19}$ C.

Table 4.1 Examples of different types of conductors

Electronic conductors	Ionic conductors	Mixed types
Metals	Seawater: $Na^+_{(aq)}$, $SO_4^{2-}_{(aq)}$	Plasma: e^-, gas ions
Graphite	$ZrO_{2(s)}$: O^{2-}	$e^-(NH_3) + Na^+(NH_3)$
Semiconductors	$RbAg_4I_5$: Ag^+	H_2 in Pd: H^+, e^-
PbO_2	Pure water: $H_3O^+_{(aq)}$, $OH^-_{(aq)}$	
Polypyrrole		

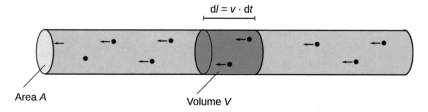

Fig. 4.2 Mobile charges passing through a cylindrical volume element V. The cylinder has a cross-section area A. The length dl of the volume element is given by the speed v of the charges multiplied with the time interval dt

If matter possesses only fixed charges, then it is called an insulator. In contrast, the existence of mobile charges makes matter a conductor. In case of an electronic conductor, the mobile charges are electrons, in ionic conductors, the mobile charges are ions. Some examples are given in Table 4.1. In solutions, ions represent the moving charges and are thus responsible for conducting electricity.

The principle of electroneutrality states that there can be no significant net charge in any macroscopic volume within a conductor. This is a consequence of the work required to separate opposite charges, or to bring like charges into close contact. This work raises the free energy change of the underlying process that may lead to unbalanced charges, thus making it less spontaneous. The amount of unbalanced charges that is allowed is due to different concentrations of oppositely charged species that are not chemically significant (and thus results in differences in the electric potential of no more than a few volts; see Sects. 4.1.4 and 4.2.3).

4.1.3 Electric Current

Current (I) is the flow of charge dQ in a particular time interval dt through a defined volume:

$$I = \left(\frac{dQ}{dt}\right)_{volume} \tag{4.1}$$

If we consider charges moving through a cylindrical wire (Fig. 4.2), we can calculate the amount of charge dQ passing through a cylindrical volume element during the time interval dt by counting the number of passing charges (N_{passing}) and multiply with the unit charge e as well as the charge state z:

$$I = \left(\frac{dQ}{dt}\right)_{volume} = \frac{N_{\text{passing}} \cdot (z \cdot e)}{dt} \tag{4.2}$$

By multiplying and diving with the volume V of the volume element and substituting

$$N_{\text{passing}} = n \cdot N_A \quad \text{and} \quad V = A \cdot dl$$

we obtain:

$$I = \frac{\frac{N_{\text{passing}}}{V} \cdot V \cdot (z \cdot e)}{dt} = \frac{\frac{n \cdot N_A}{V} \cdot A \cdot dl \cdot (z \cdot e)}{dt}$$

We can further substitute

$$c = \frac{n}{V} \quad \text{and} \quad dl = v \cdot dt$$

which yields

$$I = \frac{\frac{N_{\text{passing}}}{V} \cdot V \cdot (z \cdot e)}{dt} = \frac{c \cdot N_A \cdot A \cdot v \cdot dt \cdot (z \cdot e)}{dt}$$

This simplifies to:

$$I = \frac{c \cdot N_A \cdot A \cdot v \cdot dt \cdot (z \cdot e)}{dt} = z \cdot c \cdot N_A \cdot A \cdot v \cdot e = z \cdot c \cdot A \cdot v \cdot (e \cdot N_A)$$
$$= z \cdot c \cdot A \cdot v \cdot F \tag{4.3}$$

F is the Faraday constant and has the value of:

$$F = N_A \cdot e = 6.022 \cdot 10^{23} \, \text{mol}^{-1} \cdot 1.602 \cdot 10^{-19} \, C = 96485 \, C \, \text{mol}^{-1}$$

Equation 4.3 provides a relation between the current and the molar concentration c of charged particles with the charge state z that move with the speed v.

4.1.4   Electric Potential

Figure 4.3 shows the scheme of a simple electric circuit where a direct current power supply gives rise to electrons moving from the cathode (excess of electrons) to the anode (shortage of electrons) through a cylindrical wire. The fact that electrons move

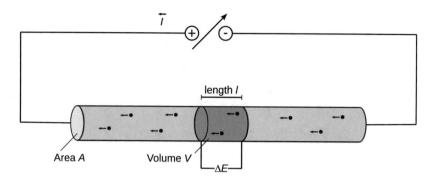

Fig. 4.3 Illustration to derive the relationship between the potential difference ΔE, current I and the travelling parameters (length, cross-sectional area) of the moving charges

from the cathode to the anode can be conceptualised by the existence of a potential difference ΔE between the cathode and the anode. The potential difference ΔE is measured in units of volts:

$$[\Delta E] = 1\,V.$$

Whereas the physical movement of electrons has the direction cathode $\rightarrow$ anode, the current I is defined to flow in the opposite direction.

As in the previous section, we consider a volume element V in the cylindrical wire of Fig. 4.3; the volume of this element can be calculated from the cross-sectional area A and the length l, as per $V = A \cdot l$.

It is obvious that there will be more current flowing, if the potential difference of the power supply is higher; I and ΔE will thus be proportional to each other. Because the potential difference between the anode and the cathode is taken as positive, and the current flows from the anode to the cathode, the directions of the two phenomena are opposite, hence:

$$\Delta E \sim -I \tag{4.4}$$

If we increase the length l of the volume element, we will need to increase the potential difference ΔE to have the same amount of charges (i.e. the same current) flowing through that volume element. Therefore:

$$\Delta E \sim l \tag{4.5}$$

The opposite is true for the cross-sectional area A. If we widen the area A, but require the same amount of charges to flow through, the potential difference ΔE needs to be decreased. It thus appears that

$$\Delta E \sim \frac{1}{A} \qquad (4.6)$$

It follows from Eqs. 4.4–4.6:

$$\Delta E \sim \frac{-I \cdot l}{A} \qquad (4.7)$$

If we define a new quantity j called the current density as

$$j = \frac{I}{A} \qquad (4.8)$$

then we obtain from Eq. 4.7:

$$j = \frac{I}{A} \sim -\frac{\Delta E}{l} \qquad (4.9)$$

As proportionality factor we introduce κ as the electric conductivity:

$$j = \frac{I}{A} = -\kappa \cdot \frac{\Delta E}{l} = -\kappa \cdot \frac{d\phi}{dx} \quad \text{with } [j] = 1 \text{ A m}^{-2} \qquad (4.10)$$

The potential difference ΔE over a distance l is called the electric field. The electric field is the gradient of the electric potential ϕ over a distance:

$$\frac{d\phi}{dx} = \frac{\Delta E}{l} \quad \text{with } \left[\frac{d\phi}{dx}\right] = 1 \text{ V m}^{-1} \qquad (4.11)$$

It follows that the conductivity κ is measured in units of siemens per metre:

$$[\kappa] = 1 \frac{A \cdot m}{m^2 \cdot V} = 1 \frac{A}{V \cdot m} = \frac{1}{\Omega \cdot m} = 1 \text{ S m}^{-1} \qquad (4.12)$$

Since conductivity is a specific property of a substance, it lends itself as a quantity to distinguish conducting from insulating matter (see Table 4.2).

4.1.5 Resistance

An important relationship for electric circuits (see Fig. 4.3) is that between the electric potential and the current flowing through the circuit. From Eq. 4.10 we can resolve an expression for the potential ΔE:

$$j = \frac{I}{A} = -\kappa \cdot \frac{\Delta E}{l} \qquad (4.10)$$

Table 4.2 Conductivity of different substances

Material	κ (S m^{-1})	Charge carrier	Property
Superconductors	∞ (at low temperature)	Electron pair	$\uparrow$
Cu	$6 \cdot 10^7$	e^-	Conducting
Hg	$1 \cdot 10^6$	e^-	
Graphite	$4 \cdot 10^4$	π-electrons	
Molten KCl	220 (at $T = 1043$ K)	K^+, Cl^-	
Battery acid	80	$H_3O^+_{(aq)}$, $HSO_4^-{}_{(aq)}$	
Seawater	5.2	Cations, anions	
Ge	2.2	e^-, holes	
0.1 M KCl$_{(aq)}$	1.3	$K^+_{(aq)}$, $Cl^-_{(aq)}$	
H_2O	$6 \cdot 10^{-6}$	$H_3O^+_{(aq)}$, $OH^-_{(aq)}$	
Typical glass	$3 \cdot 10^{-10}$	Univalent cations	
Teflon	10^{-15}	Impurities	Insulating
Vacuum, most gases	0	None	$\downarrow$

$$\Delta E = -\frac{I \cdot l}{\kappa \cdot A} \tag{4.13}$$

The quotient $\frac{l}{\kappa \cdot A}$ with the length l, the area A and the conductivity κ describes constants of a particular electric circuit and as such represents a material constant for the given system. This quotient is thus defined as the electric resistance R:

$$R = \frac{l}{\kappa \cdot A} \tag{4.14}$$

and Eq. 4.13 yields:

$$\mid \Delta E \mid = I \cdot R \tag{4.15}$$

which constitutes Ohm's law. The resistance is measured in units of ohm: $[R] = 1\ \Omega$.

The electric resistance of a conductor represents the opposition to the passage of charges through that conductor. Intuitively, the willingness of a conductor to let charges pass through will be a quantity that is inversely related to the resistance. The quantity describing the ease with which charges can pass through is called the conductance G:

$$G = \frac{1}{R} = \frac{I}{\Delta E} \tag{4.16}$$

The conductance is measured in units of siemens: $[G] = 1\ \Omega^{-1} = 1$ S. Like the resistance, the conductance is a property of the conductor used in the electric circuit.

4.1.6   Conductivity and Conductance

The electric conductivity κ is the ratio between the current density and the electric field (Eq. 4.10). As such, it is a property of the conducting material. In contrast, the conductance G describes the current-carrying capacity of the electrolytic substance; in solutions of electrolytes, the conductance is thus a property of the dissolved ions. The conductance of electrolytic solutions increases with dilution for both strong and weak electrolytes, because at low concentration there is less hindrance for the migrating ions from neighbours.

The specific conductance g measures the current-carrying capacity of all ions in a specific volume:

$$g = \frac{G}{V}, \quad \text{with } [g] = 1 \text{ S m}^{-3} \tag{4.17}$$

For strong electrolytes, the specific conductance decreases with dilution.

Whereas the conductance G depends on the physical size of the conductor, the conductivity κ is independent of the conductor size. Conductivity and conductance are related as per:

$$|j| = |\frac{I}{A}| = -\kappa \cdot \frac{\Delta E}{l} \tag{4.10}$$

$$\frac{|I| \cdot l}{A \cdot \Delta E} = \kappa$$

With $G = \frac{1}{R} = \frac{I}{\Delta E}$ it follows:

$$\frac{G \cdot l}{A} = \kappa$$

$$G = \kappa \cdot \frac{A}{l} \tag{4.18}$$

The conductance of a system is therefore equal to the conductivity of this system, multiplied by the area through which the migrating charges pass, and divided by the distance travelled.

4.2   Electrochemical Reactions

After having introduced some fundamental concepts and quantities of electricity, we will now consider their applications in a chemical context. Electrochemical reactions can proceed in systems when a suitable electric potential difference ΔE is applied. In this case, electric energy is converted into chemical energy. Such cells are called electrolytic cells.

Alternatively, chemical reactions may result in the build-up of an electric potential difference, i.e. chemical energy is converted into electric energy. These cells are called galvanic cells.

Conceptually, an electrochemical cell is separated into two half-cells, separating the oxidation and the reduction processes. The two electrochemical half reactions oxidation and reduction are combined into a net reaction, called Redox reaction. For example:

Fe^{2+} ions in aqueous solution are unstable and oxidised readily to Fe^{3+} by oxygen from ambient air:

$$
\begin{array}{lll}
O_{2(g)} + 4\,H^+_{(aq)} + 4\,e^- & \rightarrow & 2\,H_2O_{(l)} & \text{Red.} \\
Fe^{2+}_{(aq)} & \rightarrow & Fe^{3+}_{(aq)} + e^- & \text{Ox.} \\
4\,Fe^{2+}_{(aq)} + O_{2(g)} + 4\,H^+_{(aq)} & \rightarrow & 4\,Fe^{3+}_{(aq)} + 2\,H_2O_{(l)} & \text{Redox}
\end{array}
$$

Bromo-naphthalene is a component of battery paste and used as an insulating liquid in graphite paste electrodes. It can undergo a Redox reaction with iodide in lithium-iodide batteries:

$$
\begin{array}{lll}
C_{10}H_7Br_{(l)} + 2\,e^- & \rightarrow & Br^-_{(soln)} + C_{10}H_7^-_{(soln)} & \text{Red.} \\
3\,I^-_{(aq)} & \rightarrow & 2\,e^- + I_3^-_{(aq)} & \text{Ox.} \\
C_{10}H_7Br_{(l)} + 3\,I^-_{(aq)} & \rightarrow & C_{10}H_7^-_{(soln)} + Br^-_{(soln)} + I_3^-_{(aq)} & \text{Redox}
\end{array}
$$

The lead acid battery (invented by Gaston Planté in 1859, and still in use today as car battery) may produce H_2 gas when the battery is deeply or rapidly discharged:

$$
\begin{array}{lll}
2\,H^+_{(aq)} + 2\,e^- & \rightarrow & H_{2(g)} & \text{Red.} \\
Pb_{(s)} + SO_4^{2-}_{(aq)} & \rightarrow & 2\,e^- + PbSO_{4(s)} & \text{Ox.} \\
Pb_{(s)} + SO_4^{2-}_{(aq)} + 2\,H^+_{(aq)} & \rightarrow & PbSO_{4(s)} + H_{2(g)} & \text{Redox}
\end{array}
$$

4.2.1 Galvanic and Electrolytic Cells

If two half-cells are combined where the different chemical reactions give rise to an electric potential difference, the resulting electrochemical cell is called a galvanic cell. A prominent example is the so-called Daniell element shown in Fig. 4.4, left panel. Here, the spontaneously proceeding Redox reaction converts chemical into electric energy; this constitutes the working principle of batteries.

According to Faraday, the definition of anode and cathode depend on charge and discharge. The anode is the electrode to which anions flow. The anode possesses positive potential and thus gives rise to an oxidation reaction, where suitable species (elementary zinc in Fig. 4.4, left panel) deposit their electrons onto the electrode. In contrast, the cathode is the electrode attracting cations; it is the electrode with negative potential. Here, suitable species (Cu^{2+} ions in Fig. 4.4, left panel) are reduced by receiving electrons from the electrode.

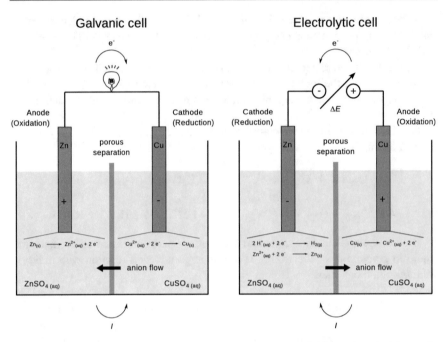

Fig. 4.4 Illustration of the processes in a galvanic cell (left panel) and electrolytic cell (right panel), using the Daniell element which consists of Zn/ZnSO₄ and Cu/CuSO₄ half-cells. At the cathode of the electrolytic cell, one would expect hydrogen gas evolving, as the H^+/H_2 Redox potential is 0.76 V in favour as compared to Zn^{2+}/Zn (see Table 4.3). However due to the hydrogen overpotential, H_2 generation is hindered and thus elementary zinc is deposited

In an electrolytic cell, a chemical reaction is forced to occur due to current flowing through a cell. An electric potential needs to be applied externally (Fig. 4.4, right panel), causing conversion of electric to chemical energy. Both electrodes are placed in a container that contains the solution of molten or dissolved electrolyte. The external power supply provides the electrons. They enter the electrolyte solution through the cathode (negative potential) and leave through the anode (positive potential).

At the anode, which has a lack of electrons due to the positive potential, electrons are removed from suitable substances at the anode surface—an oxidation occurs. The cathode, in contrast, possesses a surplus of electrons due to its negative potential. Here, electrons will be transferred onto suitable substances—this is a reduction.

Figure 4.4 contrasts the two different electrochemical cell types by using the same element, i.e. a Zn/ZnSO₄ and a Cu/CuSO₄ half-cell (the Daniell element). Of course, a galvanic cell is not supposed to be used as an electrolytic cell, hence the right panel of Fig. 4.4 is for illustration of the concept only. Note that when the Daniell element is inverted to become an electrolytic cell, the reduction at the anode does not deposit elementary zinc at the electrode, but rather produces hydrogen gas from water, since the reduction of hydrogen is electrochemically more favourable (see Sect. 4.2.8).

In technical applications, electrolytic cells are not constructed as two physical half-cells, but one cell which comprises the molten or dissolved electrolyte (e.g. $CuSO_4$) and electrodes made of the corresponding metal (Cu). In such cells, solid metal (Cu) will be deposited on the cathode; impurities either remain in solution or collect as an insoluble sludge. This process is known as electrolytic refinement and used to obtain metals of highest purity.

Rechargeable Batteries

Rechargeable batteries, such as e.g. NiMH cells or lead-acid batteries, act as galvanic cells when discharging, i.e. they convert chemical energy to electrical energy, and as electrolytic cells when being charged. In the charging process, electrical energy is converted to chemical energy.

In summary, the reaction happening at the anode is oxidation and that at the cathode is reduction. Electrons are supplied by the chemical species getting oxidised at the anode, leave the electrolyte solution through the anode and enter the electrolyte solution in the other half-cell through the cathode. In this circuit, the anode therefore has a positive potential, and the cathode a negative potential.

Overpotential

In the electrolytic cell in Fig. 4.4, the electrodeposition of zinc will occur in competition with the generation of hydrogen. Whereas thermodynamically the reduction of hydrogen is favoured (the Redox potential of the H^+/H_2 element is higher than that of the Zn^{2+}/Zn element, so less energy is required to push electrons into the H^+/H_2 process), the Zn^{2+} reduction process is kinetically favoured. By way of steric hindrance, it is difficult for hydrogen atoms to move around on the surface of the zinc electrode to eventually form H_2 molecules. This difficulty varies for different metal surfaces. In electrochemical cells, this phenomenon gives rise to an overpotential which needs to be considered when designing galvanic and, importantly, electrolytic cells. Processes such as electrolytic refinement of metals, which require large quantities of electricity, need to be optimised for power consumption and thus consider overpotentials.

4.2.2   The Faraday Laws

Of great importance for industrial applications of electrochemical processes as well as for the development of electrochemistry in general have been the Faraday laws, which Michael Faraday developed in 1834 (Ehl and Ihde 1954).

Consider the electrolytic cell illustrated in the right panel of Fig. 4.4 in the previous section. At the cathode, hydrogen gas is produced in an electrochemical reduction, powered by the application of an electrochemical potential to the element. The gas leaving the half-cell can be captured and thus its volume can be determined. The quantity of gas produced will be dependent on the current flowing through the cell, i.e. the quantity of electrical charge passing through.

This is summarised in Faraday's first law of electrolysis:

▶   The mass of a substance altered at an electrode during electrolysis is directly proportional to the quantity of electricity transferred at that electrode.

$$m \sim I \cdot t = Q \tag{4.19}$$

If one was to serially combine several different electrolytic cells in each of which a different elementary substance is produced at the cathode, there will be a particular mass of that element generated in the individual cells.

Faraday's second law of electrolysis states:

▶   For a given quantity of electric charge, the mass of a deposited/ generated elementary substance is proportional to the molar mass of that substance divided by the change in oxidation state (i.e. in most cases the charge of the cation in the electrolyte).

$$m \sim \frac{M}{z} \quad \text{with} \quad Q = \frac{I}{t} = \text{const.} \tag{4.20}$$

Applying Eq. 4.20 to two different electrolytic cells 1 and 2 through which the same amount of charge Q is passed, one obtains the following mass ratio of generated elementary substances:

$$\frac{m_1}{m_2} = \frac{\frac{M_1}{z_1}}{\frac{M_2}{z_2}} = \frac{M_1 \cdot z_2}{M_2 \cdot z_1} \tag{4.21}$$

4.2.3   The Electromotive Force

We consider a galvanic cell where two half-cells are combined to generate an electric potential difference ΔE. This potential difference arises from the different chemical potentials in the two half-cells that causes electrons to flow from one half-cell to the other. It is thus called the electromotive force (e.m.f.) of the cell (see Fig. 4.5). Importantly, the value of the e.m.f. can only be established if a negligible current is drawn from the cell (which is a requirement for any Volt meter used to determine the

Fig. 4.5 The convention for determining the electromotive force of an electrochemical cell; $e.m.f. = \Delta E = E_{\text{right}} - E_{\text{left}}$

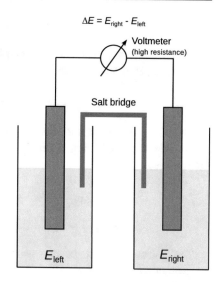

$\Delta E = E_{\text{right}} - E_{\text{left}}$

Voltmeter (high resistance)

Salt bridge

E_{left} E_{right}

voltage when placed in parallel to a resistor). If the two half-cells were connected by a short circuit (i.e. the external resistance would be zero, $R_{\text{ext}} = 0$), then current flow would be maximised and there was no potential difference ($\Delta E = 0$). If the two half-cells are connected by a Volt meter which possesses a very large resistance, then the current I is negligible and the potential difference ΔE is maximised. This is the desired configuration, as any current flowing between the two half-cells is the consequence of a proceeding Redox reaction and thus inevitably results in a lowering of the potential difference.

The combination of two half-cells is denoted by separating the electrode material and electrolyte solutions of the left and right half-cells by vertical lines. For example, the Daniell element, which we have introduced in Sect. 4.2.1, is written as:

$$\text{Zn}_{(s)} \big| \ 1 \text{ M ZnSO}_{4(\text{aq})} \big| \ 1 \text{ M CuSO}_{4(\text{aq})} \ \big| \ \text{Cu}_{(s)} \quad e.m.f. = \Delta E = +1.103 \text{ V}$$

The above cell denotes a zinc electrode immersed in an electrolyte solution of 1 M aqueous ZnSO_4 as the left cell and a copper electrode immersed in 1 M aqueous CuSO_4 solution as the right cell.

Importantly, the *e.m.f.* (or electric potential difference) is by convention defined as

$$e.m.f. = \Delta E = E_{\text{right}} - E_{\text{left}} \tag{4.22}$$

with E_{right} and E_{left} being the electric potentials of the right and left half-cells, respectively. In the example above, the right half-cell consisting of the system:

$$Cu^{2+}_{(aq)} + 2\,e^- \rightarrow Cu_{(s)} \quad E_{right} = 0.340\ V$$

possesses the higher potential and therefore pulls electrons from the left half-cell which consists of the system:

$$Zn^{2+}_{(aq)} + 2\,e^- \rightarrow Zn_{(s)} \quad E_{left} = -0.763\ V$$

Therefore, the spontaneously proceeding reactions in this electrochemical cell will consist of the reduction of Cu^{2+} ions and the oxidation of Zn, with the following net reaction:

$$Zn_{(s)} + Cu^{2+}_{(aq)} \rightarrow Zn^{2+}_{(aq)} + Cu_{(s)}$$

$$\Delta E = E_{right} - E_{left} = 0.340\ V - (-0.763\ V) = 1.103\ V \qquad (4.23)$$

The potential difference between the right and the left half-cell yields the electromotive force of that cell.

Importantly, when the number of electrons consumed or produced in the two half-cells differ, the Redox potentials must not be multiplied. In manipulating potentials, one can only change the signs of the values, not the magnitude. For example, a combination of the hydrogen electrode (see Sect. 4.2.7) with the silver/silver chloride electrode comprises the following reactions:

$$Ag^+_{(aq)} + e^- \rightarrow Ag_{(s)} \mid \cdot 2 \quad E_{right} = 0.222\ V$$
$$2\,H^+_{(aq)} + 2\,e^- \rightarrow H_{2(g)} \mid \cdot 1 \quad E_{left} = 0\ V$$

and yields the net spontaneous reaction:

$$H_{2(g)} + 2\,Ag^+_{(aq)} \rightarrow 2\,H^+_{(aq)} + 2\,Ag_{(s)}$$

$$\Delta E = E_{right} - E_{left} = 0.222\ V - 0\ V = 0.222\ V$$

The multiplication with factor 2 in above Ag^+/Ag reaction only applies to the chemical reaction, not to the Redox potential.

Sacrificial Anodes

The above example has an important application in the protection of active metals that may be subject to corrosion. Consider large metal containers such as hulls of ships, water heaters, pipelines, distribution systems, or metal tanks, all of which are made of metals that could potentially be oxidised (e.g. copper). By combination with a less valuable metal (such as zinc), an electrochemical cell is generated which uses zinc (or any metal alloy with a more negative electrochemical potential than the other metal) as the sacrificial anode. When

(continued)

exposed to environmental processes, the sacrificial anode will be consumed in place of the metal it is protecting. Obviously, these anodes must be periodically inspected and replaced when consumed, in order for the protection to continue.

4.2.4   Concentration Dependence of the Electromotive Force

The electromotive force of a cell arises from spontaneously proceeding reactions in the half-cells. In the half-cell where the oxidation occurs we can formulate the following reaction:

$$Ox + z\,e^- \rightarrow Red$$

where 'Red' denotes a substance to be oxidised (a reducing agent), and 'Ox' denotes a substance to be reduced (an oxidation agent). The electromotive force arising from this reaction is also called the Redox potential of this 'Red'-'Ox' pair.

The change in the free energy ΔG of the above reaction represents the electric work the system can provide to the environment:

$$\Delta W_{el} = \Delta G \tag{4.24}$$

We have established earlier that

$$\Delta W_{el} = R \cdot I^2 \cdot \Delta t \tag{2.10}$$

$$I = \frac{N_{passing} \cdot z \cdot e}{\Delta t} \tag{4.3}$$

If 1 mol charges are passing through, then $N_{passing} = n \cdot N_A$, and thus

$$I = \frac{N_A \cdot z \cdot e}{\Delta t} = \frac{z \cdot F}{\Delta t} \quad \text{for 1 mol charges} \tag{4.25}$$

We can then calculate the electric work based on 1 mol passing charges:

$$\Delta W_{m,el} = R \cdot I^2 \cdot \Delta t = (R \cdot I) \cdot I \cdot \Delta t$$

From Eq. 4.15 we know that $R \cdot I = \Delta E$. ΔE describes the potential difference between two electrodes or half-cells. Since we are here considering just one half-cell (the one where oxidation occurs), we describe an absolute potential $E = R \cdot I$, albeit this absolute value will be impossible to determine (see Sect. 4.2.7). Furthermore, using Eq. 4.25 yields:

$$\Delta W_{m,el} = (R \cdot I) \cdot I \cdot \Delta t = E \cdot \frac{z \cdot F}{\Delta t} \cdot \Delta t$$

$$\Delta W_{m,el} = z \cdot F \cdot E \tag{4.26}$$

Considering the molar Gibbs free energy in Eq. 4.24 allows the following conclusion:

$$\Delta W_{m,el} = z \cdot F \cdot E = \Delta G_m$$

and we remember that the difference of molar Gibbs free energy is the difference of chemical potential (Eq. 2.67):

$$z \cdot F \cdot E = \Delta \mu$$

and this difference is the difference between the chemical potential of the oxidised and reduced states of the Redox pair

$$z \cdot F \cdot E = (\mu_{Ox} - \mu_{Red})$$

Using Eq. 3.12 ($\mu = \mu^{\emptyset} + R \cdot T \cdot \ln a$, and assuming an activity coefficient of $\gamma_c = 1 \, 1\mathrm{mol}^{-1}$ such that $a = |c|$), this resolves to:

$$z \cdot F \cdot E = \mu_{ox}^{\emptyset} + R \cdot T \cdot \ln \left[\frac{c(Ox)}{c^{\emptyset}} \right] - \mu_{Red}^{\emptyset} + R \cdot T \cdot \ln \left[\frac{c(Red)}{c^{\emptyset}} \right]$$

$$z \cdot F \cdot E = \Delta \mu^{\emptyset} + R \cdot T \cdot \ln \frac{c(Ox)}{c(Red)}$$

$$E = \frac{\Delta \mu^{\emptyset}}{z \cdot F} + \frac{R \cdot T}{z \cdot F} \cdot \ln \frac{c(Ox)}{c(Red)}$$

When combining the quotient of constants $\frac{\Delta \mu^{\emptyset}}{z \cdot F}$ into a new constant $E^{\emptyset}$ (called the standard e.m.f.) and inverting the argument of the logarithm, we obtain:

$$E = E^{\emptyset} - \frac{R \cdot T}{z \cdot F} \cdot \ln \frac{c(Red)}{c(Ox)} \tag{4.27}$$

This equation is known as the Nernst equation and describes the concentration dependence of the electromotive force, and thus the Redox potential.

The standard electromotive force $E^{\emptyset}$ is the standard electrode potential (standard Redox potential) of the 'Red'-'Ox' pair.

4.2.5   Combination of Two Half-Cells

In Sect. 4.2.3, we have introduced the combination of two half-cells as illustrated by the Daniell element:

$$Zn_{(s)} \mid 1 \text{ M } ZnSO_{4(aq)} \mid 1 \text{ M } CuSO_{4(aq)} \mid Cu_{(s)} \quad e.m.f. = \Delta E = +1.103 \text{ V}$$

In this formalism, an electrochemical cell is denoted by a series of compartments. It is understood by convention that the electrode potential of the right-hand electrode is higher than that of the left-hand electrode, because the e.m.f. is reported positive in above example, and calculated as

$$e.m.f. = \Delta E = E_{\text{right}} - E_{\text{left}} \tag{4.22}$$

so if $e.m.f. > 0$ then $E_{\text{right}} > E_{\text{left}}$.

If the cell above had been written in reverse order, then the *e.m.f.* would be negative:

$$Cu_{(s)} \mid 1 \text{ M } CuSO_{4(aq)} \mid 1 \text{ M } ZnSO_{4(aq)} \mid Zn_{(s)} \quad e.m.f. = \Delta E = -1.103 \text{ V}$$

Importantly, the electrode with the higher potential is always the one where reduction occurs; the electrode with the lower potential is where oxidation occurs.

We can re-formulate the above example in a more general fashion, combining a Redox pair in the left half-cell (Red_{left}, Ox_{left}) with one in the right half-cell (Red_{right}, Ox_{right}):

$$Red_{\text{left}} \mid Ox_{\text{left}} \mid Ox_{\text{right}} \mid Red_{\text{right}} \quad e.m.f. = \Delta E > 0$$

with the following chemical reactions:

$$
\begin{array}{llll}
Red_{\text{left}} & \longrightarrow & Ox_{\text{left}} + z\,e^- & E_{\text{left}} \quad \text{oxidation} \\
Ox_{\text{right}} + z\,e^- & \longrightarrow & Red_{\text{right}} & E_{\text{right}} \quad \text{reduction} \\
Red_{\text{left}} + Ox_{\text{right}} & \longrightarrow & Ox_{\text{left}} + Red_{\text{right}} & \Delta E \quad \text{Redox}
\end{array}
$$

If $E_{\text{right}} > E_{\text{left}}$, then electrons will be flowing from the left half-cell to the right half-cell. That means ΔE is positive, and all reactions above will proceed from the left to right.

$$\Delta E = E_{\text{right}} - E_{\text{left}} \tag{4.22}$$

$$\Delta E = E_{\text{right}}^{\varnothing} - \frac{R \cdot T}{z \cdot F} \cdot \ln \frac{c(Red_{\text{right}})}{c(Ox_{\text{right}})} - E_{\text{left}}^{\varnothing} + \frac{R \cdot T}{z \cdot F} \cdot \ln \frac{c(Red_{\text{left}})}{c(Ox_{\text{left}})}$$

$$\Delta E = E_{right}^{\varnothing} - E_{left}^{\varnothing} + \frac{R \cdot T}{z \cdot F} \cdot \left(\ln \frac{c(Red_{left})}{c(Ox_{left})} - \ln \frac{c(Red_{right})}{c(Ox_{right})} \right)$$

$$\Delta E = \Delta E^{\varnothing} + \frac{R \cdot T}{z \cdot F} \cdot \left(\ln \frac{c(Red_{left})}{c(Ox_{left})} + \ln \frac{c(Ox_{right})}{c(Red_{right})} \right)$$

$$\Delta E = \Delta E^{\varnothing} + \frac{R \cdot T}{z \cdot F} \cdot \ln \frac{c(Red_{left}) \cdot c(Ox_{right})}{c(Ox_{left}) \cdot c(Red_{right})}$$

$$\Delta E = \Delta E^{\varnothing} + \frac{R \cdot T}{z \cdot F} \cdot \ln \frac{1}{K}$$

$$\Delta E = \Delta E^{\varnothing} - \frac{R \cdot T}{z \cdot F} \cdot \ln K \tag{4.28}$$

Electromotive Force Under Non-standard Concentrations

We can now consider the Daniell element at varying electrolyte concentrations, and calculate the electromotive force for example under the following conditions:

$$Zn_{(s)} \mid 0.01\ M\ ZnSO_{4(aq)} \mid 0.2\ M\ CuSO_{4(aq)} \mid Cu_{(s)} \quad e.m.f. = \Delta E = ?$$

This cell denotes a zinc electrode immersed in an electrolyte solution of 0.01 M aqueous $ZnSO_4$ as the left cell and a copper electrode immersed in 0.2 M aqueous $CuSO_4$ solution as the right cell. The potential difference in this electrochemical cell is now calculated using the Nernst equation 4.28, considering the different concentrations of the electrolytes:

$$\Delta E = \Delta E^{\varnothing} - \frac{R \cdot T}{z \cdot F} \cdot \ln K \tag{4.28}$$

We have previously established the net reaction for this cell in Eq. 4.23, and thus obtain for the standard potential difference of this cell:

$$\Delta E^{\varnothing} = 1.103\ V$$

The equilibrium constant K is calculated from the reaction in 4.23 by:

$$K = \frac{c(Zn^{2+})}{c(Cu^{2+})}$$

Therefore:

(continued)

$$\Delta E = 1.103 \text{V} - \frac{8.3144 \cdot 298 \cdot \text{J} \cdot \text{K} \cdot \text{mol}}{2 \cdot 96485 \cdot \text{K} \cdot \text{mol} \cdot \text{C}} \cdot \ln \frac{c(\text{Zn}^{2+})}{c(\text{Cu}^{2+})}$$

$$\Delta E = 1.103 \text{V} - \frac{8.3144 \cdot 298 \cdot \text{J}}{2 \cdot 96485 \cdot \text{C}} \cdot \ln \frac{0.01 \text{ M}}{0.2 \text{ M}}$$

$$\Delta E = 1.103 \text{V} - \frac{8.3144 \cdot 298 \cdot \text{V} \cdot \text{C}}{2 \cdot 96485 \cdot \text{C}} \cdot \ln 0.05$$

$$\Delta E = 1.103 \text{ V} - \frac{8.3144 \cdot 298 \cdot \text{V}}{2 \cdot 96485} \cdot (-2.996)$$

$$\Delta E = 1.103 \text{V} + \frac{2.996 \cdot 8.3144 \cdot 298}{2 \cdot 96485} \text{ V}$$

$$\Delta E = 1.114 \text{ V}$$

When the reactants of the cell have reached their equilibrium concentrations, there is no electric current flowing between the two half-cells (such as for example in a flat battery), and $\Delta E = 0$. It then follows that:

$$\Delta E = 0 = \Delta E^{\varnothing} - \frac{\text{R} \cdot T}{z \cdot \text{F}} \cdot \ln K$$

$$\Delta E^{\varnothing} = \frac{\text{R} \cdot T}{z \cdot \text{F}} \cdot \ln K \tag{4.29}$$

Hence, the standard electrode potential difference $\Delta E^{\varnothing}$ of an electrochemical cell can be used to determine the equilibrium constant K of the system. The importance of this conclusion is that from a tabulation of standard electrode potentials for individual half-cells (see Sect. 4.2.8), one can derive the standard electrode potentials for the electrochemical cell (i.e. a combination of two half-cells) and predict the value of the equilibrium constant for that system.

Flat Battery
A battery that is exhausted ('flat') no longer generates an electric current when connected to an external circuit. In this condition, there is no chemical reaction occurring in the combination of the two half-cells, which means that the overall cell reaction is at equilibrium. In equilibrium state, all chemical components are present in their equilibrium concentrations. Therefore, we can state that when the reactants of an electrochemical cell have reached their equilibrium concentrations, the *e.m.f.* of the cell is zero:
$\Delta E = 0$ for a chemical reaction at equilibrium.

4.2.6   The Thermodynamics of the Electromotive Force

From Sect. 2.2.5, when we discussed the Gibbs free energy of a reaction, we know that

$$\Delta G_m^\varnothing = -R \cdot T \cdot \ln K \tag{2.59}$$

Also, from Eq. 4.29 in the previous section, we can derive for one half-cell that

$$z \cdot F \cdot E^\varnothing = R \cdot T \cdot \ln K \tag{4.29}$$

It thus follows that

$$\Delta G_m^\varnothing = -z \cdot F \cdot E^\varnothing \tag{4.30}$$

Therefore, by measuring the standard electromotive force of an electrochemical half-cell, $E^\varnothing$, we can determine the change in the molar Gibbs free energy G_m of the underlying reaction.

4.2.7   Reference Electrodes

The electric potential of a half-cell is a potential difference itself, namely between a solid metal (electrode) and the electrolyte solution in which it is immersed. This potential difference is called the electrode potential, and it is physically impossible to measure its value. However, as we have already introduced above, it is possible to measure the difference between the electrode potential of one half-cell when combining it with another half-cell. If one chooses a particular half-cell for reasons of comparison, i.e. a reference cell, then standardised electrode potentials can be measured and tabulated.

In this context, the standard hydrogen electrode has been introduced as the reference standard, whereby the electric potential of a platinum electrode which is exposed to H_2 gas with a pressure of $p_{normal} = 1.013$ bar at $\theta_{normal} = 25\ °C$ and immersed into a solution with $a(H^+) = 1$ M is arbitrarily set to $E^\varnothing = 0$:

$$H_{2(g)} \rightarrow 2\,H^+{}_{(aq)} + 2\,e^- \quad E^\varnothing = 0$$

In order to determine standard electrode (or Redox) potentials, one thus measures the *e.m.f.* of a cell, in which the concentration of solutions are all 1 M. When determining the standard electrode potentials of any electrode, the standard hydrogen electrode is chosen as the left electrode (see Fig. 4.6).

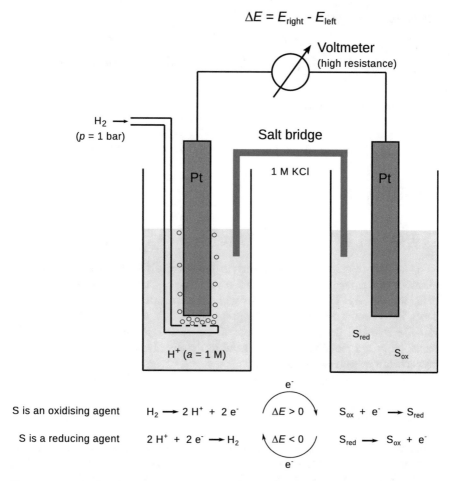

Fig. 4.6 The standard hydrogen electrode and the principle of determining standard electrode potentials

Since the standard hydrogen electrode requires a rather elaborate experimental setup, which proves impractical for many routine laboratory applications, other, more convenient reference electrodes are frequently being used. In principle, any half-cell can be employed which maintains a potential that remains practically unchanged during the course of an electrochemical measurement. One needs to evaluate the standard potential difference of the chosen electrode to the standard hydrogen electrode (see Sects. 4.2.8 and 4.3.1) and correct for this difference, when deploying the chosen electrode in a measurement. The most commonly used reference electrodes in this context are:

Standard hydrogen electrode

$$H_{2(g)} \mid H^{+}{}_{(aq)} \quad E^{\varnothing} = 0 \quad 2\,H^{+}{}_{(aq)} + 2\,e^{-} \quad \rightarrow \quad H_{2(g)}$$

Silver/silver chloride electrode

$$Ag_{(s)} \mid AgCl_{(s)} \quad E^{\varnothing} = 0.22\ \text{V} \quad AgCl_{(s)} + e^{-} \quad \rightarrow \quad Ag_{(s)} + Cl^{-}{}_{(aq)}$$

Calomel electrode

$$Hg_2Cl_{2(s)} \mid Hg_{(l)} \quad E^{\varnothing} = 0.26\ \text{V} \quad Hg_2Cl_{2(s)} + 2e^{-} \quad \rightarrow \quad 2Hg_{(l)} + 2\,Cl^{-}$$

Secondary reference electrodes such as the silver/silver chloride or the calomel electrode that are required to maintain a constant potential employ a rather insoluble metal ion salt ($AgCl$, Hg_2Cl_2) together with the elementary metal (Ag, Hg). In these cases, the potential difference with respect to the standard hydrogen electrode is not just the standard potential difference of the Redox reaction. Additionally, the transition of the metal ion from the solid to the dissolved state needs to be taken into account as well (see Sect. 4.3.1).

4.2.8   Standard Electrode Potentials

Standard electrode potentials are determined for any electrode by measuring the electromotive force or electric potential difference between that particular electrode and the standard hydrogen electrode. This enables tabulation of standard reduction potentials $E^{\varnothing}$ for individual electrodes or half cells. By convention, for such measurements to be conducted under standard conditions, the following parameters need to be ensured:

- The molar concentration of each ion needs to be $c = 1$ M
- The pressure of gases needs to be $p = p_{normal} = 1.013$ bar
- The temperature needs to be $\theta_{normal} = 25\,°C$.

Note that the standard reduction potentials indeed refer to 'normal' conditions. Selected standard electrode (Redox) potentials are given in Table 4.3. When predicting the direction of Redox reactions arising from the combination of half-cells, one needs to remember that the electron flow is from the lower to the higher potential. The electrode with the higher potential is where the reduction occurs ('oxidising'); the electrode with lower potential is where the oxidation occurs ('reducing').

Table 4.3 Standard reduction potentials for selected half-cells

Half-cell	$E^{\varnothing}$ (V)
Most electropositive, most reducing	
$Li^{+}_{(aq)} + e^{-} \rightarrow Li_{(s)}$	−3.04
$K^{+}_{(aq)} + e^{-} \rightarrow K_{(s)}$	−2.92
$Ca^{2+}_{(aq)} + 2\,e^{-} \rightarrow Ca_{(s)}$	−2.76
$Na^{+}_{(aq)} + e^{-} \rightarrow Na_{(s)}$	−2.71
$Mg^{2+}_{(aq)} + 2\,e^{-} \rightarrow Mg_{(s)}$	−2.38
$Al^{3+}_{(aq)} + 3\,e^{-} \rightarrow Al_{(s)}$	−1.66
$Zn^{2+}_{(aq)} + 2\,e^{-} \rightarrow Zn_{(s)}$	−0.76
$Fe^{2+}_{(aq)} + 2\,e^{-} \rightarrow Fe_{(s)}$	−0.41
$Cd^{2+}_{(aq)} + 2\,e^{-} \rightarrow Cd_{(s)}$	−0.40
$Ni^{2+}_{(aq)} + 2\,e^{-} \rightarrow Ni_{(s)}$	−0.23
$Pb^{2+}_{(aq)} + 2\,e^{-} \rightarrow Pb_{(s)}$	−0.13
$Fe^{3+}_{(aq)} + 3\,e^{-} \rightarrow Fe_{(s)}$	−0.04
$2\,H^{+}_{(aq)} + 2\,e^{-} \rightarrow H_{2(g)}$	0.00
$Sn^{4+}_{(aq)} + 2\,e^{-} \rightarrow Sn^{2+}_{(aq)}$	0.15
$Cu^{2+}_{(aq)} + e^{-} \rightarrow Cu^{+}_{(aq)}$	0.16
$AgCl_{(s)} + e^{-} \rightarrow Ag_{(s)} + Cl^{-}_{(aq)}$	0.22
$Hg_2Cl_{2(s)} + 2\,e^{-} \rightarrow 2\,Hg_{(l)} + 2\,Cl^{-}_{(aq)}$	0.26
$Cu^{2+}_{(aq)} + 2\,e^{-} \rightarrow Cu_{(s)}$	0.34
$MnO_4^{-}{}_{(aq)} + 8\,H^{+}_{(aq)} + 5\,e^{-} \rightarrow Mn^{2+}_{(aq)} + 4\,H_2O_{(l)}$	1.49
$H_2O_{2(aq)} + 2\,H^{+}_{(aq)} + 2\,e^{-} \rightarrow 2\,H_2O_{(l)}$	1.78
$Co^{3+}_{(aq)} + e^{-} \rightarrow Co^{2+}_{(aq)}$	1.82
$S_2O_8^{2-}{}_{(aq)} + 2\,e^{-} \rightarrow 2\,SO_4^{2-}{}_{(aq)}$	2.01
$O_{3(g)} + 2\,H^{+}_{(aq)} + 2\,e^{-} \rightarrow O_{2(g)} + H_2O_{(l)}$	2.07
$F_{2(g)} + 2\,e^{-} \rightarrow 2\,F^{-}_{(aq)}$	2.87
Most electronegative, most oxidising	

The potential difference of a galvanic cell (one that produces electric power) is calculated as the difference between the electrode with the higher potential (reduction) and that with the lower potential (oxidation): $\Delta E_{galv} = E_{red} - E_{ox}$. Importantly, when the number of electrons consumed or produced in the two half-cells differ, the reduction potentials must not be multiplied

4.2.9 Absolute Electrode Potentials

The determination of absolute electrochemical reduction potentials of isolated half-cells in solution is a challenging task. Therefore, most commonly, reduction potentials are measured relative to other half-cells which establishes a series of reduction potential values that can be ordered from lowest to highest potential. The anchor for this series is the standard hydrogen electrode which has arbitrarily been assigned a value of 0 V.

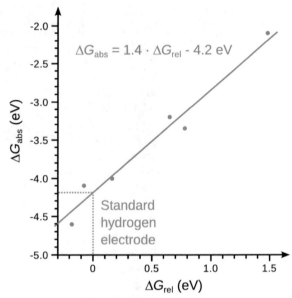

Fig. 4.7 Absolute electrode
potentials as determined by
Evan Williams and colleagues
(Donald et al. 2008)

The availability of an absolute scale would help to bridge a current divide in the electrochemical characterisation of solids, in particular semiconductors, and solutions. When dealing with solid/gas interfaces, it is typically the energy of a free electron *in vacuo* that is taken as a reference energy. There have thus been numerous efforts to estimate the potential of the standard hydrogen electrode versus a free electron.

A recent approach in this context comprises of the experimental measurement of the energy gained by hydrated metal ions of the type $[M(H_2O)_{32}]^{2+}$ or $[M(NH_3)_6(H_2O)_{55}]^{3+}$ when capturing an electron, by means of FT/ICR mass spectrometry (Donald et al. 2008). The reduction of a cluster of hydrated metal ions is accompanied by the loss of water molecules, and the sum of the binding energies of these molecules is correlated with the energy deposited onto the cluster by the gained electron. The energies determined in this fashion represent absolute free energy changes for the reduction of a metal cluster, ΔG_{abs}. These absolute values can be compared to the relative free energy changes ΔG_{rel} measured for the metal clusters by using the conventional reference half-cell methodology (Fig. 4.7). The linear correlation obtained allows for conversion of relative into absolute reduction potentials, and also allows provides an estimate of the absolute reduction potential of the standard hydrogen electrode which is about 4.2 ± 0.4 V (Donald et al. 2008).

4.2.10 External Potential Difference

As we have introduced in Sect. 4.2.1, when combining two half-cells into an electrochemical cell, that cell can be set up as a galvanic or electrolytic cell by

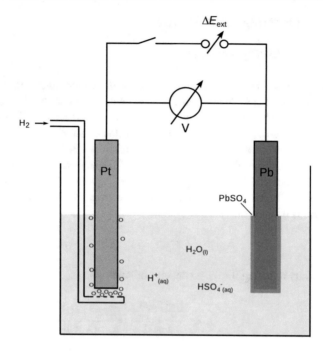

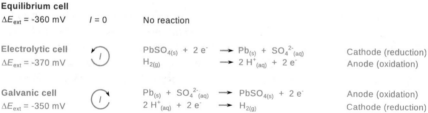

Equilibrium cell
$\Delta E_{ext} = -360$ mV $\quad I = 0 \quad$ No reaction

Electrolytic cell
$\Delta E_{ext} = -370$ mV

$PbSO_{4(s)} + 2e^- \longrightarrow Pb_{(s)} + SO_4{}^{2-}{}_{(aq)}$ Cathode (reduction)
$H_{2(g)} \longrightarrow 2H^+{}_{(aq)} + 2e^-$ Anode (oxidation)

Galvanic cell
$\Delta E_{ext} = -350$ mV

$Pb_{(s)} + SO_4{}^{2-}{}_{(aq)} \longrightarrow PbSO_{4(s)} + 2e^-$ Anode (oxidation)
$2H^+{}_{(aq)} + 2e^- \longrightarrow H_{2(g)}$ Cathode (reduction)

Fig. 4.8 Depending on the external potential difference, an electrochemical cell can either run as electrolytic cell, galvanic cell or rest at equilibrium

choosing an appropriate external potential difference ΔE_{ext} (see Fig. 4.8). If the electric potential difference of a cell is ΔE_{cell}, then any external potential difference that is more positive will result in the cell working as a galvanic cell:

$$\Delta E_{ext} > \Delta E_{cell} \text{ galvanic cell } Pb_{(s)} + H_2SO_{4(aq)} \rightarrow PbSO_{4(aq)} + H_{2(g)}$$

An external potential difference that is more negative than ΔE_{cell} will operate the cell as an electrolytic cell with the reverse chemical reaction:

$$\Delta E_{ext} < \Delta E_{cell} \text{ electrolytic cell } PbSO_{4(aq)} + H_{2(g)} \rightarrow Pb_{(s)} + H_2SO_{4(aq)}$$

4.3   Electrolytes in Solution

4.3.1   Solubility Product

In the previous section, we have introduced the silver/silver chloride electrode as a frequently used reference electrode. The electrode is made of a silver wire coated with silver chloride that is immersed into a solution containing chloride ions. The step which determines the potential of this electrode (half-cell) is the transition of Ag^+ ions from the electrode into solution. Here, the metal ions ($Ag^+_{(s)}$) are in a heterogeneous equilibrium between the solid electrode material ($AgCl_{(s)}$) and the dissolved ions ($Ag^+_{(aq)}$, $Cl^-_{(aq)}$). A saturated aqueous AgCl solution contains only 13 µM Ag^+ and 13 µM Cl^- ions; the equilibrium of the following disassociation reaction is thus on the left hand side

$$AgCl_{(s)} + H_2O \quad \rightarrow \quad Ag^+_{(aq)} + Cl^-_{(aq)}$$

and results in an equilibrium constant of small value:

$$K = \frac{\frac{c(Ag^+)}{c^{\emptyset}} \frac{c(Cl^-)}{c^{\emptyset}}}{\frac{c(AgCl)}{c^{\emptyset}}} = \frac{[Ag^+] \cdot [Cl^-]}{[AgCl]} \tag{4.31}$$

Remember that H_2O is omitted from calculation of the equilibrium constant, since it is neither consumed nor produced. The concentration of any solid substances are considered to be constant in solvation reactions, hence $c(AgCl) = $ const.

The silver chloride concentration can thus be multiplied into the equilibrium constant, yielding a new constant, the solubility product, which is solely a function of the concentration of dissolved ions:

$$K_{sp} = K \cdot [AgCl] = [Ag^+] \cdot [Cl] = \frac{c(Ag^+)}{c^{\emptyset}} \frac{c(Cl^-)}{c^{\emptyset}} \tag{4.32}$$

The electrode (half-cell) potentials are themselves potential differences (see Fig. 4.9). They arise from the transition of ions from the solid to the dissolved state. To stay consistent with the nomenclature used in previous sections, we will denote these differences as E and $E^{\emptyset}$, respectively, to indicate that we are referring to an electrochemical half-cell. In analogy to the way in which we derived Eq. 4.27, we can formulate an expression for the potentials arising from Ag^+ transitions in the silver/silver chloride electrode, by using Eq. 4.32:

$$E(Cl^-|AgCl|Ag^+) = E^{\emptyset}(Ag^+|Ag) + \frac{R \cdot T}{z \cdot F} \cdot \ln \frac{\left[Ag^+_{(aq)}\right]}{\left[Ag^+_{(s)}\right]}$$

$$\Rightarrow E(Cl^-|AgCl|Ag^+) = E^{\emptyset}(Ag^+|Ag) + \frac{R \cdot T}{F} \cdot \ln \left[Ag^+_{(aq)}\right] \tag{4.33}$$

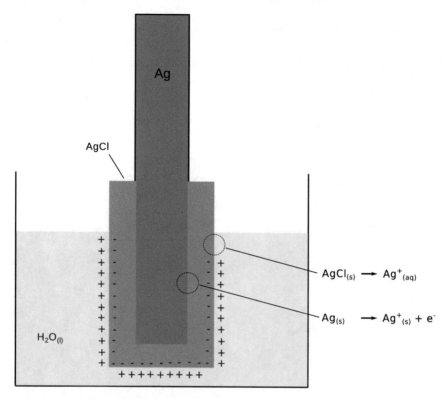

Fig. 4.9 Illustration of transition processes at electrodes and the generation of an electrode potential. Positively charged ions transition from the solid to the dissolved state. This changes the electroneutrality and leads to generation of an electrostatic double layer. The double layer is responsible for the electrode potential

From Eq. 4.32 we obtain

$$c_{aq}(Ag^+) = \frac{K_{sp}}{c_{aq}(Cl^-)} \cdot \left(c^\varnothing\right)^2$$

which we can substitute into Eq. 4.33:

$$E(Cl^-|AgCl|Ag) = E^\varnothing(Ag^+|Ag) + \frac{R \cdot T}{F} \cdot \ln \frac{K_{sp}}{\left[\frac{c_{aq}(Cl^-)}{c^\varnothing}\right]} \tag{4.34}$$

Using the logarithm rule of $\frac{K_{sp}}{\left[\frac{c_{aq}(Cl^-)}{c^\varnothing}\right]} = \ln K_{sp} - \ln \frac{c_{aq}(Cl^-)}{c^\varnothing}$, we obtain:

$$E(\text{Cl}^-|\text{AgCl}|\text{Ag}) = E^{\varnothing}(\text{Ag}^+|\text{Ag}) + \frac{R \cdot T}{F} \cdot \ln K_{sp} - \frac{R \cdot T}{F} \ln \frac{c_{aq}(\text{Cl}^-)}{c^{\varnothing}}$$

As $\frac{R \cdot T}{F} \cdot \ln K_{sp}$ is a constant, it can be combined with the standard potential of the $\text{Ag}^+ | \text{Ag}$ transition:

$$E(\text{Cl}^-|\text{AgCl}|\text{Ag}) = E^{\varnothing}(\text{Cl}^-|\text{AgCl}|\text{Ag}) - \frac{R \cdot T}{F} \cdot \ln \frac{c_{aq}(\text{Cl}^-)}{c^{\varnothing}} \qquad (4.35)$$

Determining the Solubility Product from Standard Electrode Potentials
Since the solubility product of a poorly soluble salt forms part of the standard electrode potential if that salt is part of the solid electrode material, one can determine the solubility product from these (tabulated) standard electrode potentials. Comparison of Eqs. 4.34 and 4.35 shows that the right hand sides of both equations can be set equal:

$$E^{\varnothing}(\text{Ag}^+|\text{Ag}) + \frac{R \cdot T}{F} \cdot \ln \frac{K_{sp}}{\left[\frac{c_{aq}(\text{Cl}^-)}{c^{\varnothing}}\right]} = E^{\varnothing}(\text{Cl}^-|\text{AgCl}|\text{Ag}) - \frac{R \cdot T}{F} \cdot \ln \frac{c_{aq}(\text{Cl}^-)}{c^{\varnothing}}$$

Using the logarithm rule of $\ln \frac{K_{sp}}{\left[\frac{c_{aq}(\text{Cl}^-)}{c^{\varnothing}}\right]} = \ln K_{sp} - \ln \frac{c_{aq}(\text{Cl}^-)}{c^{\varnothing}}$, we obtain:

$$E^{\varnothing}(\text{Ag}^+|\text{Ag}) + \frac{R \cdot T}{F} \cdot \ln K_{sp} - \frac{R \cdot T}{F} \cdot \ln \frac{c_{aq}(\text{Cl}^-)}{c^{\varnothing}}$$

$$= E^{\varnothing}(\text{Cl}^-|\text{AgCl}|\text{Ag}) - \frac{R \cdot T}{F} \cdot \ln \frac{c_{aq}(\text{Cl}^-)}{c^{\varnothing}}$$

$$E^{\varnothing}(\text{Ag}^+|\text{Ag}) + \frac{R \cdot T}{F} \cdot \ln K_{sp} = E^{\varnothing}(\text{Cl}^-|\text{AgCl}|\text{Ag})$$

$$\frac{R \cdot T}{F} \cdot \ln K_{sp} = E^{\varnothing}(\text{Cl}^-|\text{AgCl}|\text{Ag}) - E^{\varnothing}(\text{Ag}^+|\text{Ag})$$

$$\ln K_{sp} = \frac{F}{R \cdot T} \cdot \left[E^{\varnothing}(\text{Cl}^-|\text{AgCl}|\text{Ag}) - E^{\varnothing}(\text{Ag}^+|\text{Ag})\right] \qquad (4.36)$$

The tabulated standard electrode potentials for the two reactions are:

$$E^{\varnothing}(\text{Cl}^-|\text{AgCl}|\text{Ag}) = 0.800 \text{ V}$$

$$E^{\varnothing}(\text{Ag}^+|\text{Ag}) = 0.222 \text{ V}$$

This yields for Eq. 4.36:

(continued)

$$\ln K_{sp} = \frac{96485 C \cdot K \cdot mol}{8.3144 \cdot 298 \cdot mol \cdot J \cdot K} \cdot (0.800\ V - 0.222\ V)$$

With $1\ C = 1\ J\ V^{-1}$ this resolves to:

$$\ln K_{sp} = \frac{96485 \cdot (-0.57.8) J \cdot K \cdot mol \cdot V}{8.3144298 V \cdot mol \cdot J \cdot K}$$

$$\ln K_{sp} = -22.5$$

$$K_{sp} = e^{-22.5} = 1.7 \cdot 10^{-10}$$

From Eq. 4.32, we know that the units of the solubility product for silver chloride are:

$$\left[K_{sp}\right] = 1\ mol^2\ l^{-2}$$

and therefore:

$$K_{sp}(AgCl) = 1.7 \cdot 10^{-10}\ mol^2\ l^{-2}$$

4.3.2 Colligative Properties and the van't Hoff factor

As we have seen in Sect. 3.2.2, colligative properties result from the reduction of the chemical potential $\mu^*_{solvent}$ of the pure liquid solvent, as a result of the presence of the solute. For an ideal solvent/solution we can state:

$$\text{In the absence of solute} : \mu_{solvent} = \mu^*_{solvent}$$
$$\text{In the presence of solute} : \mu_{solvent} = \mu^*_{solvent} + R \cdot T \cdot \ln x_{solute} \qquad (3.13)$$

Because x_{solute} is the mole fraction, its value in case of a solution is larger than 0 and less than 1:

$$0 < x_{solute} < 1 \Rightarrow \ln x_{solute} < 0$$

Hence the second term in Eq. 3.13 is always negative and thus $\mu_{solvent} < \mu^*_{solvent}$.

These properties depend on the particular solvent and on the concentration of solute, but not on the nature of the solute. If we assume ideal solutions, then the solute and the solvent have identical intermolecular forces. Therefore, the enthalpy of solution is zero ($\Delta H_{sol} = 0$), since the potential energy of solvent and solute molecules is not affected. However, the intermixing of solute and solvent will raise the entropy of the solution: $\Delta S_{sol} > 0$. Therefore, the lowering of the chemical potential of the liquid solvent is an entropic effect.

In summary, colligative properties:

Table 4.4 Colligative properties

Effect	Equation	Parameters	Concentration
Lowering of vapour pressure (Raoult's law)	$p = x \cdot p^*$(3.12)	p^*: vapour pressure of pure liquid	Mole fraction x
Melting point depression	$\Delta T_f = K_f \cdot b$(4.37)	K_f: cryoscopic constant	Molality b
Boiling point elevation	$\Delta T_b = K_b \cdot b$(4.38)	K_b: ebullioscopic constant	Molality b
Osmosis	$\Pi = c \cdot R \cdot T$(4.39)	Π: osmotic pressure	Molar concentration c

Table 4.5 Commonly used measures of concentration

Mole fraction	$x_j = \frac{n_j}{\Sigma_i n_i}$	$[x] = 1$	(2.74)
Molality	$b = \frac{n_{solute}}{m_{solvent}}$	$[b] = 1 \text{ mol kg}^{-1}$	(4.40)
Molar concentration	$c = \frac{n_{solute}}{V_{solution}}$	$[c] = 1 \text{ mol l}^{-1}$	(2.73)
Mass ratio	$w = \frac{m_{solute}}{m_{solvent}}$	$[w] = 1$	(4.41)

- Do not depend on the chemical nature of the solute
- Depend on the concentration of the solute
- Depend on the nature of the solvent
- Are entropic effects, thus depend on the number of dissolved particles.

There are four important colligative properties (see Table 4.4), all of which are related to the concentration of the solute.

For historical reasons, different types of concentrations are used for the different phenomena (Table 4.5). The concentrations are:

Raoult's law, which describes the lowering of the vapour pressure of a liquid in the presence of solutes has already been described in an earlier Sect. 3.2.1. Intriguingly, the fact that the vapour pressure of a solution is lower than that of the pure solvent results in the observation that the boiling point of a solution is higher than that of the pure solvent.

In order to consider the entropic effects of solutes, a factor that considers the number of dissolved particles in the solution needs to be introduced. This is especially important for electrolytes, as the number N of particles always increases upon dissociation:

$$NaCl_{(s)} \quad \rightarrow \quad Na^+_{(aq)} + Cl^-_{(aq)}$$
$$N = 1 \qquad \qquad N = 2 = v$$

Therefore, the van't Hoff equation for the osmotic pressure is corrected by a factor i, called the van't Hoff factor:

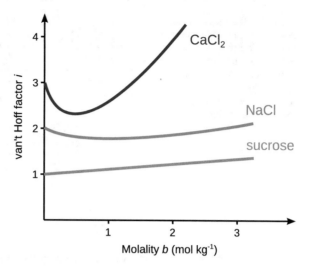

Fig. 4.10 The van't Hoff factor i approaches the theoretical number of ions into which an electrolyte dissociates (ν) only at infinite dilution. For non-electrolytes, the van't Hoff factor may also differ from the theoretical value of $\nu = 1$ at increasing concentrations

$$\Pi = i \cdot c \cdot R \cdot T \tag{4.39}$$

Similarly, for boiling point elevation and melting point depression one obtains:

$$\Delta T_b = i \cdot K_b \cdot b \tag{4.38}$$

$$\Delta T_f = i \cdot K_f \cdot b \tag{4.37}$$

For an ideal (i.e. infinite) dilution, the van't Hoff correction factor i approaches the theoretical number of ions ν into which the solute molecule dissociates (see Fig. 4.10). The variation of the van't Hoff factor from the theoretically expected dissociation numbers for both electrolytes and non-electrolytes may arise from the following phenomena:

- Difference in internal pressure of solute and solvent
- Polarity
- Compound formation or complexation
- Association of either solute or solvent

In case of electrolytes, additional effects arise from:

- The dissociation of weak/strong electrolytes
- Interaction of the ions of strong electrolytes.

4.3.3   Degree of Dissociation

If the deviation of the van't Hoff factor i from the theoretically expected value ν is due to dissociation only, then the degree of dissociation α can be determined as per:

Fig. 4.11 Illustration of ion
association. In solution, ions
of opposite electrical charge
may come together to form a
distinct chemical entity. Top
panel: fully solvated ions;
bottom panel: an ion triplet
with solvent sharing

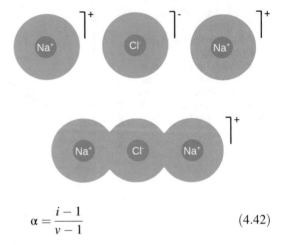

$$\alpha = \frac{i-1}{v-1} \tag{4.42}$$

where v is the theoretical number of ions yielded per solute molecule (e.g. for $CaCl_2$:
$v = 3$).

The degree of dissociation α can be reliably estimated for weak electrolytes, as
there are only few ions at all concentrations, and therefore other solution effects are
negligible.

For strong electrolytes, Eq. 4.42 works well at low concentrations, but substantially deviates from experimental observations at moderate or strong concentrations,
due to interactions between ions. With increasing concentrations, attractive
interactions between dissolved electrolytes become important, and ion pairs or
triplets may form (Fig. 4.11). Since ions involved in those pairings no longer account
as individual ions, the degree of dissociation is less than expected.

4.3.4   Activities and Ionic Strength

In cases such as the one described in the previous section, where significant
deviations from the ideal behaviour need to be considered, it is appropriate to use
an 'effective concentration'. Therefore, for strong electrolytes as well as weak
electrolytes in the presence of salts (e.g. buffer systems), activities instead of
concentrations should be used in order to quantify the deviation from the completed
dissociation behaviour.

We have previously defined the activity a such that the following equation is true
for a real liquid (see Sect. 3.2.3):

$$\mu = \mu^* + R \cdot T \cdot \ln a \tag{3.12}$$

where the activity a (non-ideal solution) is related to the mole fraction x (ideal
solution) by a scaling factor called the activity coefficient γ:

Table 4.6 Definition of activity in terms of frequently used concentration measures

Activity in terms of the mole fraction x	$a = \gamma_x \cdot x$	(4.43)
Activity for a volatile solvent (Raoult's law: $p = x \cdot p^*$)	$a = \frac{p}{p^*}$	(4.44)
Activity in terms of molality b	$a = \gamma_b \cdot b$	(4.45)
Activity in terms of molar concentration c	$a = \gamma_c \cdot c$	(4.46)

$$a = \gamma \cdot x \tag{3.13}$$

Importantly, activity can be defined in terms of any other concentration measure (see Table 4.6). The activity of solutes is usually less than the concentration, but it approaches the same value as the concentration at high dilution. Activities may be different for cations and anions of an electrolyte. For ideal solvents, the activity coefficient $\gamma = 1$, for non-ideal solvents, the activity coefficient approaches 1 at very high concentration of the solvent.

When comparing activities of electrolytes, experimental conditions should be chosen to ensure constant ionic strength. If a solution contains N different types of ions with the charge states $z_1, z_2, \ldots, z_N$ and at concentrations $c_1, c_2, \ldots, c_N$, it is appropriate to use an average concentration. The ionic strength I is such an averaged concentration, weighted by the square of the charge state:

$$I = \frac{1}{2} \cdot \Sigma_{i=1}^{N} c_i \cdot z_i^2 \tag{4.47}$$

Here, c_i is the molar concentration and z_i is the charge of the ith ion; N is the number of different ions in the solution.

The ionic strength is important in many biochemical applications. For instance, when evaluating the effect of pH on an enzymatic reaction, the effect of the salt concentration in the buffer may obscure the results, unless the buffer is adjusted to ionic strength in each experiment.

4.3.5 pH Buffers

In analogy to the solubility product of electrolytes (see Sect. 4.3.1), one can consider the following dissociation of water

$$2\,H_2O_{(l)} \rightleftharpoons H_3O^+{}_{(aq)} + OH^-{}_{(aq)} \quad K = \frac{[H_3O^+] \cdot [OH^-]}{[H_2O]^2} \tag{4.48}$$

and define the ion product of water as

$$K_w = K \cdot [H_2O]^2 = [H_3O^+] \cdot [OH^-] \tag{4.49}$$

The equilibrium constant K has a very small numerical value, since the equilibrium of reaction 4.48 resides on the left hand side. Since the molar concentration of water can be assumed constant ($c = 55.56$ mol l^{-1}), it is combined with the equilibrium constant K to yield the ion product of water: $K_w = 1.0 \cdot 10^{-14}$.

Since in pure water, the amounts of hydronium (H_3O^+) and hydroxide (OH^-) ions resulting from the dissociation reaction above are equal, the concentration of hydronium ions (for convenience frequently replaced by protons) can be calculated from Eq. 4.49:

$$c(H^+) = c(H_3O^+) = [H_3O^+] \cdot c^\emptyset = \sqrt{K_w} \cdot c^\emptyset = 1.0 \cdot 10^{-7} \text{ mol l}^{-1} \quad (4.50)$$

This gives rise to the definition of the pH and pOH, which are calculated as the negative decadic logarithms of the H^+ and OH^- concentrations, respectively:

$$pH = -\lg \frac{c(H^+)}{1\,M} \quad (4.51)$$

$$pOH = -\lg \frac{c(OH^-)}{1\,M} \quad (4.52)$$

From Eq. 4.50 follows that:

$$pH + pOH = 14 \quad (4.53)$$

The pH of aqueous solutions can be affected by:

- Neutral salts, which may modify the ionic strength
- Water addition, which may either change activity coefficients or act as a weak acid or base
- Temperature, which changes K_w and K_a or K_b in case of acids/bases (see below).

Therefore, pH buffers are of high practical importance in many (bio-)chemical formulations (e.g. drug preparations, creams, etc) as well as living cells where, for example, buffers in blood keep the pH at ~7.4 (blood pH outside the range 6.9 and 7.8 puts life in danger).

Buffers are compounds or mixtures of compounds that withstand a change in pH upon addition/generation of acid or base (in a reasonable window). Buffer compounds are typically weak acids with their conjugate bases, or weak bases and their conjugate acids. For example, on addition of a strong acid to a solution containing equal quantities of acetic acid and sodium acetate, the hydrogen ions react with the acetate according to

$$H_3O^+_{(aq)} + H_3C - COO^-_{(aq)} \rightleftharpoons H_3C - COOH_{(aq)} + H_2O$$

thus converting the hydronium ion to water and eliminating the acidic property.

Obviously, the capacity of the buffer system to 'eliminate' additional hydronium ions will be exhausted, when all available molecules of the weak base have been consumed. The amount of either strong acid or base that can be added before a significant change in the pH will occur is a matter of stoichiometry. The maximum amount of strong acid that can be buffered is equal to the amount of conjugate base present in the buffer. Similarly, the maximum amount of base that can be tolerated is equal to the amount of weak acid present in the buffer.

The pH of a solution that contains a buffer system consisting of a weak acid and conjugate base can be calculated based on the equilibrium of the dissociation reaction

$$HA_{(aq)} + H_2O \rightleftharpoons H_3O^+{}_{(aq)} + A^-{}_{(aq)}$$

where HA denotes the acid (e.g. H_3CCOOH) and A^- the conjugate base (e.g. H_3CCOO^-). The equilibrium constant K_a for the dissociation reaction is given by:

$$K_a = \frac{[H_3O^+] \cdot [A^-]}{[HA]}$$

This yields for the concentration of hydronium ions:

$$c(H_3O^+) = [H_3O^+] \cdot c^\emptyset = K_a \cdot \frac{[HA]}{[A^-]} c^\emptyset$$

We divide the left and right hand side of this equation by the unit of the molar concentration (1 M), and then subject both sides to a logarithmic operation, with $lg = \log_{10}$:

$$lg\frac{c(H_3O^+)}{c^\emptyset} = lg\left(K_a \cdot \frac{[HA]}{[A^-]}\right)$$

Using the logarithm rule of $\log(a \cdot b) = \log a + \log b$, we obtain:

$$lg\frac{c(H_3O^+)}{c^\emptyset} = lgK_a + lg\frac{[HA]}{[A^-]}$$

After multiplying the equation with (-1):

$$-lg\frac{c(H_3O^+)}{c^\emptyset} = -lgK_a - lg\frac{[HA]}{[A^-]}$$

we recognise that the left hand side is the definition of the pH: $pH = -lg\frac{c(H_3O^+)}{c^\emptyset}$ $= -lg[H_3O^+]$ and the expression $-lgK_a$ defines the pK_a of the acid HA: $pK_a = -lgK_a$

This yields:

$$pH = pK_a - \lg\frac{[HA]}{[A^-]} \quad \text{or} \quad pH = pK_a + \lg\frac{[A^-]}{[HA]} \qquad (4.54)$$

which is known as the Henderson-Hasselbalch equation.

For real (non-ideal) situations, the equilibrium constant needs to be set up with activities instead of concentrations. If we consider a buffer system consisting of acetic acid (HAc) and acetate (Ac$^-$) for illustration, this yields:

$$K_a = \frac{a(H_3O^+) \cdot a(Ac^-)}{a(HAc)} = \frac{a(H_3O^+) \cdot \gamma_c(.Ac^-) \cdot c(Ac^-)}{\gamma_c(HAc) \cdot c(HAc)}$$

Since HAc constitutes the solvent, and is thus present at a large concentration (i.e. its mole fraction approaches 1), we know from Sect. 4.3.4 that:
$\gamma_x \to 1$ as $x \to 1$, therefore $\gamma_c(HAc) \approx 1$, which yields:

$$K_a = a(H_3O^+) \cdot \gamma_c(Ac^-) \cdot \frac{c(Ac^-)}{c(HAc)}$$

Resolving for $a(H_3O^+)$ yields:

$$a(H_3O^+) = K_a \cdot \frac{c(HAc)}{\gamma_c(Ac^-) \cdot c(Ac^-)}$$

In order to the logarithm of both sides, we need to divide by the units of activity which in this case are 1 M:

$$-\lg a(H_3O^+) = -\lg\left[K_a \cdot \frac{c(HAc)}{\gamma_c(Ac^-) \cdot c(Ac^-)}\right]$$

Using the logarithm rule $\log(a \cdot b) = \log a + \log b$, we obtain:

$$-\lg a(H_3O^+) = -\lg K_a - \lg\frac{c(HAc)}{\gamma_c(Ac^-) \cdot c(Ac^-)}$$

$$-\lg a(H_3O^+) = -\lg K_a - \lg\frac{1}{\gamma_c(Ac^-)} - \lg\frac{c(HAc)}{c(Ac^-)}$$

With the rule $\log\left(\frac{1}{a}\right) = -\log a$, this yields:

$$-\lg a(H_3O^+) = -\lg K_a + \lg\gamma(Ac^-) + \lg\frac{c(Ac^-)}{c(HAc)}$$

and therefore:

$$pH = pK_a + lg \left[\frac{c(Ac^-)}{c(HAc)}\right] + lg\, \gamma(Ac^-) \tag{4.55}$$

which is the Henderson-Hasselbalch equation considering activities. Comparison with Eq. 4.54 shows that the additional term of $lg\, \gamma(Ac^-)$ needs to be taken into account.

4.3.6 Applications of Conductivity Measurements

When titrating weak acids (bases) with strong bases (acids), the determination of the end point of titration may be difficult due to the buffering properties described in the previous section. In such cases, conductivity measurements (also called conductometry) can be used to detect the end points of titrations. Conductometric titrations are also a useful alternative for acid-base titrations where indicators can not be used, because the sample or titrant are coloured. The observable parameter in these titrations is the conductivity κ (see Eq. 4.10).

Figure 4.12 compares observations in conductometric titrations with different acid/base pairings. When a strong base (e.g. NaOH, the analyte) is titrated with a strong acid (e.g. HCl, the titrant), the solution initially contains $Na^+_{(aq)}$ and $OH^-_{(aq)}$ ions, therefore high conductivity is observed (Fig. 4.12, left panel). As the titration proceeds, the conductivity falls sharply since $OH^-_{(aq)}$ ions react with $H^+_{(aq)}$ to neutral H_2O and the added $Cl^-_{(aq)}$ ions replacing $OH^-_{(aq)}$ have a lower conductivity than $OH^-_{(aq)}$ (see Table 5.5). At the end point, the exact amount of $H^+_{(aq)}$ has been added to match the amount of $OH^-_{(aq)}$, and the solution shows a conductivity expected of a solution of NaCl in water. As more titrant is added, the conductivity begins to rise sharply, due to $H^+_{(aq)}$ having higher conductivity than $Na^+_{(aq)}$. Determination of the end point is conveniently done by extrapolating the two branches of the conductometric plot.

When a weak base (e.g. NaAc) is titrated with a strong acid (e.g. HCl), the initial conductivity of the solution is low, since the salt of the weak base is only partially

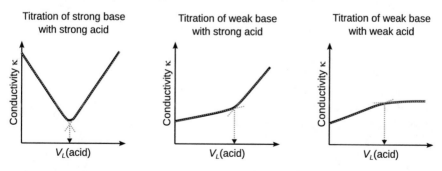

Fig. 4.12 Schematic plots of conductometric titration data for strong acid/strong base (left), strong acid/weak base (middle), weak acid/weak base (right). Typical experimental conductivity values for such titrations are in the range of $\mu S\ cm^{-1}$

dissociated (Fig. 4.12, middle panel). Addition of $H^+_{(aq)}$ and $Cl^-_{(aq)}$ leads to a rise in conductivity, as the acid-base reaction causes dissociation of the salt and thus releases $Na^+_{(aq)}$. After the amount of base is been matched by $H^+_{(aq)}$ at the endpoint, the conductivity rises sharply, as now excess $H^+_{(aq)}$ become available which possess substantially higher conductivity than any other ions (see Table 5.5).

4.3.7 The Debye-Hückel Theory

So far, any deviation from ideal behaviour has been treated by an empirical approach. In order to be able to account for non-ideal behaviour, the concentration of an electrolyte has been replaced with the activity, and we demanded that all underlying thermodynamic relations such as the variation of the chemical potential (Eq. 3.12) should then remain valid even for non-ideal solutions. The value of the activity may be found by experimentally determining the activity coefficient and using the known concentration. With these activity values, fundamental thermodynamic quantities such as the chemical potential can then be calculated.

Experimental Determination of Activity Coefficients
The activity a and the activity coefficient γ are intimately linked to the mole fraction or concentration whenever a system deviates from ideal behaviour (see Table 4.6). Any such phenomena can thus be used, in principle, to determine activity coefficients.

For example, the colligative property of vapour pressure lowering by a solute may be used to determine the activity coefficient of the solvent. From Raoult's law, we know that

$$x = \frac{p}{p^*} \tag{3.12}$$

where x is the mole fraction of the solute, p the vapour pressure of the solution and p^* the vapour pressure of the pure solvent. In Table 4.6, we have thus defined the activity of a non-ideal solvent as

$$a = \frac{p}{p^*} \tag{4.44}$$

Remembering that the activity in terms of the mole fraction is generally defined as

$$a = \gamma_x \cdot x \tag{4.43}$$

we obtain:

(continued)

$$\gamma_x \cdot x = \frac{p}{p^*} \quad \Rightarrow \quad \gamma_x = \frac{p}{p^* \cdot x} \tag{4.56}$$

which allows for the experimental determination of the activity coefficient γ of the solvent directly from observation of the colligative property of vapour pressure lowering by a solute. The only requirement is that the vapour of the solution/solute behaves as an ideal gas. If that is not the case under the chosen conditions, fugacities instead of pressures need to be used.

Debye and Hückel developed a theoretical basis to quantify the non-ideal behaviour of electrolyte solutions. Their theory is based on the following assumptions:

- The electrolyte solution needs to be dilute (i.e. $c < 0.01$ mol l^{-1})
- The electrolytes are completely dissociated into ions
- On average, each ion is surrounded by ions of opposite charge (see Fig. 4.13).

Based on these assumptions, Debye and Hückel calculated the chemical potential of a completely dissociated electrolyte from the chemical potentials of the cat- and anions. These calculations result in a relationship between the mean activity coefficient $\gamma_\pm$ and the ionic strength I of the solution, called the Debye-Hückel limiting law:

$$\lg \gamma_\pm = z_+ \cdot z_- \cdot A \cdot \sqrt{I} \tag{4.57}$$

where z_+ and z_- are the charge numbers of the cat- and anion of the electrolyte, and I is the ionic strength as defined in Eq. 4.47. A is a constant for the solvent of interest; in case of aqueous solutions at $\theta = 25\,°C$, $A = 0.5099\ dm^{3/2}\ mol^{-1/2}$.

Due to the principle of electroneutrality (see Sect. 4.1.2), one can only prepare solutions that contain cat- as well as anions. Therefore, the Debye-Hückel theory provides the above relationship only for a mean activity coefficient. Notably,

Fig. 4.13 The Debye-Hückel theory assumes an ionic atmosphere: on average, each ion is surrounded by a sphere of counter-ions

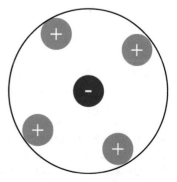

electrolytes that possess the same charge numbers also possess the same mean activity coefficients.

Using the Debye-Hückel limiting law, we can now return to the Henderson-Hasselbalch equation for the non-ideal solutions, illustrated by the acetic acid/acetate buffer system (Eq. 4.55):

$$pH = pK_a + \lg \frac{c(Ac^-)}{c(HAc)} + \lg \gamma(Ac^-) \tag{4.55}$$

The last term in this equation can be evaluated from Eq. 4.57 as follows:

$$\lg \gamma_\pm = z(Ac^-) \cdot A \cdot \sqrt{I} = (-1) \cdot 0.51 \frac{1^{3/2}}{mol^{1/2}} \cdot \sqrt{I}$$

Therefore:

$$pH = pK_a + \lg \frac{c(Ac^-)}{c(HAc)} - 0.51 \frac{1^{3/2}}{mol^{1/2}} \cdot \sqrt{I} \tag{4.58}$$

4.3.8   Tonicity

As well as controlling the pH of a solution, it is often necessary to control the osmotic pressure Π of a solution, especially when preparing biochemical buffers. For example, in cases where membranes are involved, a difference in the osmotic pressure between the two sides of the membrane may have devastating effects and cause swelling, contraction or even rupture of the membrane.

The measure of the osmotic pressure gradient of two solutions that are separated by a semi-permeable membrane is called tonicity. It is thus a qualitative description of the relative concentrations of solutes in the two solutions and determines the direction of diffusion. Three different cases can be distinguished: hypertonic, hypotonic and isotonic solutions. Commonly, tonicity refers to the osmotic pressure, but it can also be defined for any other colligative property (melting point, boiling point). Isotonic solutions thus describe solutions that possess the same colligative properties.

In biological settings, a hypertonic solution is one with a higher concentration of solutes outside the cell than inside the cell. Therefore, hypertonic solutions will cause diffusion of water out of the cell in order to balance the concentration of the solutes. Hypotonic solutions, in contrast, possess a lower concentration of solutes outside the cell than is found inside the cell. The direction of diffusion in this case is thus such that water flows into the cell, causing swelling and possible bursting. If an outside solution provides the same concentration of solutes as found inside the cell, then this solution is called isotonic. Water molecules diffuse through the membrane in both directions at equal rates. Isotonic solutions have great importance for biochemical and medical applications; they are for example used as intravenously infused fluids with patients or as washing solutions for the eye in laboratories.

The Concentration of an Isotonic NaCl Solution

The melting point of human blood and tears is $-0.52\,°C$. Determine the mass ratio w of an isotonic NaCl solution in water ($K_f = 1.86$ K kg mol^{-1}).

The colligative property of melting point depression is given by:

$$\Delta T_f = i \cdot K_f \cdot b \tag{4.37}$$

The molality b is given by:

$$b = \frac{n(\text{solute})}{m(\text{solvent})} = \frac{n(\text{NaCl})}{m(\text{H}_2\text{O})} \tag{4.40}$$

The concentration of solute is expressed here as the mass ratio w, which is defined as:

$$w = \frac{m(\text{solute})}{m(\text{solvent})} = \frac{m(\text{NaCl})}{m(\text{H}_2\text{O})} \tag{4.41}$$

We thus obtain from Eq. 4.40:

$$b = \frac{n(\text{NaCl})}{m(\text{H}_2\text{O})} = \frac{m(\text{NaCl})}{m(\text{H}_2\text{O}) \cdot M(\text{NaCl})} = \frac{w(\text{NaCl})}{M(\text{NaCl})}$$

This yields for the mass ratio w:

$$w(\text{NaCl}) = b(\text{NaCl}) \cdot M(\text{NaCl})$$

From Eq. 4.37, we can substitute for the molality:

$$w(\text{NaCl}) = \frac{\Delta T_f}{i \cdot K_f} \cdot M(\text{NaCl})$$

$$w(\text{NaCl}) = \frac{0.56 \cdot 58 \cdot \text{K} \cdot \text{g} \cdot \text{mol}}{2 \cdot 1.86 \cdot \text{mol} \cdot \text{K} \cdot \text{kg}} = \frac{0.56 \cdot 58 \cdot 10^{-3} \cdot \text{K} \cdot \text{kg} \cdot \text{mol}}{2 \cdot 1.86 \cdot \text{mol} \cdot \text{K} \cdot \text{kg}}$$

$$w(\text{NaCl}) = 0.009 = 0.9\%$$

4.4   Exercises

1. Nickel-cadmium batteries are based on the following half-cell reactions:

$$NiO(OH)_{(s)} + H_2O_{(l)} + e^- \quad \rightarrow \quad Ni(OH)_{2(s)} + OH^-_{(aq)}$$
$$Cd_{(s)} + 2\,OH^-_{(aq)} \quad \rightarrow \quad Cd(OH)_{2(s)} + 2\,e^-$$

   (a) The *e.m.f.* of a nickel-cadmium cell is 1.4 V, and the standard Redox
         potential of $Cd(OH)_{2(s)}$ is -0.809 V. What is the standard Redox
         potential of $NiO(OH)_{(s)}$?
   (b) What is the standard free energy of formation of $NiO(OH)_{(s)}$?

2. Calculate the standard *e.m.f.* and the equilibrium constant under standard
   conditions for the following cell:

$$Pt_{(s)}, H_{2(g)} \,\big|\, HCl_{(aq)} \,\big|\, AgCl_{(s)}, Ag_{(s)}$$

3. Calculate the pH of a solution with a formal concentration of $5 \cdot 10^{-7}$ M of the
   strong acid HI at 25 °C.

4. What *e.m.f.* would be generated by the the following cell at 25 °C, assuming ideal
   behaviour:

$$Pt_{(s)}, H_{2(g)}(1\text{ bar}) \,\big|\, H^+_{(aq)}(0.03\text{ M}) \,\big|\, Cl^-_{(aq)}(0.004\text{ M}) \,\big|\, AgCl_{(s)}, Ag_{(s)}$$

5. The lactate/pyruvate Redox system can be described as per:

$$pyruvate + 2\,H^+ + 2\,e^- \rightleftharpoons lactate$$

   The standard reduction potential is measured at $c(\text{lactate}) = c(\text{pyruvate}) = c(H^+)$
   $= 1$ M and has a value of $E^{\varnothing} = 0.21$ V. Calculate and plot the pH dependency of
   the reduction potential for this system at 298 K, assuming $c(\text{lactate}) = c(\text{pyruvate})$.

Molecules in Motion

5

5.1 Transport Processes

We have already seen that physico-chemical processes do not always require a chemical reaction to proceed, such as for example when considering solutions or phase changes. Irreversible processes that arise from non-equilibrium conditions may also include the spatial translocation of objects and properties. In particular, one can observe the transfer of

- matter
- energy
- any other property.

Such transport processes are fundamental processes in biological settings (molecular transport in the cell) and engineering (e.g. liquid flow, thermo devices, etc). Four important instances of transport processes are:

- diffusion: migration of matter along a concentration gradient
- thermal conduction: migration of energy along a temperature gradient
- electric conduction: migration of charges along an electric potential gradient
- viscosity: migration of a linear momentum along a velocity gradient.

We have already considered electric conduction in Sect. 4.1, and will consider these properties again in more detail in Sects. 5.1.4–5.1.6. However, to introduce some fundamental concepts and parameters, we shall start with an introduction to the kinetic molecular theory of gases.

© Springer International Publishing AG, part of Springer Nature 2018
A. Hofmann, *Physical Chemistry Essentials*,
https://doi.org/10.1007/978-3-319-74167-3_5

5.1.1   The Kinetic Molecular Theory of Gases

In the kinetic molecular theory of gases, we only consider energy contributions that arise from the kinetic energy of the individual gas molecules. We assume an ideal gas, and therefore contend that

- the gas consists of molecules of mass m in random motion
- the molecules have negligible size
- the molecules interact only through brief, infrequent, elastic collisions.

Elastic collisions are those where the total translational kinetic energy of molecules is conserved.

From this kinetic molecular theory, one can derive an equation that relates pressure and volume of an ideal gas with the speed of the individual gas molecules:

$$p \cdot V = \frac{1}{3} \cdot n \cdot M \cdot c^2 \tag{5.1}$$

with c being the root mean square speed of the molecules, i.e. a speed averaged over the entire population of gas molecules:

$$c = \sqrt{\langle v^2 \rangle} \tag{5.2}$$

In the above equation, $\langle v^2 \rangle$ is the arithmetic mean of the squared speeds:

$$\langle v^2 \rangle = \frac{1}{n} \Sigma_{i=1}^{n} v_i^2$$

The speed of the individual gas molecules only depends on the temperature (see Eq. 2.1 which defines the relationship between temperature and average kinetic energy of molecules). Therefore, at constant temperature, the root mean square speed c will also be constant. It follows straight from Eq. 5.1 that at constant root mean square speed c (i.e. constant temperature) the product $p \cdot V$ is constant. This is otherwise known as Boyle's law, which states that

▶   the absolute pressure exerted by a given mass of an ideal gas is inversely proportional to the volume it occupies, if the temperature and amount of gas remain unchanged within a closed system:

$$p \sim \frac{1}{V} \quad \text{or} \quad p_1 \cdot V_1 = p_2 \cdot V_2 \tag{5.3}$$

Since we are considering an ideal gas, we can further develop Eq. 5.1:

$$p \cdot V = \frac{1}{3} \cdot n \cdot M \cdot c^2 = n \cdot R \cdot T \tag{5.4}$$

and derive an expression for the root mean square speed of the gas molecules:

$$c = \sqrt{\frac{3 \cdot R \cdot T}{M}} \tag{5.5}$$

Analysis of Eq. 5.5 shows that

- the higher the temperature, the higher the speed of the molecules, and
- heavy molecules travel slower than light molecules.

5.1.2 The Maxwell–Boltzmann Distribution

The root mean square speed c of the gas molecules is an averaged speed over the entire population of gas molecules in the system. Obviously, the speeds of individual gas molecules vary and span a range of different values. A molecule may be travelling rapidly, but then collide and travel slower. It may then accelerate again, only to be slowed down by the next collision.

The distribution of speeds (velocities) can be calculated based on the Boltzmann distribution, which is a probability distribution over various possible states of a system frequently used in statistical mechanics (Maxwell 1860a, 1860b). The resulting function is called the Maxwell–Boltzmann distribution:

$$f(v) = 4 \cdot \pi \cdot \sqrt{\left(\frac{M}{2 \cdot \pi \cdot R \cdot T}\right)^3} \cdot v^2 \cdot e^{-\frac{M \cdot v^2}{2 \cdot R \cdot T}} \tag{5.6}$$

In Eq. 5.6, the exponential factor $e^{-\frac{M \cdot v^2}{2 \cdot R \cdot T}}$ resembles the well-known Boltzmann factor that describes the ratio of two states which only depends on the energy difference between the two states:

$$\frac{f_{state\ 1}}{f_{state\ 2}} = e^{-\frac{E_{state\ 2} - E_{state\ 1}}{k_B \cdot T}} \tag{5.7}$$

A graphical representation of the Maxwell–Boltzmann distribution for either varying molecular masses or temperatures is shown in Fig. 5.1. For a particular gas (Fig. 5.1 right panel, molecular mass is constant), the average speed of the molecules, as indicated by the position of the peak, increases with increasing temperature. At the same time, the distribution of different speeds becomes lees uniform within the population of gas molecules when the temperature increases; this is obvious from the distribution curves spreading out. Notably, the areas underneath the curves remain the same, since the number of gas molecules in the system is considered to be constant.

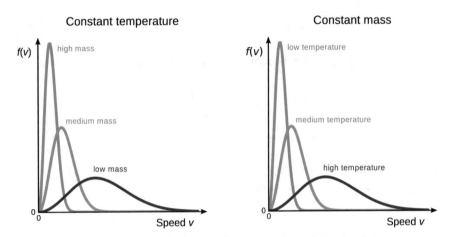

Fig. 5.1 Graphical representation of the Maxwell–Boltzmann distribution (Eq. 5.6) for either varying temperatures or molecular masses

When comparing different gases with different molecular masses (Fig. 5.1 left panel) at the same temperature (and assuming the same number of molecules), the Maxwell–Boltzmann distribution shows that lighter molecules possess higher average speeds than heavier molecules. Also, the distribution of individual speeds is less uniform in populations of lighter molecules.

5.1.3   Transport Properties

The generally accessible concept of flux as a stream of moving bodies (objects that have a mass) can be readily expanded to any object, including those that have no mass, such as physical properties. The migration of a property, i.e. property transport, can phenomenologically be described by its flux:

$$\text{flux } J = \frac{\text{quantity of property}}{\text{area passed} \cdot \text{time interval}} \tag{5.8}$$

In order to calculate the total quantity of a property that is migrating, one can re-arrange above equation to obtain:

$$\text{quantity of property} = J \cdot A \cdot \Delta t$$

In general, we can distinguish two types of properties migrating, matter and energy (Table 5.1):

Table 5.1 Migration of properties

Property migrating	Process	Flux	Units of flux
Matter	Diffusion	Number of molecules per area per time	$\frac{1}{m^2 s}$
Energy	Thermal conduction	Energy per area per time = power per area	$1\frac{J}{m^2 s} = 1\frac{W}{m^2}$
Momentum	Laminar flow	Force per area	$1\frac{N}{m^2}$

5.1.4 Diffusion: Flux of Matter

In most cases, it is found that the flux J of one property is proportional to the first derivative of another. This fundamental observation is called the general transport equation.

For example, the flux of molecules (*property₁* = matter) that migrate by diffusion along a particular direction x is proportional to the first derivative of the concentration of these molecules (*property₂* = concentration) along that direction. If we choose to work with the molar concentration c, then the first derivative of c with respect to the travel coordinate x is $\frac{dc}{dx}$, and we can denote the above rule mathematically as:

$$J \sim \frac{dc}{dx} \tag{5.9}$$

This relationship is known as Fick's first law of diffusion.

Since matter migrates from areas of high concentration to low concentration, the direction of migration is opposite to the concentration gradient. The opposite direction between flux and concentration gradient is expressed by a minus sign, when the proportionality relationship in Eq. 5.9 is converted to an equation:

$$J = -\text{const.} \cdot \frac{dc}{dx}$$

In this equation, the constant factor will become D', a coefficient for the diffusion process:

$$J = -D' \cdot \frac{dc}{dx} \tag{5.10}$$

Practically, it is not the molar concentration c that is used for the particle flux, but rather the concentration expressed as number of particles per volume, $\frac{N}{V}$. Therefore, Eq. 5.10 is transformed as per:

Table 5.2 Units of the parameters for particle flux (diffusion)

Parameter	Units	Explanation
Concentration gradient	$\left[\frac{\mathrm{d\,IN}}{\mathrm{d}x}\right] = \frac{\frac{1}{m^3}}{m} = 1\ m^{-4}$	Number of molecules per volume travelling over a distance
Flux	$[J] = 1\ m^{-2}\ s^{-1}$	Number of molecules passing an area per time interval
Diffusion coefficient	$[D] = 1\ m^2\ s^{-1}$	

Table 5.3 Units of the parameters for thermal conduction

Parameter	Units	Comment
Temperature gradient	$\left[\frac{\mathrm{d}T}{\mathrm{d}x}\right] = 1\mathrm{Km}^{-1}$	
Flux	$[J] = 1\mathrm{Jm}^{-2}\mathrm{s}^{-1} = 1\mathrm{Wm}^{-2}$	Also called the heat flux density
Thermal conductivity	$[\kappa] = 1\mathrm{JK}^{-1}\mathrm{m}^{-1}\mathrm{s}^{-1} = 1\mathrm{WK}^{-1}\mathrm{m}^{-1}$	

$$J = -D' \cdot \frac{\mathrm{d}c}{\mathrm{d}x} = -D' \cdot \frac{\mathrm{d}\left(\frac{n}{V}\right)}{\mathrm{d}x} = -D' \cdot \frac{\mathrm{d}\left(\frac{N}{N_A \cdot V}\right)}{\mathrm{d}x}$$

$$J = -D' \cdot \frac{1}{N_A} \cdot \frac{\mathrm{d}\left(\frac{N}{V}\right)}{\mathrm{d}x} \tag{5.11}$$

$$J = -D \cdot \frac{\mathrm{d\,IN}}{\mathrm{d}x}$$

Equation 5.11 contains the diffusion coefficient D that is commonly used. (According to the above transformations, D and D' are related as per $D = \frac{10^{-3}}{N_A} \cdot D'$). The SI units of the different components in eq. 5.11 are given in Table 5.2.

5.1.5 Thermal Conduction: Flux of Energy

Energy migrates along a temperature gradient; this transport is called thermal conduction. Similar to the diffusion process, energy migrates from high to low temperature, i.e. opposite the temperature gradient. When applying the general transport equation for thermal conduction, we therefore need to include a minus sign to account for the opposite direction of flux and gradient:

$$J = -\kappa \cdot \frac{\mathrm{d}T}{\mathrm{d}x} \tag{5.12}$$

The proportionality constant κ is called the thermal conductivity. The SI units of the different components in Eq. 5.12 are given in Table 5.3.

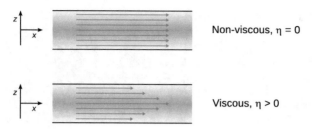

Fig. 5.2 Top: If a substance flow in a tube has negligible resistance, the speed is the same all across the tube. Bottom: When a viscous substance flows through a tube, its speed at the walls is substantially less than in the centre of the tube

5.1.6 Viscosity: Flux of Momentum

A substance streaming through a tube (Fig. 5.2) can be considered as consisting of laminar layers that flow in the same direction. If the substance is very fluid ('non-viscous') and flows through the tube with negligible resistance, then all layers move with the same speed. In contrast, the speed of the individual layers in a viscous substance are different: at the walls, the liquid moves with a substantially lesser speed than in the centre of the tube, giving rise to a 'U'- or 'V'-shaped profile.

When molecules from an outer layer (which moves at slow speed) switch to a neighbouring layer that moves faster, the neighbouring layer will be retarded because of the lower momentum of the switching molecules. The opposite will happen when molecules switch from a faster (inner) to a slower moving (outer) layer.

When molecules switch from one layer to another (a movement perpendicular to the flow direction z), then a momentum of $(m \cdot v_x)$ or $(m \cdot v_y)$ migrates from one layer to another. The flux of momentum is described by:

$$J = -\text{const.} \cdot \frac{\mathrm{d}(m \cdot v_x)}{\mathrm{d}x}$$
$$J = -(\text{const.} \cdot m) \cdot \frac{\mathrm{d}v_x}{\mathrm{d}x} \tag{5.13}$$

$$J = -\eta \cdot \frac{\mathrm{d}v_x}{\mathrm{d}x} \tag{5.14}$$

The constant of proportionality, η, is called the viscosity. If all layers move at the same velocity, the gradient $\frac{\mathrm{d}(m \cdot v_x)}{\mathrm{d}x}$ is zero, and there is no flux of momentum; the substance may still have a viscosity, though! The SI units of the different components in Eq. 5.14 are given in Table 5.4.

When discussing the flow of a substance through a tube in the above paragraph, most likely liquids came to mind; for example, water as a liquid with negligible viscosity, and glycerol as a liquid with considerable viscosity. However, these phenomena are not limited to liquids. For example, the shape of the Bunsen burner flame is due to the velocity profile across the tube.

Table 5.4 Units of the parameters for the momentum flux

Parameter	Units	Comment
Momentum gradient	$\left[\frac{d(m \cdot v_x)}{dx}\right] = 1 \text{ kg s}^{-1} = 1(\text{N s})\text{m}^{-1}$	
Flux	$[J] = 1 \text{ kg m}^{-1}\text{s}^{-2} = 1 \text{ N m}^{-2}$	
Viscosity	$[\eta] = 1 \text{ kg m}^{-1}\text{s}^{-1} = 1 \text{ N s m}^{-2} = 10 \text{ P}$	The unit poise is named after Jean Léonard Marie Poiseuille

5.1.7 The Transport Parameters of the Ideal Gas

Earlier in this section, we introduced the kinetic molecular theory of gases (Sect. 5.1.1), and started to link the behaviour of particular gas molecules to macroscopic laws. We then learned about transport properties and can now apply these to an ideal gas. Using the kinetic theory, expressions for the different transport parameters can be derived. We will not derive these relationships rigorously, but rather discuss their impact on the gas molecules.

The diffusion coefficient D of the ideal gas is obtained as per

$$D = \frac{1}{3} \cdot \lambda \cdot c \tag{5.15}$$

Here, λ is the mean free path length, i.e. the average distance a molecule travels without collision; c is the mean speed of the molecules.

We can predict the following effects:

- Since the mean free path λ of gas molecules decreases with increasing pressure (more collisions), Eq. 5.15 tells us that the diffusion coefficient D also decreases with increasing pressure. This means that at higher pressure, molecules diffuse more slowly.
- The mean speed c increases with increasing temperature, and according to Eq. 5.15 so does the diffusion coefficient D. This means that at higher temperatures, molecules diffuse more quickly.
- The mean free path λ (and thus the diffusion coefficient D) increases when the collision cross-section of molecules decrease. Smaller molecules therefore diffuse quicker than large molecules.

The thermal conductivity κ of an ideal gas A is given by

$$\kappa = \frac{1}{3} \cdot \lambda \cdot c \cdot C_{V,\text{m}} \cdot c(\text{A}) \tag{5.16}$$

where λ and c are the mean free path and mean speed as before. $C_{V,\text{m}}$ is the molar heat capacity at constant volume and $c(\text{A})$ the molar concentration of the gas.

This allows the following predictions:

- The mean free path λ is inversely proportional to the concentration (high concentration means more molecules which make the occurrence of collisions more likely, thus decreasing the mean free path). The thermal conductivity κ would thus be expected to decrease with increasing concentration, but since the concentration itself features as a factor in Eq. 5.16, the two effects balance and κ is thus independent of the concentration. Since pressure and concentration are 'two sides of the same medal' with gases (a high concentration of gas is accompanied by high pressure), we can conclude that the thermal conductivity κ is independent of the pressure.
- The thermal conductivity is larger for gases with a larger heat capacity.

For the viscosity η of an ideal gas A, the following relationship is obtained:

$$\eta = \frac{1}{3} \cdot M \cdot \lambda \cdot c \cdot c(A) \qquad (5.17)$$

We can thus predict that:

- Since the mean free path λ is inversely proportional to the concentration, and the concentration itself features as a factor in Eq. 5.17, the viscosity η is independent of the concentration, and thus also of the pressure.
- The mean speed c increases with increasing temperature, and so does the viscosity η. At higher temperatures, gases have a higher viscosity.

   This behaviour is in contrary to observations with a liquid: for a molecule in a liquid to move it must overcome intermolecular interactions. With increasing temperature, more molecules acquire this energy and can move; the viscosity of a liquid thus decreases with increasing temperature.

5.2   Molecular Motion in Liquid Solutions

We have seen in Sect. 4.1.4 that ions can be dragged through a liquid solvent by applying a potential difference ΔE between two opposing electrodes. From the potential difference ΔE over a distance l, we defined the electric field $\frac{d\phi}{dx}$ as

$$\frac{d\phi}{dx} = \frac{\Delta E}{l} \qquad (4.11)$$

   The fundamental property that characterises the ability of ions to move through the solution is the resistance R (see Sect. 4.1.5); the higher the resistance, the harder it is for the ions to migrate. The conductance G of a solution is the inverse of the resistance. Therefore, the lower the resistance, the higher becomes the conductance and the easier it gets for ions to migrate through the solution:

$$G = \frac{1}{R} \tag{4.16}$$

The conductance G increases with the cross-sectional area A and decreases with the length l; as the constant of proportionality we introduced the conductivity κ:

$$G = \kappa \cdot \frac{A}{l} \tag{4.18}$$

If we measure the conductance of an electrolyte solution in suitable container and two fixed electrodes, then the cross-sectional area A and the length l are invariant parameters of the experimental system. The measured conductance will thus depend only on the number of charged species present solution. If we normalise the proportionality factor κ (the conductivity) with respect to the molar concentration of ions (c), we obtain a property that is a characteristic constant for a particular ion. This constant is called the molar conductivity Λ_m:

$$\Lambda_m = \frac{\kappa}{c} \quad \text{with units of } [\Lambda_m] = 1 \text{ S m}^2 \text{ mol}^{-1} \tag{5.18}$$

5.2.1 Conductivities of Electrolyte Solutions

In extensive measurements with strong electrolytes (substances that are fully dissociated in solution), Friedrich Kohlrausch found in the nineteenth century that at low concentrations, the molar conductivities vary with the square root of the concentration:

$$\Lambda_m = \Lambda_{0m} - \kappa \cdot \sqrt{c} \tag{5.19}$$

which is known as Kohlrausch's law and introduces the limiting molar conductivity Λ_{0m} as a maximum molar conductivity that is attained only at indefinite dilution. The dependence on $\sqrt{c}$, rather than c, is due to inter-ionic interactions. While migrating, ions of a particular charge may pass ions of opposite charge and thus retard their migration. Kohlrausch's law is only valid for strong electrolytes and describes the non-linear decrease of the molar conductivity with increasing concentration.

For weak electrolytes (which only dissociate partially in solution), the molar conductivity depends strongly on concentration. The more dilute a solution, the greater its molar conductivity, due to the increased ionic dissociation (see Fig. 5.3). The degree of dissociation α (see Sect. 4.3.3) can be obtained from the ratio of the molar and limiting molar conductivities:

Fig. 5.3 The molar
conductivities of strong and
weak electrolytes increase
with decreasing
concentrations, but to
different extents. The molar
conductivity at infinite
dilution is the limiting molar
conductivity Λ_{0m}

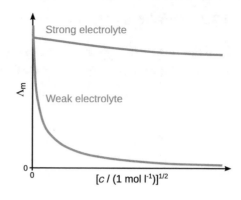

$$\alpha = \frac{\Lambda_m}{\Lambda_{0m}}. \tag{5.20}$$

The dissociation constant K_d of weak electrolytes is related to the degree of dissociation by Ostwald's dilution law:

$$K_d = \frac{[A^+] \cdot [B^-]}{[AB]} = \frac{\alpha^2}{1 - \alpha} \cdot [AB]_0 \tag{5.21}$$

where $[A^+]$, $[B^-]$ and $[AB]$ are the equilibrium concentrations of the cation, anion and nondissociated electrolyte, respectively. $[AB]_0$ is the total concentration of electrolyte. Considering Eq. 5.20, Ostwald's dilution law can also be expressed in the form of molar conductivities:

$$K_d = \frac{\Lambda_m^2}{(\Lambda_{0m} - \Lambda_m) \cdot \Lambda_{0m}} [AB]_0 \tag{5.22}$$

Experimentally, it could also be shown that the limiting molar conductivity (Λ_{0m}) is comprised of the two independent limiting molar conductivities of the anions (λ_-) and cations (λ_+):

$$\Lambda_{0m} = \nu_+ \cdot \lambda_+ + \nu_- \cdot \lambda_- \tag{5.23}$$

This finding is also known as the law of the independent migration of ions. In Eq. 5.23, ν denotes the number of ions per formula. For example:

$$NaCl \quad \nu_+ = 1, \nu_- = 1$$
$$CaCl_2 \quad \nu_+ = 1, \nu_- = 2$$

5.2.2 Mobilities of Ions

In an electric field $E = \frac{d\phi}{dx}$, where the two electrodes are at a distance $dx = l$, a particle with the charge $Q = z \cdot e$ experiences the force:

$$F_{el} = z \cdot e \cdot E = z \cdot e \cdot \frac{d\phi}{l} \tag{5.24}$$

The particle is thus accelerated towards the appropriate electrode, but it has to overcome friction in the liquid medium. According to Stokes' law of friction, the frictional force F_{fr} of a spherical particle with radius r is related to its velocity v by

$$F_{fr} = f \cdot v \text{ with the frictional constant } f = 6 \cdot \pi \cdot \eta \cdot r. \tag{5.25}$$

Here, η is the viscosity of the solvent, which we have introduced in Sect. 5.1.6.

Since both forces, the electrostatic attraction and the frictional force, act in opposite direction, the final drift speed of the moving particle is established, when both forces balance each other (i.e. no net force):

$$F_{el} = F_{fr} \tag{5.26}$$

In the above equation, we substitute the expressions from 5.24 and the left, and 5.25 on the right hand side:

$$z \cdot e \cdot E = f \cdot v$$

Resolving for the drift speed yields:

$$v = \left(z \cdot \frac{e}{f} \right) \cdot E = u \cdot E \tag{5.27}$$

where u is the mobility of the ion and E the electric field the ion is exposed to.

Resolving Eq. 5.27 for u and substituting the expression for the frictional constant from above delivers an expression that links the ion mobility to the charge of the ion as well as the viscosity of the liquid medium:

$$u = z \cdot \frac{e}{f} = \frac{z \cdot e}{6 \cdot \pi \cdot \eta \cdot r} \quad \text{with the units} [u] = 1\,\text{m}^2\,\text{V}^{-1}\,\text{s}^{-1}. \tag{5.28}$$

The ion mobility u provides a link between the charge z of an ion and its conductivity λ:

$$\lambda = z \cdot u \cdot F \tag{5.29}$$

In this equation, the constant $F = N_A \cdot e$ is the Faraday constant. Earlier, we have introduced the limiting molar conductivity Λ_{0m} as

Table 5.5 Limiting molar conductivity of selected ions in water at 25 °C. Note the extraordinarily high limiting conductivity of the proton and the hydroxyl ion in aqueous solution

Cation	λ_{0+} $mS\ m^2\ mol^{-1}$	Anion	λ_{0-} $mS\ m^2\ mol^{-1}$
H^+	35.0	OH^-	19.9
Li^+	3.87	Cl^-	7.63
Na^+	5.01	Br^-	7.84
K^+	7.35	I^-	7.68
Mg^{2+}	10.6	SO_4^{2-}	16.0
Ca^{2+}	11.9	NO_3^-	7.14
Ba^{2+}	12.7	$H_3C\text{-}COO^-$	4.09

$$\Lambda_{0m} = \nu_+ \cdot \lambda_+ + \nu_- \cdot \lambda_- \tag{5.23}$$

where ν_+ and ν_- denote the stoichiometric number of cations and anions in the formula of the electrolyte, and λ_+ and λ_- are the conductivities of the cation and anion, respectively.

By combining Eqs. 5.23 and 5.29, we can now conclude for the limiting molar conductivity:

$$\Lambda_{0m} = \nu_+ \cdot z_+ \cdot u_+ \cdot F + \nu_- \cdot z_- \cdot u_- \cdot F \tag{5.30}$$

In the absence of inter-ionic interactions (i.e. at low concentrations) for a symmetrical $z{:}z$ electrolyte, such as $CuSO_4$ ($\nu_+ = \nu_- = 1, z_+ = z_- = z = 2$) this equation simplifies to:

$$\Lambda_{0m} = (u_+ + u_-) \cdot z \cdot F \tag{5.31}$$

Therefore, ion mobilities are highly useful parameters in determining the limiting molar conductivities of electrolytes (Table 5.5).

5.2.3 Ion–Ion Interactions

In the absence of an electric field, the atmosphere surrounding a particular ion is spherical. When an ion migrates through an electric field, its atmosphere is no longer spherical but distorted, since the centres of gravity do no longer coincide. This gives rise to a reduced mobility (relaxation). The opposite direction of the movement of ionic atmosphere and the central ion also causes a viscous drag called the electrophoretic effect (Fig. 5.4).

The quantitative treatment of these phenomena is complicated, but has been achieved with the Debye–Hückel–Onsager theory. This theory allows calculation of numerical values for limiting molar conductivities which are in good agreement with experimental data at low molar concentrations (mM and below). A rigorous treatment of the Debye–Hückel–Onsager theory is beyond the scope of this introduction, but we will summarise the main finding which culminates in a linear

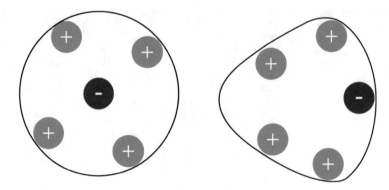

Fig. 5.4 In the absence of an electric field, the atmosphere surrounding an ion is spherical (left). However, when an ion experiences an external electric field, its surrounding atmosphere gets distorted, giving rise to reduced mobility and a viscous drag (right)

relationship between the conductivity κ and the limiting molar conductivity of an electrolyte Λ_{0m}:

$$\kappa = A + B \cdot \Lambda_{0m} \quad \text{or} \quad \Lambda_{0m} = -\frac{A}{B} + \frac{1}{B} \cdot \kappa \tag{5.32}$$

We remember that Kohlrausch's law delivered a relationship between Λ_{0m} and κ that satisfies the generic expression in Eq. 5.32:

$$\Lambda_{0m} = \Lambda_m + \sqrt{c} \cdot \kappa \quad \text{therefore} \quad -\frac{A}{B} = \Lambda_m \quad \text{and} \quad \frac{1}{B} = \sqrt{c} \tag{5.19}$$

which can be resolved to $A = -\frac{\Lambda_m}{\sqrt{c}}$ and $B = \frac{1}{\sqrt{c}}$

The theory by Debye, Hückel and Onsager yields the following expressions for the two coefficients A and B:

$$A \sim \frac{z^2}{\eta \cdot \sqrt{T}} \tag{5.33}$$

and

$$B \sim \frac{z^3}{\sqrt{T^3}} \tag{5.34}$$

5.2.4   Diffusion

Earlier in this chapter, we have introduced diffusion as flux of matter. Since this phenomenon is of utmost importance for any chemical and biochemical process, we will extend the discussion of this particle transport and also consider some thermo-dynamic aspects that are linked to this transport property.

Thermodynamically, we know that if a molecule is moving from a location with the chemical potential $\mu_1 = \mu$ to another location with the chemical potential $\mu_2 = \mu + d\mu$, the work done by the system is $dW = d\mu$.

In a system, where the chemical potential depends on the spatial position x, this results in a chemical potential gradient, i.e. the differential of μ with respect to $x : \left(\frac{\delta\mu}{\delta x}\right)$. For this process, we assume constant pressure and temperature, and there-fore obtain for the work dW done by the system during the diffusion:

$$dW = d\mu = \left(\frac{\delta\mu}{\delta x}\right)_{p,T} \cdot dx \qquad (5.35)$$

According to classical Newton mechanics, translocation work can generally be expressed in terms of a force opposing the direction of translocation:

$$dW = -F \cdot dx \qquad (5.36)$$

The fact that the force F opposes the spatial translocation x results in a minus sign in the above equation. By comparison of Eqs. 5.35 and 5.36, we conclude that the force F is defined by

$$F = -\left(\frac{\delta\mu}{\delta x}\right)_{p,T} \qquad (5.37)$$

and is thus called the thermodynamic force. The thermodynamic force does not necessarily represent a real force that is pushing the particles down the slope of the chemical potential. It rather represents the spontaneous tendency of molecules to disperse, and thus provides a link between classical mechanics and the 2nd law of thermodynamics.

5.2.5 Fick's First Law of Diffusion

If we now consider a solution that contains a solute of activity a, we can calculate the chemical potential of the solute as per:

$$\mu = \mu^{\varnothing} + R \cdot T \cdot \ln a \qquad (3.12)$$

If the solution is not uniform and the activity depends on the position x, then there exists a gradient of the chemical potential (as discussed in the previous section). From Eq. 5.37, we know that this gradient is the thermodynamic force:

$$F = -\left(\frac{\delta\mu}{\delta x}\right)_{p,T} = -\left(\frac{\delta\mu^{\varnothing}}{\delta x}\right)_{p,T} - R \cdot T \cdot \left(\frac{\delta \ln a}{\delta x}\right)_{p,T} \qquad (5.38)$$

Using Eq. 3.12, the gradient $\left(\frac{\delta\mu}{\delta x}\right)_{p,T}$ can be separated into two terms, namely $\left(\frac{\delta\mu^{\varnothing}}{\delta x}\right)_{p,T}$ and $\left(\frac{\delta \ln a}{\delta x}\right)_{p,T}$. Since $\mu^{\varnothing}$ is the standard chemical potential it is a constant parameter and thus not dependent on the location. Its differential with respect to x is thus zero. This results in the following expression for the thermodynamic force:

$$F = -R \cdot T \cdot \left(\frac{\delta \ln a}{\delta x}\right)_{p,T} \qquad (5.39)$$

which we can simplify under the assumption that the solution is ideal; in that case, $a = c$:

$$F = -R \cdot T \cdot \left(\frac{\delta \ln c}{\delta x}\right)_{p,T} \qquad (5.40)$$

Using the fact that $d\ln y = \frac{1}{y}$ (see A.2.4), we can use the transformation of $\frac{d \ln y}{d x} = \frac{1}{y} \cdot \left(\frac{dy}{dx}\right)$ to obtain:

$$F = -\frac{R \cdot T}{c} \cdot \left(\frac{\delta c}{\delta x}\right)_{p,T} \qquad (5.41)$$

This equation describes the flux of diffusing particles as motion in response to a thermodynamic force which arises from the concentration gradient.

The particles reach a steady drift speed v_{dr}, when the thermodynamic force is matched by the viscous drag. We can then construct the following chain of conclusions:

The drift speed v_{dr} is proportional to the force F : $v_{dr} \sim F$

The particle flux J is also proportional to the drift speed: $J \sim v_{dr}$

The force F is proportional to the concentration gradient: $F \sim \left(\frac{dc}{dx}\right)$

$$\text{Therefore} : J \sim v_{dr} \sim F \sim \left(\frac{dc}{dx}\right)$$

This means that the flux J is proportional to the concentration gradient; we have already come across this relationship in Sect. 5.1.4: Fick's first law of diffusion.

From the introduction of Fick's first law in that section, we remember that we can express the flux in terms of the concentration gradient by using the diffusion coefficient D as a constant of proportionality:

$$J = -D \cdot \frac{dc}{dx} \tag{5.10}$$

The flux is also related to the drift speed v_{dr} by:

$$J = v_{dr} \cdot c \tag{5.42}$$

Note the equality of units: $[J] = \frac{1}{m^2 \cdot s}$

$$[v_{dr} \cdot c] = \left(\frac{m}{s} \cdot \frac{6.022 \cdot 10^{23}}{mol} \right) \cdot \left(\frac{mol}{10^{-3} m^3} \right)$$

v_{dr} includes the Avogadro constant N_A.
By combining Eqs. 5.10 and 5.42, one obtains:

$$v_{dr} \cdot c = -D \cdot \frac{dc}{dx} \tag{5.43}$$

and from Eq. 5.41 we know that $\left(\frac{dc}{dx} \right) = -\frac{F \cdot c}{R \cdot T}$, which then yields the numerical relationship between the particle drift speed v_{dr} and the thermodynamic force:

$$v_{dr} = -\frac{D}{c} \cdot \frac{dc}{dx} = \frac{D \cdot F}{R \cdot T} \tag{5.44}$$

The constant of proportionality between the drift speed and the thermodynamic force is therefore $\frac{D}{R \cdot T}$.

5.2.6 The Einstein Relation

In Sect. 5.2.2, we discussed the mobilities of ions and derived the following expression for the ion drift speed in an electric field E:

$$v_{dr} = u \cdot E \tag{5.44}$$

In this equation, u is the mobility and $E = \frac{d\phi}{dx}$ the electric field.

In the case of 1 mol of ions moving through an electric field, the driving force F is the electrostatic force F_{el}, which is calculated according to

$$F_{el} = N_A \cdot z \cdot e \cdot E = z \cdot F \cdot E \tag{5.24}$$

Note: 'F' on the right hand side is the Faraday constant
Combining Eqs. 5.44 and 5.24 yields:

$$u \cdot E = \frac{z \cdot F \cdot E \cdot D}{R \cdot T}$$

which simplifies to

$$u = \frac{z \cdot F \cdot D}{R \cdot T} \tag{5.45}$$

and thus delivers an expression for the diffusion constant D as per:

$$D = \frac{u \cdot R \cdot T}{z \cdot F} \tag{5.46}$$

Note: 'F' on the right hand side is the Faraday constant

This important relationship between the diffusion coefficient D and the ion mobility u is known as the Einstein relation.

Typical ion mobilities take values at the order of $u = 5 \cdot 10^{-8}\,\mathrm{m^2\,s^{-1}\,V^{-1}}$ for which Eq. 5.46 yields a value of $D = 1 \cdot 10^{-9}\,\mathrm{m^2\,s^{-1}}$ as a typical value of the diffusion coefficient of an ion in water.

5.2.7 The Stokes-Einstein Equation

We can now take advantage of the relationship between the ion mobility u and the diffusion coefficient D given by the Einstein relation and combine it with Eq. 5.28, which we derived when introducing the ion mobility in Sect. 5.2.2:

$$\text{Einstein relation}: u = \frac{z \cdot F \cdot D}{R \cdot T} \tag{5.45}$$

$$u = z \cdot \frac{e}{f} = \frac{z \cdot e}{6 \cdot \pi \cdot \eta \cdot r} \tag{5.28}$$

Mobility u of an ion with radius r and charge z in a medium of viscosity η
The combination of both equations yields:

$$u = \frac{z \cdot e}{6 \cdot \pi \cdot \eta \cdot r} = \frac{z \cdot F \cdot D}{R \cdot T} \tag{5.47}$$

We then substitute the Faraday constant $F = N_A \cdot e$ and obtain:

$$\frac{z \cdot e}{6 \cdot \pi \cdot \eta \cdot r} = \frac{z \cdot N_A \cdot e \cdot D}{R \cdot T}$$

The product of charge state and elementary charge, $(z \cdot e)$, can be eliminated, and with $k_B = \frac{R}{N_A}$ the above equation becomes:

$$\frac{1}{6 \cdot \pi \cdot \eta \cdot r} = \frac{D}{k_B \cdot T}$$

which resolves for the diffusion coefficient D as:

$$\text{Stokes-Einstein equation} : D = \frac{k_B \cdot T}{6 \cdot \pi \cdot \eta \cdot r} \qquad (5.48)$$

This equation relates the diffusion coefficient D of a solute of radius r migrating through a medium with the viscosity of that medium, η.

This important equation is called the Stokes-Einstein equation and is the basis of determination of the diffusion coefficient D by viscosity measurements. Importantly, there is no reference to the charge state of the solute, so all molecules (not just ions) can be assessed this way.

5.3   Exercises

1. At 25 °C, the molar ionic conductivities Λ_m of the alkali ions Li^+, Na^+ and K^+ are 3.87, 5.01 and 7.35 mS m^2 mol^{-1}, respectively. What are their mobilities?

2. Fullerene (C_{60}) has a diameter of 10.2 Å. Estimate the diffusion coefficient of fullerene in benzonitrile at 25 °C. The viscosity of benzonitrile at that temperature is 12.4 mP. How large is the error of your estimate when comparing the result to the experimental value of $4.1 \cdot 10^{-10}$ m^2 s^{-1}?

3. The mean free pathlength of gas molecules can be calculated according to Maxwell as $\lambda = \frac{1}{\sqrt{2} \cdot IN \cdot \sigma}$, where the particle density is $IN = \frac{p}{k_B \cdot T}$. Calculate the diffusion coefficient of argon at 298 K and a pressure of 1 bar. The collisional cross-section of argon is $\sigma = 0.41$ nm^2.

4. The diffusion coefficient of sucrose in water is $5.2 \cdot 10^{-6}$ cm^2 s^{-1} at room temperature. Estimate the effective radius of a sucrose molecule, if water has a viscosity of 10 mP.

Kinetics

6

6.1 Introduction

6.1.1 From Thermodynamics to Reaction Kinetics

The thermodynamic principles we have introduced in Sect. 2, and subsequently applied to a variety of systems, all had a common point of focus: they were targeting systems in equilibrium. A chemical reaction that has reached equilibrium still exhibits a forward and a reverse reaction, but the rates of both processes are equal, and since the reactions are of opposing direction, there is no net change.

In this part, we want to address the question of what happens to reactants in the course of time. This will also lead us to investigate what particular reaction pathways are engaged during a chemical reaction. The fundamental kinetics concepts will thus characterise a reaction with respect to

- time
- concentration
- temperature.

6.1.2 Spontaneous and Non-Spontaneous Reactions, Processes at Equilibrium

We have seen earlier (see Sect. 2.2.1) that the Gibbs free energy G provides a single criterion of spontaneity and equilibrium:

$$G = H - T \cdot S \tag{2.44}$$

$$\Delta G = \Delta H - T \cdot \Delta S \quad (T = \text{const.}, p = \text{const.}) \tag{2.42}$$

© Springer International Publishing AG, part of Springer Nature 2018
A. Hofmann, *Physical Chemistry Essentials*,
https://doi.org/10.1007/978-3-319-74167-3_6

Importantly, the function G is based on state functions, and hence a state function itself. It was derived that:

- $\Delta G > 0$ for a non-spontaneous process
- $\Delta G = 0$ for a process at equilibrium
- $\Delta G < 0$ for a spontaneous process.

With knowledge of the enthalpy of reaction and by calculating the entropy change ΔS_{sys} (see Sect. 2.1.10), we can therefore predict whether or not a reaction will be spontaneous at a selected temperature.

For processes that are at equilibrium, the principle of Le Châtelier (Le Chatelier and Boudouard 1898) allows us to predict what happens to the position of the equilibrium when the conditions are changed:

▶   A system at equilibrium, when subject to a disturbance, responds in a way that tends to minimise the effect of the disturbance.

Energy changes during a reaction
Consider the following reaction which describes the synthesis of ammonia from its elements:

$$N_2 + 3\,H_2 \rightleftharpoons 2\,NH_3 \quad \Delta G = -32\,kJ\,mol^{-1}$$
$$\Delta H = -92.4\,kJ\,mol^{-1}$$

In this example, ΔG is the free energy change of the system, as 1 mol of nitrogen gas reacts with hydrogen to form ammonia. ΔG is a difference in free energy, i.e. the system has a lower free energy after the reaction proceeded to the right-hand side of the chemical equation. The fact that the free energy change is negative indicates that the reaction will proceed from the left to the right (albeit it requires pressures of 150–250 bar, temperatures of 300–550 °C and a catalyst—a technologically important process known as the Haber-Bosch process).

We see that the change of enthalpy during the reaction takes a much more negative value than the change of the free energy. This makes for two observations. First, the fact that the enthalpy change is negative indicates that the reaction is exothermic, i.e. 92.4 kJ mol^{-1} of heat are produced for every mol of nitrogen being transformed. Second, the question is what happens to the remaining energy of 60.4 kJ mol^{-1}. This is energy "consumed" as the reaction proceeds from the left to right, because the disorder of the system is decreased. Whereas there are four molecules of starting products, there are only two molecules of end products; the system has taken a state with

(continued)

more order. Therefore, the entropy term $T \cdot \Delta S$ accounts for 60.4 kJ for each mol of nitrogen transformed.

Note: Since the chemical reaction above describes the synthesis of ammonia from its elements, the change of enthalpy for this reaction is called the enthalpy of formation.

6.1.3 Stoichiometry and Molecularity

A stoichiometric equation is the simplest equation involving whole numbers of the molecules involved in the overall chemical reaction. For example:

$$N_2 + 3H_2 \rightarrow 2NH_3$$

$$2KMnO_4 + 16HCl \rightarrow 2KCl + 2MnCl_2 + 8H_2O + 5Cl_2$$

$$N_2 + O_2 \rightarrow 2N_2O$$

Stoichiometric equations do not represent the kinetic equations! The molecularity of a reaction as calculated by adding the stoichiometric coefficients of the reactants does therefore in most cases not agree with the experimentally observed order of a reaction (see Sect. 6.2.2). However:

▶ For an elementary reaction, the molecularity calculated as the number of reacting species in the stoichiometric equation, is the same as the order of the reaction.

▶ An elementary reaction is defined as a single step reaction with a single transition state and no intermediates.

Therefore, more complex chemical reactions consist of a sequence of reactions each known as an elementary reaction. The concept of the transition state will be discussed in more detail in Sect. 6.5.2.

Reactions that involve

- One molecule are called unimolecular,
- Two molecules are called bimolecular, and
- Three molecules are called termolecular.

6.2   Reaction Rates, Rate Constants and Orders of Reaction

6.2.1   The Rate of Reaction

What is the rate of a reaction? Studies in the mid-nineteenth century by Cato M. Guldberg and Peter Waage led them to put forward the law of mass action (Guldberg 1864; Waage 1864; Waage and Guldberg 1864) which states that

▶   the rate of any chemical reaction is proportional to the product of the masses of the reacting substances, with each mass raised to a power equal to the coefficient that occurs in the chemical equation:

$$\nu_A A + \nu_B B \rightarrow \nu_C C + \nu_D D$$

$$v \sim m(A)^{\nu_A} \cdot m(B)^{\nu_B} \tag{6.1}$$

After being re-discovered by van't Hoff (1877), this law is now of historical interest, since the rate expressions derived from this law are now known to apply only to elementary reactions. However, it is still useful for obtaining the correct equilibrium equation for a reaction

$$K = \frac{\prod\limits_{i=1}^{N_{products}} \left(\frac{c_i}{c^{\varnothing}}\right)^{\nu_i}}{\prod\limits_{j=1}^{N_{reactants}} \left(\frac{c_j}{c^{\varnothing}}\right)^{\nu_j}} = \frac{\left(\frac{c(C)}{c^{\varnothing}}\right)^{\nu_C} \cdot \left(\frac{c(D)}{c^{\varnothing}}\right)^{\nu_D}}{\left(\frac{c(A)}{c^{\varnothing}}\right)^{\nu_A} \cdot \left(\frac{c(B)}{c^{\varnothing}}\right)^{\nu_B}} = \frac{[C]^{\nu_C} \cdot [D]^{\nu_D}}{[A]^{\nu_A} \cdot [B]^{\nu_B}} \tag{6.2}$$

since Guldberg and Waage recognised that chemical equilibrium is a dynamic process in which the rates of reaction for the forward and backward processes must be equal at chemical equilibrium.

In Physical Chemistry, the rate of reaction is defined as the change of concentration over time, divided by the appropriate stoichiometric coefficient. Convention describes a positive rate as a formation of products (=loss of reactants). So for the following reaction,

$$\nu_A A + \nu_B B \rightarrow \nu_C C + \nu_D D$$

where A is consumed and C is produced, the reaction rate is defined as:

$$v = -\frac{1}{\nu_A} \cdot \frac{dc(A)}{dt} = +\frac{1}{\nu_C} \cdot \frac{dc(C)}{dt} \tag{6.3}$$

The rate of a reaction can be expressed with respect to different components
When considering the following reaction

$$2\,NOBr_{(g)} \rightarrow 2\,NO_{(g)} + Br_{2(g)}$$

with the formation of NO proceeding with $\frac{dc(NO)}{dt} = 0.16\,\text{mmol dm}^{-3}\text{s}^{-1}$, what is the rate of formation of NO in this reaction?

$$v = \frac{1}{v(NO)} \cdot \frac{dc(NO)}{dt} = \frac{1}{2} \cdot 0.16\,\text{mmol dm}^{-3}\text{s}^{-1} = 0.080\,\text{mmol dm}^{-3}\text{s}^{-1}$$

For the consumption of NOBr one obtains the rate:

$$v = -\frac{1}{v(NOBr)} \cdot \frac{dc(NOBr)}{dt}$$

Since $-\frac{dc(NOBr)}{dt} = \frac{dc(NO)}{dt}$ and $v(NOBr) = 2$, this yields:

$$v = -\frac{1}{2} \cdot \left(-0.16\,\text{mmol dm}^{-3}\text{s}^{-1}\right) = 0.080\,\text{mmol dm}^{-3}\text{s}^{-1}$$

This illustrates that the rate for a reaction can be calculated with respect to any component, but always yields the same numerical value.

6.2.2   Rate Laws, Rate Constants and Reaction Order

The rate of a reaction is often proportional to the concentrations of the reactants, raised to a power:

$$v_A A + v_B B \rightarrow v_C C + v_D D$$

$$v \sim c(A)^m, v \sim c(B)^n$$

This proportionality relationship can be transformed into an equation:

$$v = \text{const.} \cdot c(A)^m \cdot c(B)^n$$

Importantly, the exponents m and n cannot be predicted *a priori*. In some instances, they agree with the stoichiometric coefficients (i.e. $m = v_A$, $n = v_B$), but in many instances this is not the case. The constant factor is called the rate constant k:

$$v = k \cdot c(A)^m \cdot c(B)^n \quad \text{(rate law)} \tag{6.4}$$

If $m = n = 1$, the units of k for this equation are:

$$[k] = 1 \, \text{dm}^3 \, \text{mol}^{-1} \, \text{s}^{-1}$$

The rate constant is independent of the concentrations of the reactants, but dependent on the temperature (see Sect. 6.5). Since the exponents m and n cannot be predicted, their values need to be determined experimentally. Equation 6.4 is thus subject to experimental verification and is called the rate law of the reaction.

The exponents m and n define the order of a reaction. A reaction with a rate law as given in Eq. 6.4 is then said to be of

- m-th order in A
- n-th order in B
- the overall order of $(m + n)$

The exponents m, n, ... do not need to be integers, they can also be fractional, e.g. $\frac{1}{2}$.

Some reactions obey a zero-order rate law and therefore proceed with a rate that is independent of the concentration of the reactant (as long as some is present). For example, phosphine (PH_3) is catalytically decomposed on a tungsten (W) surface

$$4 \, PH_{3(g)} \rightarrow P_{4(g)} + 6 \, H_{2(g)}$$

and follows a zero-order reaction. In that case, the generic rate law 6.4 simplifies to:

$$v = k$$

Reactions of zero-th order are typically found when there is a limiting parameter in the reaction, such as e.g. the surface of a catalyst.

Some reactions have complicated rate laws with no overall order. For example, the formation of HBr from its elements

$$H_{2(g)} + Br_{2(g)} \rightarrow 2 \, HBr_{(g)}$$

has the following rate equation:

$$v = \frac{k_1 \cdot c(H_2) \cdot c(Br_2)^{3/2}}{c(Br_2) - k_{-1} \cdot c(HBr)}$$

This reaction is of first order with respect to H_2, but of indefinite order with respect to Br_2 and HBr (see also Sect. 6.4.2).

6.2.3   Reaction Profile

Since the concentrations of reactants and products change with time during a chemical reaction, the reaction rates are often monitored experimentally by continuously measuring the concentration of a particular species during the reaction. The resulting plot that represents the change of the concentration of a species with time is called a reaction profile.

The term 'continuous' needs to be taken with some caution. In some cases, samples are taken at various time points during a reaction and then analysed as to the concentration of a particular species (e.g. by quenching methods). Such measurements are taken at select time intervals and thus yield discrete data points. In other cases, the proceeding reaction may be monitored with a sensor (e.g. thermocouple, photodetector of a UV/Vis absorption spectrometer, etc) that delivers an electric signal which is acquired as digital data by a computer. Despite such data are often taken as continuous data, the data acquisition is not continuous but depends on the sampling rate of the analog–digital converter (i.e. consists of—a large number of—discrete data points).

The choice of the sampling rate in the experimental setting has repercussions in the mathematical treatment of kinetics (see Fig. 6.1). If a low sampling rate is applied (such as e.g. in the offline analysis/quenching method), data points are acquired spaced by long time intervals. This is called a reaction profile with discrete data points. The rates measured from various pairs of points then correspond to a mathematical Δ. Note that when graphically depicting data from discrete measurements, the appropriate choice is a scatter plot where each data point is depicted as a discrete point with a symbol (and possibly error bars, if multiple measurements are available). Discrete data points should never be connected by lines. If a mathematical model, such as a rate equation, is available, discrete data points are fitted with that equation and the fit may be superimposed as a continuous line graph on the discrete data points. The rate may then be extracted by differentiation of the underlying mathematical (fit) equation. Since the fit line is based on a mathematical equation and thus data are known for every point on the ordinate (x-axis), this constitutes truly continuous data and thus represented by connected dots (=line).

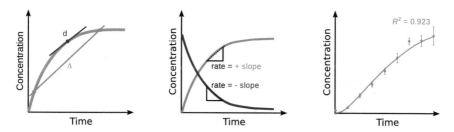

Fig. 6.1 Reaction profiles. Left: effects of low sampling rates. Centre: reaction profiles appear different, depending on whether reactant or product concentrations are followed; however, the reaction rate is the same. Right: illustration of a scatter plot and data fit

At high sampling rates, a near-continuous profile is acquired and the differences between the different data points approaches the mathematical 'd', which is the slope of the curve at a particular point.

The left panel of Fig. 6.1 shows that a low sampling rate (turquoise) results in acquisition of discrete data points; the rates determined from discrete data points are approximations, since they represent macroscopic differences taken between two clearly distinct points ('Δ'). High sampling rates enable a near-continuous data acquisition and the rates may be determined as the slopes of tangents at any point ('d'). The centre panel of Fig. 6.1 illustrates different reaction profiles for the same reaction. Depending on whether product (turquoise) or reactant (magenta) concentrations are followed, the rate is obtained either as the positive or the negative slope from the reaction profile. The right panel of Fig. 6.1 illustrates a scatter plot of discrete data for a chemical reaction. Multiple independent measurements at individual time points allow calculation of the mean (black dots) and the estimated standard deviation (shown as positive and negative error bars). Data fitting with an appropriate rate equation results in a truly continuous reaction profile from which rates can be determined by differentiation of the underlying rate law. The goodness of fit needs to be indicated (here illustrated as an R^2 evaluation).

6.3   Rate Equations

A summary of the differential and integrated rate equations for the different reaction orders, along with their derived half-lives is given in Table 1.11.

6.3.1   Differential Rate Equation

For the reaction

$$\nu_A A \rightarrow \nu_B B + \nu_C C$$

The rate law of the form

$$v = k \cdot c(A)^m$$

can be expressed as

$$-\frac{1}{\nu_A} \cdot \frac{dc(A)}{dt} = k \cdot c(A)^m \tag{6.5}$$

which constitutes the differential form of the rate equation.

The exponent in the rate law, m, determines the order of the reaction; it is said to proceed with the m-th order. The differential rate equations for the different reaction orders are:

$$-\frac{1}{\nu_A} \cdot \frac{dc(A)}{dt} = k \cdot c(A)^0 = k \quad 0^{th}\,order \tag{6.6}$$

$$-\frac{1}{\nu_A} \cdot \frac{dc(A)}{dt} = k \cdot c(A)^1 = k \cdot c(A) \quad 1^{st}\,order \tag{6.7}$$

$$-\frac{1}{\nu_A} \cdot \frac{dc(A)}{dt} = k \cdot c(A)^2 \quad 2^{nd}\,order \tag{6.8}$$

$$-\frac{1}{\nu_A} \cdot \frac{dc(A)}{dt} = k \cdot c(A)^3 \quad 3^{rd}\,order \tag{6.9}$$

6.3.2   Integrated Rate Expression for the First Order Rate Law

In order to derive an expression for a given rate law that allows calculation of numerical values for the concentration of a reactant over time, the differential rate expression needs to be integrated:

$$v = -\frac{1}{\nu_A} \cdot \frac{dc(A)}{dt} \Rightarrow c_t(A) = ?$$

To integrate the differential 1st-order rate law

$$-\frac{1}{\nu_A} \cdot \frac{dc(A)}{dt} = k \cdot c(A) \tag{6.7}$$

one first identifies the two variables that are of interest for the integration; here, these are $c(A)$ and t. The equation is then re-arranged such that the two variables get isolated on opposite sides:

$$-\frac{1}{\nu_A} \cdot \frac{1}{c(A)} \cdot dc(A) = k \cdot dt$$

The integration has to be carried out with respect to the two differential variables, i.e. $\frac{1}{c(A)} \cdot dc(A)$ on the left side and dt on the right side. The factors $-\frac{1}{\nu_A}$ and k are constants with respect to the integration and thus can be taken outside the integral. In order to calculate determined integrals, one also needs to specify the boundaries within which the integration is supposed to happen. We start at time t_0 where the concentration of A is the initial concentration $c_0(A)$, and proceed up to time t, where the concentration of A is $c_t(A)$:

$$-\frac{1}{\nu_A} \cdot \int_{c_0}^{c_t} \frac{1}{c(A)} \cdot dc(A) = k \cdot \int_{t_0}^{t} dt \qquad (6.10)$$

The integral on the left side, $\int \frac{1}{c(A)} \cdot dc(A)$, corresponds to the type $\int x^{-1} dx$ and is resolved to $\ln x$. On the right hand side, $\int dt$ is of the type $\int dx$ which resolves to x (see Table A.2). Equation 6.10 with determined integrals thus yields:

$$-\frac{1}{\nu_A} \cdot [\ln c(A)]_{c_0}^{c_t} = k \cdot [t]_{t_0}^{t} \qquad (6.11)$$

The expressions of the resolved integrals $[x]_{x_{\text{start}}}^{x_{\text{end}}}$ need to be calculated as per:

$$[x]_{x_{\text{start}}}^{x_{\text{end}}} = x_{\text{end}} - x_{\text{start}}$$

Equation 6.11 then yields:

$$-\frac{1}{\nu_A} \cdot [\ln c_t(A) - \ln c_0(A)] = k \cdot (t - t_0)$$

With $t_0 = 0$ one obtains:

$$-\frac{1}{\nu_A} \cdot [\ln c_t(A) - \ln c_0(A)] = k \cdot t$$

In order to calculate a numerical value for the only unknown, $c_t(A)$, we isolate the expression for $c_t(A)$ on one side:

$$\ln c_t(A) - \ln c_0(A) = -\nu_A \cdot k \cdot t \qquad (6.12)$$

$$\ln c_t(A) = -\nu_A \cdot k \cdot t + \ln c_0(A) \qquad (6.13)$$

Equation 6.13 shows that if we measure the concentration of A, $c_t(A)$, at various time points t, and plot $\ln c_t(A)$ versus t, we obtain a line with a negative slope, $(-\nu_A \cdot k)$, and a y-intercept of $\ln c_0(A)$ (Fig. 6.2). Linear relationships in x–y plots are a convenient way of analysing data, and historically (before the advent of the computer) the only way to determine numerical values of fitting constants.

With modern fitting software, it is, of course, possible to fit non-linear relationships. If Eq. 6.12 is resolved for $c_t(A)$, one obtains:

Fig. 6.2 Analysis of 1st-order kinetics by means of linear regression

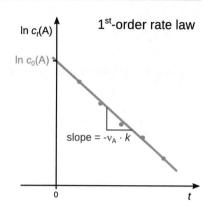

$$\ln \frac{c_t(A)}{c_0(A)} = -\nu_A \cdot k \cdot t \tag{6.14}$$

$$\frac{c_t(A)}{c_0(A)} = e^{-\nu_A \cdot k \cdot t}$$

$$c_t(A) = c_0(A) \cdot e^{-\nu_A \cdot k \cdot t} \tag{6.15}$$

A closer look at Eqs. 6.13 and 6.15 reveals that the experimentally determined rate constant is indeed ($\nu_A \cdot k$) and includes the stoichiometric coefficient! This needs to be taken into account when calculating the rate constant k from experimental data.

6.3.3 Half-lives and Time Constants

Half-lives and time constants are secondary parameters derived from the rate equations and can be calculated for reactions of any order.

The half-life $t_{1/2}$ describes the time required to reduce the concentration of a reactant to 1/2 of its initial value. The time constant τ describes the time required to reduce the concentration of a reactant to $\frac{1}{e}$ of its initial value.

For 1st-order chemical reactions, the half-life and time constant are particularly useful indicators for the rate of the reaction, since with 1st-order reactions the two parameters do not depend on the initial concentration of reactants is the half-life $t_{1/2}$ of a reactant.

In order to derive an expression for half-life of a 1st-order reaction, we consider its definition:

$$t = 0: \quad c_t(A) = c_0(A)$$

$$t = t_{1/2}: \quad c_t(A) = \frac{1}{2} \cdot c_0(A)$$

Using this relationship in the 1st-order rate Eq. 4.13 yields:

$$\ln \frac{c_t(A)}{c_0(A)} = -k \cdot t$$

$$\ln \frac{\frac{1}{2} \cdot c_0(A)}{c_0(A)} = -k \cdot t_{1/2}$$

$$-\ln \frac{1}{2} = k \cdot t_{1/2}$$

Making use of a logarithm rule that multiplies (-1) into the argument of the logarithm yields:

$$\ln 2 = k \cdot t_{1/2}$$

and thus

$$t_{1/2} = \frac{\ln 2}{k} \quad \text{Half-life of a 1st-order reaction}$$

Irrespective of the initial concentration, the time it takes in a 1st-order reaction to reduce the initial concentration to half its value is always $\frac{0.693}{k}$.

Earlier, we have introduced the time constant τ of a reaction as the time it takes for the concentration of a reactant to fall to $\frac{1}{e}$ ($e \approx 2.71828$) of its initial value. As above in the case of the half-life, we use this relationship and apply it to the 1st-order rate law 4.13 which yields:

$$\ln \frac{\frac{1}{e} \cdot c_0(A)}{c_0(A)} = -k \cdot \tau$$

$$-\ln \frac{1}{e} = k \cdot \tau$$

$$\ln e = k \cdot \tau$$

$$1 = k \cdot \tau$$

$$\tau = \frac{1}{k} \quad \text{Time constant of a } 1^{\text{st}}\text{-order reaction} \qquad (6.17)$$

6.3.4 Integrated Rate Expression for the Second Order Rate Law

The differential 2nd-order rate law is given by Eq. 6.8:

$$-\frac{1}{\nu_A} \cdot \frac{dc(A)}{dt} = k \cdot c(A)^2 \qquad (6.8)$$

We re-arrange the equation such that the two variables $c(A)$ and t get isolated on opposite sides:

$$-\frac{1}{\nu_A} \cdot \frac{1}{c(A)^2} \cdot dc(A) = k \cdot dt$$

and integrate from time t_0 where the concentration of A is $c_0(A)$ to time t, where the concentration of A is $c_t(A)$:

$$-\frac{1}{\nu_A} \cdot \int_{c_0}^{c_t} \frac{dc(A)}{c(A)^2} = k \cdot \int_{t_0}^{t} dt \qquad (6.18)$$

The integral on the left side, $\int \frac{dc(A)}{c(A)^2}$ corresponds to the type $\int x^{-2} dx$ and is resolved to $\frac{1}{-2+1} \cdot x^{-2+1} = -x^{-1}$. On the right hand side, $\int dt$ is of the type $\int dx$ which resolves to x (see Table A.2). Equation 6.18 with determined integrals thus yields:

$$\frac{1}{\nu_A} \cdot \left[\frac{1}{c(A)} \right]_{c_0}^{c_t} = k \cdot [t]_{t_0}^{t}$$

$$\frac{1}{\nu_A} \cdot \left(\frac{1}{c_t(A)} - \frac{1}{c_0(A)} \right) = k \cdot (t - t_0)$$

With $t_0 = 0$, this yields:

$$\left(\frac{1}{c_t(A)} - \frac{1}{c_0(A)} \right) = \nu_A \cdot k \cdot t$$

Fig. 6.3 Analysis of
2nd-order kinetics by means
of linear regression

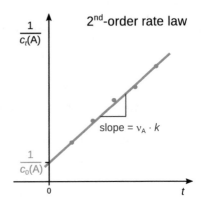

$$\frac{1}{c_t(A)} = v_A \cdot k \cdot t + \frac{1}{c_0(A)} \tag{6.19}$$

or, when resolving for $c_t(A)$:

$$c_t(A) = \frac{c_0(A)}{1 + v_A \cdot k \cdot t \cdot c_0(A)} \tag{6.20}$$

Equation 6.19 shows that for the second order rate law, a linear correlation is obtained when plotting $\frac{1}{c_t(A)}$ versus t, where the slope is $v_A \cdot k$ and the y-intercept is $\frac{1}{c_0(A)}$ (Fig. 6.3).

6.3.5   Integrated Rate Expression for the Third Order Rate Law

The differential 3rd-order rate law is given by Eq. 6.9:

$$-\frac{1}{v_A} \cdot \frac{dc(A)}{dt} = k \cdot c(A)^3 \tag{6.9}$$

For integration, we again isolate the two variables, $c(A)$ and t, on opposite sides:

$$-\frac{1}{v_A} \cdot \frac{1}{c(A)^3} \cdot dc(A) = k \cdot dt$$

and integrate from $[t_0, c_0(A)]$ to $[t, c_t(A)]$:

$$-\frac{1}{\nu_A} \cdot \int_{c_0}^{c_t} \frac{dc(A)}{c(A)^3} = k \cdot \int_{t_0}^{t} dt \qquad (6.21)$$

The integral on the left side, $\displaystyle\int_{c_0}^{c_t} \frac{dc(A)}{c(A)^3}$ corresponds to the type $\int x^{-3} dx$ and is

resolved to $\frac{1}{-3+1} \cdot x^{-3+1} = -\frac{1}{2}x^{-2}$. On the right hand side, $\displaystyle\int_{t_0}^{t} dt$ is of the type $\int dx$

which resolves to x (see Table A.2). Equation 6.21 with determined integrals thus yields:

$$-\frac{1}{\nu_A} \cdot \left[-\frac{1}{2 \cdot c(A)^2} \right]_{c_0}^{c_t} = k \cdot [t]_{t_0}^{t}$$

$$\frac{1}{2 \cdot \nu_A} \cdot \left(\frac{1}{c_t(A)^2} - \frac{1}{c_0(A)^2} \right) = k \cdot (t - t_0)$$

$$\frac{1}{c_t(A)^2} = 2 \cdot \nu_A \cdot k \cdot t + \frac{1}{c_0(A)^2} \qquad (6.22)$$

Therefore, a plot of $\frac{1}{c_t(A)^2}$ against t will yield a line with slope $2 \cdot \nu_A \cdot k$ and intercept $\frac{1}{c_0(A)^2}$ (Fig. 6.4).

Fig. 6.4 Analysis of 3rd-order kinetics by means of linear regression

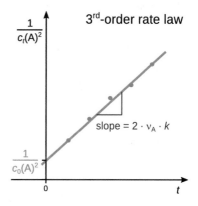

6.3.6   Integrated Rate Expression for the Zero-th Order Rate Law

In order to integrate the differential rate law of zero-th order

$$-\frac{1}{\nu_A} \cdot \frac{dc(A)}{dt} = k \tag{6.6}$$

the two variables, $c(A)$ and t, are isolated on opposite sides of the equation:

$$-\frac{1}{\nu_A} \cdot dc(A) = k \cdot dt$$

We then integrate from $[t_0, c_0(A)]$ to $[t, c_t(A)]$:

$$-\frac{1}{\nu_A} \cdot \int_{c_0}^{c_t} dc(A) = k \cdot \int_{t_0}^{t} dt \tag{6.23}$$

The integrals on both sides side, $\int_{c_0}^{c_t} dc(A)$ and $\int_{t_0}^{t} dt$, correspond to the type $\int dx$ which resolves to x (see Table A.2). Equation 6.23 with determined integrals thus yields:

$$-\frac{1}{\nu_A} \cdot [c(A)]_{c_0}^{c_t} = k \cdot [t]_{t_0}^{t}$$

which results in

$$-\frac{1}{\nu_A} \cdot [c_t(A) - c_0(A)] = k \cdot (t - t_0)$$

With $t_0 = 0$ this yields:

$$-\frac{1}{\nu_A} \cdot [c_t(A) - c_0(A)] = k \cdot t$$

$$c_t(A) - c_0(A) = -\nu_A \cdot k \cdot t$$

$$c_t(A) = -\nu_A \cdot k \cdot t + c_0(A) \tag{6.24}$$

Fig. 6.5 Analysis
of 0th-order kinetics
by means of linear regression

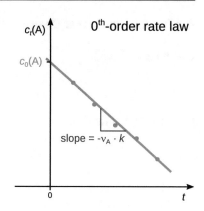

Therefore, a plot of $c_t(A)$ against t will yield a line with slope $-\nu_A \cdot k$ and intercept $c_0(A)$ (Fig. 6.5).

6.4   Determination of the Rate Law

6.4.1   Analysing Data Using the Integrated Rate Laws

If kinetic data for a chemical reaction have been obtained, there are two ways to check whether the reaction adheres to any of the standard rate laws (0th–3rd order).

If at least two data points, t and $c_t(A)$, are known, one can substitute these values into the four integrated rate laws (Eqs. 6.24, 6.15, 6.19 and 6.22) and estimate the rate constants k for the different orders. For the correct order, the two data points need to yield the same value for the rate constant.

Alternatively, the linear relationships (Eqs. 6.24, 6.13, 6.19 and 6.22) of the integrated rate equations can be plotted and subjected to a linear fit (see Figs. 6.2, 6.3, 6.4, and 6.5). The rate law that produces the best fit is likely to be the right one (Table 6.1).

Importantly, the experimentally determined rate constant k_{exp} as obtained from the slope of these linear relationships is related to the 'true' rate constant k by a proportionality factor that also depends on the stoichiometric coefficients of the reaction.

Table 6.1 The rate law may be determined by graphical representation of data and checking for a linear relationship. The rate constant is then obtained from the slope

Order	Rate law	Graph	Rate constant
0	$-\frac{1}{\nu_A} \cdot \frac{dc(A)}{dt} = k$	$c(A)$ vs t	$k = -\frac{1}{\nu_A} \cdot k_{exp}$
1	$-\frac{1}{\nu_A} \cdot \frac{dc(A)}{dt} = k \cdot c(A)$	$\ln c(A)$ vs t	$k = -\frac{1}{\nu_A} \cdot k_{exp}$
2	$-\frac{1}{\nu_A} \cdot \frac{dc(A)}{dt} = k \cdot c(A)^2$	$\frac{1}{c(A)}$ vs t	$k = \frac{1}{\nu_A} \cdot k_{exp}$
3	$-\frac{1}{\nu_A} \cdot \frac{dc(A)}{dt} = k \cdot c(A)^3$	$\frac{1}{c(A)^2}$ vs t	$k = \frac{1}{2 \cdot \nu_A} \cdot k_{exp}$

6.4.2   Experimental Design: Isolation Method

Given the reaction

$$A + B \rightarrow C + D$$

with the general rate law

$$v = k \cdot c(A)^m \cdot c(B)^n \tag{6.4}$$

we are now concerned with experimental approaches that allow determination of the exponents m, n, etc.

In the isolation method, all starting reactants are supplied in excess except for one (e.g. reactant A). One can then assume that all excessively supplied reactants remain approximately at their initial concentration; the one component with limited supply is 'isolated'. In the above example, one can then assume for B that:

$$c(B) = c_0(B) = \text{const.}$$

For the rate law (Eq. 6.4), it follows that

$$v = k \cdot c(A)^m \cdot c(B)^n = k \cdot c(A)^m \cdot c_0(B)^n$$

The rate constant k and the constant factor $c_0(B)^n$ can be combined to an apparent rate constant k':

$$v = k' \cdot c(A)^m \tag{6.25}$$

This resulting rate law only features the concentration of one component (the isolated reactant) and is thus called a pseudo-1st-order rate law.

In order to exploit the pseudo-1st-order rate law in such a way, one needs to determine the rate of a reaction at a given time t, as well as the concentration of A at that particular time t. The most frequently used approach in this context is to just consider the initial rates, i.e. the rate at time $t = 0$. This eliminates the need to re-examine the concentration of A, as it will be essentially the starting concentration of the reactant, $c_0(A)$. In the method of initial rates, the rate is thus measured at the very beginning of the reaction when the concentration of the reactants is still at their initial values. Equation 6.25 then becomes:

$$v_0 = k' \cdot c_0(A)^m$$

and by subjecting the equation to a logarithmic transformation (and using the logarithm rule of $\log(a \cdot b) = \log a + \log b$), one obtains:

Fig. 6.6 Determination of the reaction order using the isolation method on initial rates

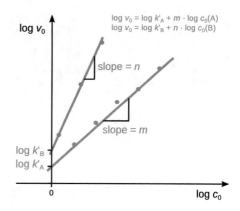

$$\log v_0 = \log k' + \log[c_0(A)^m]$$

With the rule of $\log a^m = m \cdot \log a$ this yields:

$$\log v_0 = \log k' + m \cdot \log c_0(A) \tag{6.26}$$

Experimentally, the initial rates are determined for a series of different initial concentrations of A. A plot of $\log v_0$ versus $\log c_0(A)$ then results in a line with slope m and intercept $\log k'_A$ (see Fig. 6.6).

Using this pseudo-1st-order rate law, the exponent of the isolated reactant can be obtained and thus the true order of the reaction with respect to this reactant may be established. Subsequently, all other starting reactants are isolated, one at a time, thus allowing conclusions as to the remaining reaction orders.

A caveat for the method of initial rates

The method of initial rates may not always reveal the full rate law. Reaction products may participate in the reaction and thus affect the rate. As first proposed in 1907 (Bodenstein and Lind 1907), in the reaction that synthesises HBr from H_2 and Br_2

$$H_{2(g)} + Br_{2(g)} \rightarrow 2\,HBr_{(g)}$$

the product HBr engages in the radical reaction mechanism (which consists of consecutive elementary reactions; see Sect. 6.6):

(continued)

Initiation
$$Br_2 \rightarrow Br^\bullet + Br^\bullet$$

Propagation
$$Br^\bullet + H_2 \rightarrow HBr + H^\bullet$$
$$H^\bullet + Br_2 \rightarrow HBr + Br^\bullet$$
$$H^\bullet + HBr \rightarrow Br^\bullet + H_2$$

Termination
$$Br^\bullet + Br^\bullet \rightarrow Br_2$$
$$H^\bullet + Br^\bullet \rightarrow HBr$$
$$H^\bullet + H^\bullet \rightarrow H_2$$

Based on this reaction mechanism, the following rate equation is derived:

$$v = \frac{k_1 \cdot c(H_2) \cdot c(Br_2)^{3/2}}{c(Br_2) - k_2 \cdot c(HBr)}$$

6.4.3 Reactions Approaching Equilibrium

In practice, kinetic studies often conducted immediately after initiation of a reaction, i.e. far from the equilibrium. Therefore, reverse reactions need not be considered and the method of initial rates may be applied. However, when a reaction actually approaches equilibrium, the concentrations of products have to be taken into account. As an exemplary case, we consider a chemical equilibrium

$$A \rightleftharpoons B \tag{6.27}$$

where both the forward and reverse reactions follow a 1st-order rate law (which is, for example, the case in some isomerisation reactions).

The concentration of A is thus decreased by the forward reaction, but increased by the reverse reaction, and one obtains the following rate expression:

$$\frac{dc(A)}{dt} = -k_1 \cdot c(A) + k_{-1} \cdot c(B) \tag{6.28}$$

With

$$c(A) + c(B) = c_0(A) \Rightarrow c(B) = c_0(A) - c(A)$$

we can substitute $c(B)$ in Eq. 6.28 and obtain:

$$\frac{dc(A)}{dt} = -k_1 \cdot c(A) + k_{-1} \cdot [c_0(A) - c(A)]$$

$$= -k_1 \cdot c(A) - k_{-1} \cdot c(A) - k_{-1} \cdot c_0(A)$$

$$\frac{dc(A)}{dt} = -(k_1 + k_{-1}) \cdot c(A) - k_{-1} \cdot c_0(A) \qquad (6.29)$$

The integrated form of the differential Eq. 6.29 resolves to:

$$c(A) = \frac{k_{-1} + k_1 \cdot e^{-(k_1 - k_{-1}) \cdot t}}{k_{-1} + k_1} \cdot c_0(A) \qquad (6.30)$$

From Eq. 6.30, one can calculate the equilibrium concentration of A by considering a very large value for the time t; if the reaction is allowed to continue for a long time, it will certainly have reached equilibrium. With knowledge of the function e^{-x} (see Fig. 6.7) it becomes clear that:

$$\text{if } t = \infty \Rightarrow e^{-(k_1 - k_{-1}) \cdot t} \to 0$$

Therefore, the term $k_1 \cdot e^{-(k_1 - k_{-1}) \cdot t}$ in Eq. 6.30 can be neglected and the equilibrium concentration of A resolves to:

$$c_{eq}(A) = \frac{k_{-1}}{k_{-1} + k_1} \cdot c_0(A) \qquad (6.31)$$

For the equilibrium concentration of B, one obtains:

Fig. 6.7 The function e^{-x} anneals to zero when x takes very large values

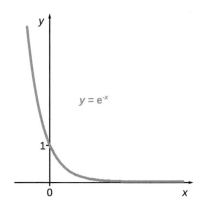

$$c_{eq}(B) = c_0(A) - c_{eq}(A) = c_0(A) - \frac{k_{-1}}{k_{-1} + k_1} \cdot c_0(A)$$

$$c_{eq}(B) = \frac{k_{-1} + k_1}{k_{-1} + k_1} \cdot c_0(A) - \frac{k_{-1}}{k_{-1} + k_1} \cdot c_0(A)$$

$$c_{eq}(B) = \frac{k_{-1} + k_1 - k_{-1}}{k_{-1} + k_1} \cdot c_0(A)$$

$$c_{eq}(B) = \frac{k_1}{k_{-1} + k_1} \cdot c_0(A) \qquad (6.32)$$

As illustrated in Fig. 6.8, in the course of the reaction, the concentration of A decreases over time and attains its equilibrium value once the equilibrium has been reached. In turn, the concentration of B increases over time up to the value of its equilibrium concentration. Importantly, at equilibrium the rates of the forward and reverse reactions are the same. We can thus conclude:

$$v_1 = v_{-1}$$

$$k_1 \cdot c_{eq}(A) = k_{-1} \cdot c_{eq}(B)$$

$$\frac{k_1}{k_{-1}} = \frac{c_{eq}(B)}{c_{eq}(A)} = K \qquad (6.33)$$

This relationship provides a link between the equilibrium constant of the reaction, K (see Eq. 6.2), a thermodynamic quantity, and the rate constants, k_1 and k_{-1}, which are kinetic properties.

Fig. 6.8 Time-dependent development of the concentrations of A and B in reaction 6.27

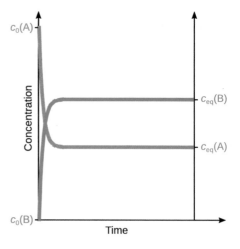

6.4.4   Relaxation Methods

When a system is disturbed and thus taken off equilibrium conditions, the term relaxation refers to the return to a state of equilibrium. When a chemical reaction is at equilibrium, and environmental parameters such as temperature or pressure are suddenly changed, the system will adjust itself to new equilibrium conditions (cf. Le Châtelier's principle; Sect. 6.1.2).

Manfred Eigen (1967 Nobel Prize in Chemistry) developed methodologies in the 1950s to experimentally determine rate constants of fast reactions based on these phenomena (Eigen et al. 1953; Eigen 1954). Technically, it is possible to elicit sudden changes in temperature (T-jump) or pressure (p-jump). If a reaction has a definite reaction enthalpy $\Delta_r H$ ($\Delta_r H \neq 0$), the equilibrium constant K of that reaction is dependent on the temperature (van't Hoff equation, Eqs. 2.60, 2.61 in Sect. 2.2.6). In response to a sudden temperature change (T-jump), the reaction will respond by a relaxation process that established the new equilibrium based on the new temperature. Similarly, if the reaction volume $\Delta_r V$ has a definite value ($\Delta_r V \neq 0$), then its equilibrium constant is dependent on the pressure. A sudden pressure change (p-jump) will result in a relaxation that adjusts the equilibrium to the new conditions.

Irrespective of the order of the underlying reaction, the relaxation process will follow a 1st-order kinetics.

Since the equilibrium composition of a reaction depends on the temperature or pressure, the concentrations of the individual components after a T- or p-jump will differ from the equilibrium concentrations $c_{eq}(A)$, $c_{eq}(B)$ and $c_{eq}(C)$ by a concentration difference $\Delta_{jump}c$. For the simple equilibrium

$$A + B \underset{k_{-1}}{\overset{k_1}{\rightleftharpoons}} C$$

this means

$$c(A) = c_{eq}(A) + \Delta_{jump}c$$

$$c(B) = c_{eq}(B) + \Delta_{jump}c$$

$$c(C) = c_{eq}(C) - \Delta_{jump}c$$

Without rigorously deriving the consequences here, it can be shown that the change of the concentration difference $\Delta_{jump}c$ with time occurs as per

$$\frac{d(\Delta_{jump}c)}{dt} = -\Delta_{jump}c \cdot \left\{ k_1 \cdot \left[c_{eq}(A) + c_{eq}(B) \right] + k_{-1} \right\}$$

After separation of variables and integration, this leads to the relationship

$$\Delta_{jump}c(t) = -\Delta_{jump}c(0) \cdot e^{-\left\{k_1 \cdot \left[c_{eq}(A)+c_{eq}(B)\right]+k_{-1}\right\} \cdot t} \tag{6.34}$$

In Eq. 6.34, $\Delta_{jump}c(t)$ is the concentration change at time t, and $\Delta_{jump}c(0)$ the concentration change immediately after the T/p-jump. By comparison, we find that this equation takes the form of equation

$$c_t(A) = c_0(A) \cdot e^{-\nu_A \cdot k \cdot t} \tag{6.15}$$

which describes a 1st-order kinetics.

The time constant τ of a first order reaction as then as per Eq. 6.17:

$$\frac{1}{\tau} = k$$

Note: k is the true rate constant from Eq. 6.15.

Applied to Eq. 6.34 for the relaxation after a T/p-jump, the time constant becomes:

$$\frac{1}{\tau} = k_1 \cdot \left[c_{eq}(A) + c_{eq}(B)\right] + k_{-1}$$

By using the relationship $K = \frac{k_1}{k_{-1}}$, this can be transformed to:

$$\frac{1}{\tau} = k_1 \cdot \left(c_{eq}(A) + c_{eq}(B) + \frac{1}{K}\right) \tag{6.35}$$

Due to the system relaxing after a T/p-jump, the time constant τ is now called relaxation time.

T-jump and p-jump experiments therefore offer a way to determine rate constants of the forward and reverse reactions of an equilibrium. If the equilibrium constant is known, a single measurement of the relaxation time τ can reveal the values for k_1 and k_{-1}.

6.4.5 Monitoring the Progress of a Reaction

The rates of chemical reactions are measured using techniques that monitor the concentrations of species present in the reaction mixture. Moderately fast reactions can be monitored by 'freezing' the state of a reaction at a certain time—a method called quenching. The quenched reaction can then be analysed offline, using a suitable experimental procedure. Fast reactions often require real-time analysis or the application of relaxation methods (see previous section). We will look at some experimental methodologies for both cases in Sects. 6.4.6 and 6.4.7.

Importantly, the time scale of a particular reaction often does not solely depend on the value of the rate constant. Only for 1st-order reactions does the half-life $t_{1/2}$ depend exclusively on the rate constant k:

$$t_{1/2} = \frac{\ln 2}{k} \quad \text{Half-life of a 1st-order reaction} \qquad (6.16)$$

For 2nd- and 3rd-order reactions, the half-lives depend on the concentration of the reactants as per:

$$t_{1/2} = \frac{1}{k \cdot c_0(A)} \quad \text{Half-life of a 2nd-order reaction} \qquad (6.36)$$

$$t_{1/2} = \frac{3}{2 \cdot k \cdot c_0(A)^2} \quad \text{Half-life of a 3rd-order reaction} \qquad (6.37)$$

and such reactions can thus be slowed down by choosing lesser initial concentrations. However, the accuracy of concentration measurements decreases as concentrations become smaller, and therefore there will be technical limits. For reactions that follow 0th-order, the half-life is

$$t_{1/2} = \frac{c_0(A)}{2 \cdot k} \quad \text{Half-life of a 0th-order reaction.} \qquad (6.38)$$

So in order to increase the half-life in such cases, larger initial concentrations are required.

A reaction where at least one component is a gas might result in an overall change of pressure in a system of constant volume. Such reactions may be followed by recording the variation of pressure with time. Spectrophotometry is a widely applicable technique; it measures the absorption of electromagnetic radiation and is particularly useful, if a component in the reaction mixture has characteristic spectral properties. Reactions where the number or type of ions change can be monitored by recording the electric conductivity. If hydrogen ions are produced or consumed during a reaction, its progress can be followed by monitoring the pH of the solution.

Examples of monitoring reaction kinetics

(1)

$$H_{2(g)} + Br_{2(g)} \rightarrow 2\,HBr_{(g)}$$

This gas phase reaction may be monitored by absorption of visual light by Br_2. The visual absorption maximum of Br_2 is at $\lambda_{max} = 420$ nm (Hubinger and Nee 1995).

(2)

$$(H_3C)_3CCl_{(aq)} + H_2O_{(l)} \rightarrow (H_3C)_3COH_{(aq)} + H^+{}_{(aq)} + Cl^-{}_{(aq)}$$

Here, the conductivity or pH of the solution may be monitored.

(3)

The kinetics of the alkaline hydrolysis of acetylsalicylic acid (aspirin) in solution can be analysed by quenching the reaction mixture at various time points and determining the amount of hydroxyl ions remaining at that time by back-titration.

Alternatively, the reaction may be monitored by UV/Vis spectroscopy. The product of the reaction, salicylic acid/salicylate, possesses an absorption maximum at $\lambda_{max} = 298$ nm; acetylsalicylic acid has no significant absorbance at this wavelength.

(4)

$$H_2O_{2(aq)} + 2\,H^+{}_{(aq)} + 2\,I^-{}_{(aq)} \rightarrow 2\,H_2O + I_{2(aq)}$$

The reaction of hydrogen peroxide with iodide in acid solution is a fast exothermic reaction. The kinetics of this process can be monitored via the released heat output, i.e. in terms of temperature rise. This can be done with a calorimeter.

6.4.6 Experimental Techniques for Moderately Fast Reactions

In order to achieve homogeneous mixing of reactants at the beginning of a reaction, the experimental challenges increase with increase of the reaction rate. With moderately fast reactions (half-lives at the order of seconds), conventional mixing techniques are quite appropriate. Typically, such reactions are started by addition of one of the reactants to a stirred solution containing the remaining reactants.

Data acquisition may either occur using a real-time technique (e.g. pH electrode, recording of absorbance) or by stopping the reaction at various time points (quenching) and analysing the concentration of reactants with a suitable protocol. There are three types of quenching methods:

Conventional Quenching
Here, the reaction is stopped after it has been allowed to progress for a certain time. Reaction intermediates may be trapped and analysis can proceed at any time. This works for reactions that are slow enough such that there is negligible reaction going on during the quenching process.

Freeze Quench Method
Here, the reaction is quenched by rapid cooling of the mixture. Concentrations of reactants, intermediates and products can then be measured.

Chemical Quench Flow Method
Reactants are mixed as in the flow method (see next section). The reaction is quenched by another reagent (e.g. acid, base) after the mixture travelled along a fixed length through the outlet tube. Once the reaction has been quenched, analysis of concentrations can proceed with any suitable "slow" method.

6.4.7 Experimental Techniques for Fast Reactions

For fast reactions, mixing of reactants at the start of the reaction needs to be achieved by flow methods. Here, reactants are mixed rapidly as they are introduced into a chamber. The reaction continues while the mixture flows through the outlet tube. Two types of flow methods are commonly in use: the continuous flow and the stopped flow (see Fig. 6.9).

With exception of the chemical quench flow method (see previous section), data acquisition in flow methods is typically by real time methods, whereby either a small sample is withdrawn or the bulk reaction mixture is monitored. Figure 6.10 shows a stopped flow instrument with an integrated absorbance spectrometer.

Enzymatic activity of carbonic anhydrase by stopped-flow kinetics
Carbonic anhydrases are essential enzymes in all organisms regulating CO_2 but also pH homeostasis. They catalyse the (de-)hydration of CO_2:

(continued)

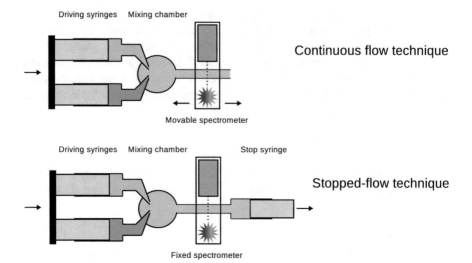

Fig. 6.9 Schematic comparison of the experimental setup for continuous flow (top) and stopped flow (bottom)

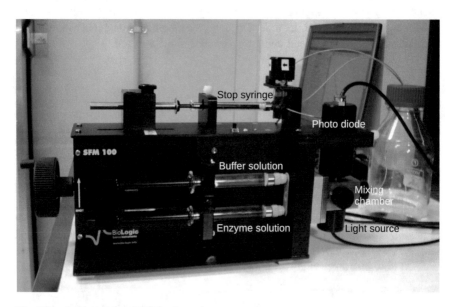

Fig. 6.10 A stopped flow UV/Vis absorption spectrometer

$$CO_{2(aq)} + H_2O_{(l)} \rightleftharpoons HCO_3^-{}_{(aq)} + H^+{}_{(aq)}$$

In the forward direction, the environmental pH is lowered, in the reverse direction, the pH is increased. This pH change, which is due to the production of 1 mol H^+ for each mol CO_2 consumed, allows monitoring of the reaction (Khalifah 1971). For stopped-flow experimentation, this is done by a pH indicator whose colour change is monitored spectrophotometrically (compared to the fast reaction, pH electrodes require too long to equilibrate for a stable read-out). The pH indicator used at mildly acidic pH is m-cresol purple; its spectral characteristics at different pH conditions are shown in Fig. 6.11. m-cresol purple has an absorption band at $\lambda_{max} = 578$ nm; the absorbance decreases with decreasing pH. The indicator in the reaction mix thus allows monitoring of H^+ production and, therefore, CO_2 consumption.

Figure 6.12 shows a reaction profile obtained by stopped-flow kinetics experiments with human carbonic anhydrase II. Note that the reaction has finished after 80 ms!

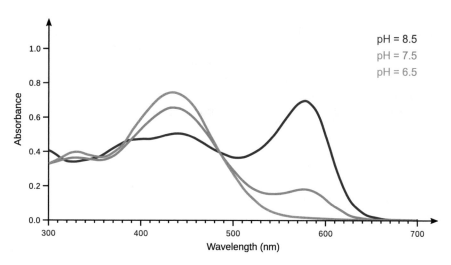

Fig. 6.11 Absorption spectra of m-cresol purple in the visual range at different pH conditions (6.5, 7.5, 8.5)

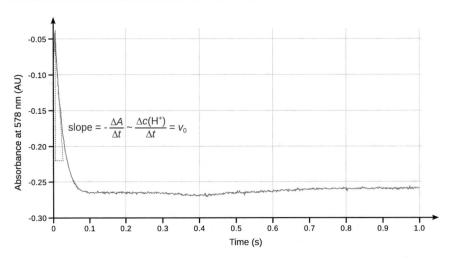

Fig. 6.12 Reaction profile of human carbonic anhydrase II from a stopped flow kinetics assay using m-cresol purple as indicator

6.5   Temperature Dependence of Reaction Rates

6.5.1   The Arrhenius Equation

We have previously discussed the temperature dependence of the equilibrium constant K (Sect. 2.2.6) which has been developed by van't Hoff in 1884. Since the equilibrium constant is the ratio of the rate constants of the forward and the reverse reaction (Sect. 6.4.3, Eq. 6.33), it is reasonable to expect that both rate constants would be dependent on the temperature as well. This concept has been developed by Svante Arrhenius in 1889, who found that the temperature dependence of most reaction rates adheres to the Arrhenius equation (Arrhenius 1889a, 1889b):

$$k = A \cdot e^{-\frac{E_a}{R \cdot T}} \tag{6.39}$$

Here, A is called the pre-exponential factor and E_a is the activation energy. When writing Eq. 6.39 in its logarithmic form

$$\ln k = \ln A - \frac{E_a}{R \cdot T} \tag{6.40}$$

it becomes clear that a plot of $\ln k$ versus $\frac{1}{T}$ yields a line with slope $-\frac{E_a}{R}$; such a plot is called an Arrhenius plot. The comparison of two reactions with different activation energies E_a shows that the higher the activation energy of a reaction, the stronger the reaction depends on the temperature. This is illustrated in Fig. 6.13, where the reaction with the higher activation energy is shown in blue. Due to the larger

Fig. 6.13 Temperature
dependence of the rate
constant according to
Arrhenius

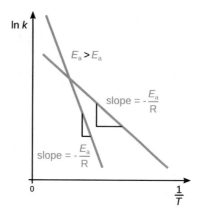

value of E_a, the slope of that line is steeper, and therefore the rate constant k varies
stronger with temperature T.

If a reaction has an activation energy of $E_a = 0$, its rate constant does not depend
on temperature. If negative activation energies are found, the rate constant decreases
with increasing temperature—an indication of a complex reaction mechanism. There
are also several cases, where the relationship between $\ln k$ and $\frac{1}{T}$ is not linear
correlation; this means that the reaction does not behave according to the Arrhenius
equation. Non-Arrhenius behaviour may be observed in explosions, enzyme
reactions, heterogeneous catalysis and in cases of pre-equilibria.

As a general case, the activation energy at any temperature can be calculated
according to

$$E_a = R \cdot T^2 \cdot \left(\frac{d \ln k}{dT}\right) \tag{6.41}$$

which applies to both Arrhenius and non-Arrhenius behaviour.

Since the variation of the rate constant k of a reaction directly affects the rate v,
there is a practical implication of the Arrhenius equation: for many reactions at room
temperature, the reaction rates double for every 10 K increase in temperature.

6.5.2 Interpretation of the Arrhenius Equation

When following the course of a reaction between two reactants A and B, the two
molecules come into contact with each other, conformations change, and atoms are
exchanged or discarded. These events are collectively summarised as the reaction
coordinate. The potential energy along the reaction coordinate has a maximum
value, and the corresponding atomic structure is called the activated complex (see
Fig. 6.14). Here, the reactants are in a state of distortion where any further distortion
will send them on the way to the products. This geometric configuration is thus
called the transition state.

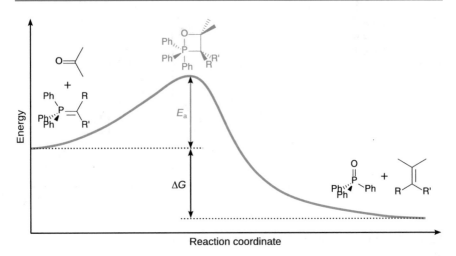

Fig. 6.14 Energy profile of a reaction without intermediate where the products have lower free energy than the reactants. E_a is the activation energy and ΔG is the free energy of the reaction. See further development of this concept in the transition state theory (Fig. 6.20)

The energy difference of the system between the start of the reaction and the activated complex is the activation energy E_a as introduced in the Arrhenius Equation (6.40). One can thus conclude that E_a is the minimum kinetic energy that the reactants must have in order to form products. The pre-exponential factor A is a measure of the rate with which collisions between molecules occur, irrespective of their energy. The product between the rate of collisions and the minimum kinetic energy required for reaction is thus the rate of successful collisions.

This fundamental concept and the exponential factor $e^{-\frac{E_a}{RT}}$ turn out to be essential components of the collision theory (Sect. 6.8) as well as the transition state theory (Sect. 6.11). Since the rate constant of a reaction is macroscopic parameter, the Arrhenius activation energy derived from Eq. 6.40 is also a macroscopic property, and should thus not be interpreted at the level of single molecules. E_a is an average energy determined from many individual collisions with varying collision parameters (collision angle, kinetic energy, internal energy); it is not simply a molecular threshold energy.

6.6 Linking the Rate Laws with Reaction Mechanisms

So far, we have focused on the explanation of kinetic observations with respect to mathematically formulated rate laws. In the following, we will try to explain the observed rate laws in terms of a postulated reaction mechanism.

6.6.1 Elementary Reactions

Most reactions occur as a sequence of steps which are called elementary reactions, each of which involves only a small number of different molecules or ions (see Sect. 6.1.3).

For example, the net reaction describing the formation of HBr from its elements

$$H_{2(g)} + Br_{2(g)} \rightarrow 2\,HBr_{(g)}$$

can be broken down into the elementary reactions arising from a radical reaction mechanism (Bodenstein and Lind 1907; Christiansen 1967):

Initiation
$$Br_2 \rightarrow Br^{\bullet} + Br^{\bullet}$$

Propagation
$$Br^{\bullet} + H_2 \rightarrow HBr + H^{\bullet}$$
$$H^{\bullet} + Br_2 \rightarrow HBr + Br^{\bullet}$$
$$H^{\bullet} + HBr \rightarrow Br^{\bullet} + H_2$$

Termination
$$Br^{\bullet} + Br^{\bullet} \rightarrow Br_2$$
$$H^{\bullet} + Br^{\bullet} \rightarrow HBr$$

In an elementary reaction, there is no phase specified; it is simply a proposed individual step of a larger reaction mechanism. For instance in the following step

$$H^{\bullet} + Br_2 \rightarrow HBr + Br^{\bullet}$$

an H atom attacks a Br_2 molecule: a bimolecular reaction. In Sect. 1.3, we introduced the molecularity of the reaction as the number of molecules that come together to react in the elementary reaction. It was also discussed that the molecularity needs to be distinguished from the reaction order which is an empirical quantity, derived from the experimental rate law. However, the reaction order of the individual elementary reactions can be derived directly from the molecularity.

Unimolecular Elementary Reactions

$$A \rightarrow P$$

follow a 1st-order rate law, because the number of molecules A that can decay is proportional to the number of molecules A initially available:

$$\frac{dc(A)}{dt} = -k \cdot c(A) \tag{6.7}$$

Bimolecular Elementary Reactions

$$A + B \rightarrow P$$

follow a 2nd-order rate law, because their rate is proportional to the rate with which the two reactants meet—which in turn depends on their concentrations:

$$\frac{dc(A)}{dt} = -k \cdot c(A) \cdot c(B) \tag{6.8}$$

6.6.2   Consecutive Elementary Reactions

When investigating a reaction mechanism, the postulated mechanism can only be explored by a detailed investigation of the system, considering that side products or intermediates may appear in the course of the reaction. Importantly, if a reaction is an elementary bimolecular reaction, then it has a 2nd-order kinetics. However, if 2nd-order kinetics are observed for a net reaction, then it may be a complex reaction.

In the following, we will need to combine a series of consecutive simple steps, i.e. elementary reactions, in order to arrive at a reaction mechanism and the corresponding rate law.

Some reactions proceed through formation of an intermediate (I), so the entire mechanism is described by consecutive unimolecular reactions:

$$A \xrightarrow{k_1} I \xrightarrow{k_2} P$$

An example is the decay of radioactive isotopes, such as

$$^{239}U \xrightarrow{\quad t_{1/2} = 23.5 \text{ min} \quad} {}^{239}Np \xrightarrow{\quad t_{1/2} = 2.35 \text{ d} \quad} {}^{239}Pu$$

When we consider the variation of concentration with time for consecutive unimolecular reactions, it becomes clear that A decays according to a 1st-order rate law to I:

$$\frac{dc(A)}{dt} = -k_1 \cdot c(A)$$

The intermediate I is formed from A, and decays to P; both processes follow 1st-order rate laws:

$$\frac{dc(I)}{dt} = k_1 \cdot c(A) - k_2 \cdot c(I)$$

P is formed from I in a 1st-order reaction:

$$\frac{dc(P)}{dt} = k_2 \cdot c(I)$$

In order to obtain expressions for the concentrations of A, I and P, the differential equations need to be integrated. This yields for the above equations:

$$c(A) = c_0(A) \cdot e^{k_1 \cdot t} \tag{6.42}$$

$$c(I) = \frac{k_1}{k_2 - k_1} \cdot \left(e^{-k_1 \cdot t} - e^{-k_2 \cdot t} \right) \cdot c_0(A) \tag{6.43}$$

$$c(P) = \left(1 + \frac{k_1 \cdot e^{-k_2 \cdot t} - k_2 \cdot e^{-k_1 \cdot t}}{k_2 - k_1} \right) \cdot c_0(A) \tag{6.44}$$

The time-dependent development of the concentrations of A, I and P is illustrated graphically in Fig. 6.15.

6.6.3 Steady-State Approximation

It is quite apparent from the algebraic expressions 6.42–6.44 in the previous section, that there is an increase in mathematical complexity when looking at reactions with more than one step. A reaction scheme that comprises many steps may become virtually unsolvable, and numerical rather than algebraic integration may be required.

Alternatively, an approximation can be introduced that yields more directly accessible results than the rigorous mathematical treatment. The (quasi-)steady-state approximation assumes that, after an initial induction period, where concentrations of intermediates rise from zero to a peak value, they remain approximately constant and the rate of change of concentrations of intermediates are negligibly small:

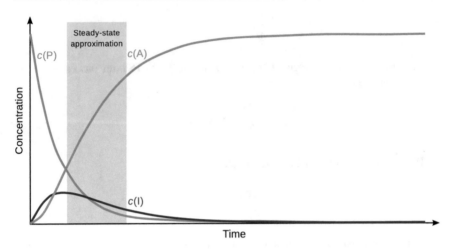

Fig. 6.15 Concentrations of reactant, intermediate and product in a set of consecutive elementary interactions

$$c(I) \approx \text{const.} \Rightarrow \frac{dc(I)}{dt} \approx 0 \tag{6.45}$$

Figure 6.15 shows that this is indeed an approximation as the curve depicting the concentration of the intermediate I does *not* possess a slope $\left(= \frac{dc(I)}{dt}\right)$ of zero after rising to a peak value. Nevertheless, we assume the approximation stated by Eq. 6.45 for the following considerations and obtain for the concentration of the intermediate:

$$\frac{dc(I)}{dt} = k_1 \cdot c(A) - k_2 \cdot c(I) = 0$$

and thus

$$c(I) = \frac{k_1}{k_2} \cdot c(A)$$

Which yields for the rate of product concentration change:

$$\frac{dc(P)}{dt} = k_2 \cdot c(I) = \frac{k_1 \cdot k_2}{k_2} \cdot c(A) = k_1 \cdot c(A)$$

This reveals a 1st-order rate law, whereby P is formed by a decay process of A:

$$c(A) = c_0(A) \cdot e^{-k_1 \cdot t} \tag{6.15}$$

For the rate of product formation, one thus obtains:

$$\frac{dc(P)}{dt} = k_1 \cdot c_0(A) \cdot e^{-k_1 \cdot t} \tag{6.46}$$

In order to integrate Eq. 6.46, we first isolate the two differential variables on opposite sides of the equation:

$$dc(P) = k_1 \cdot c_0(A) \cdot e^{-k_1 \cdot t} dt$$

and then consider the development of products in the time interval $0 \rightarrow t$, i.e. from $c_0(P)$ to $c_t(P)$:

$$\int_{c_0(P)}^{c_t(P)} dc(P) = \int_0^t k_1 \cdot c_0(A) \cdot e^{-k_1 \cdot t} dt$$

The factors k_1 and $c_0(A)$ are constants with respect to the integral and can thus be taken outside the integral:

$$\int_{c_0(P)}^{c_t(P)} dc(P) = k_1 \cdot c_0(A) \cdot \int_0^t e^{-k_1 \cdot t} dt$$

The integral on the left side of the equation is of the type $\int dx = \int x^0 dx$ and resolves to $x^{0+1} = x$. On the right hand side, the integral is of the type $\int e^{a \cdot x} dx$ which resolves to $\frac{1}{a} \cdot e^{a \cdot x}$. We thus obtain:

$$[c(P)]_{c_0(P)}^{c_t(P)} = k_1 \cdot c_0(A) \cdot \left[-\frac{1}{k_1} \cdot e^{-k_1 \cdot t} \right]_0^t$$

$$[c_t(P) - c_0(P)] = -\frac{k_1}{k_1} \cdot c_0(A) \cdot \left(e^{-k_1 \cdot t} - e^{-k_1 \cdot 0} \right)$$

Since there is no product present at time $t = 0$, it follows that $c_0(P) = 0$; for the term $e^{-k_1 \cdot 0}$, one obtains $e^0 = 1$. Therefore:

$$(c_t(P) - 0) = -c_0(A) \cdot \left(e^{-k_1 \cdot t} - 1 \right)$$

$$c_t(P) = c_0(A) \cdot \left(1 - e^{-k_1 \cdot t} \right) \tag{6.47}$$

When compared to the expression we obtained for the product concentration by rigorous kinetic analysis above

$$c(\mathrm{P}) = \left(1 + \frac{k_1 \cdot e^{-k_2 \cdot t} - k_2 \cdot e^{-k_1 \cdot t}}{k_2 - k_1}\right) \cdot c_0(\mathrm{A}) \qquad (6.44)$$

it becomes clear that expression 6.47—obtained with the steady-state approximation—constitutes a special case of Eq. 6.44. In the case when

$$k_2 \gg k_1$$

k_1 in the denominator of the quotient in 6.44can be neglected. Also, a large value of
k_2 will give rise to $e^{-k_2 \cdot t}$ approaching zero, based on the behaviour of the function e^{-x}
(see Fig. 6.16). Under these conditions, Eq. 6.44 becomes:

$$c(\mathrm{P}) = \left(1 + \frac{k_1 \cdot 0 - k_2 \cdot e^{-k_1 \cdot t}}{k_2}\right) \cdot c_0(\mathrm{A})$$

$$c(\mathrm{P}) = \left(1 + \frac{-k_2 \cdot e^{-k_1 \cdot t}}{k_2}\right) \cdot c_0(\mathrm{A})$$

$$c(\mathrm{P}) = \left(1 - e^{-k_1 \cdot t}\right) \cdot c_0(\mathrm{A})$$

and thus shows the relationship (6.47) obtained with the steady-state approximation.
   It is found that k_2 does not have to be much larger than k_1 in order to obtain
reasonably accurate results from the steady-state approximation. For example,
$k_2 = 20 \cdot k_1$ already yields a good agreement.

6.6.4 The Rate-Determining Step

In the previous section, we have considered the special case of $k_2 \gg k_1$ in the
following set of consecutive reactions

Fig. 6.16 The function e^{-x}
anneals to zero when
x assumes very large values

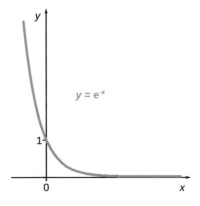

$$A \xrightarrow{k_1} I \xrightarrow{k_2} P$$

which means that the first reaction proceeds a lot slower than the second reaction. The first reaction thus becomes the rate-determining step.

In consecutive reactions, the rate-determining step is the one with the lowest rate constant. In many cases, the rate law for a consecutive reaction with a rate-determining step can thus be written in a straight-forward way: the rate of the overall reaction equals the rate of the rate-determining step.

6.6.5 Pre-Equilibria

In many instances of consecutive reactions, the first reaction needs to be treated as an equilibrium process (pre-equilibrium). We thus consider the case where an intermediate I is formed by reaction of A and B in an equilibrium reaction, and that the reaction to the product P proceeds a lot slower than the reverse reaction in the equilibrium:

$$A + B \underset{k_{-1}}{\overset{k_1}{\rightleftharpoons}} I \xrightarrow{k_2} P$$

Since we assume the product formation as rate-determining step, the rate constant for the second reaction takes a much smaller value than the rate constant for the reverse reaction of the equilibrium:

$$k_{-1} \gg k_2$$

For the first reaction, the equilibrium constant is given by:

$$K = \frac{[I]}{[A] \cdot [B]} = \frac{c(I) \cdot c^{\emptyset}}{c(A) \cdot c(B)} = \frac{k_1}{k_{-1}} \cdot c^{\emptyset}$$

Substituting this into the rate law for the product formation yields:

$$\frac{dc(P)}{dt} = k_2 \cdot c(I) = k_2 \cdot \frac{K}{c^{\emptyset}} \cdot c(A) \cdot c(B) \tag{6.48}$$

This is the form of a 2nd-order rate law with the rate constant $k = k_2 \cdot \frac{K}{c^{\emptyset}} = k_2 \cdot \frac{k_1}{k_{-1}}$.

6.6.6   Kinetic and Thermodynamic Control

A common feature of reactions in organic chemistry (e.g. nitration of mono-substitute benzenes yielding o-, m-, p-derivatives) is the formation of multiple products from one set of reactants:

$$A + B \rightarrow P_1 \quad \text{rate constant}: k_1; \quad \text{rate}: v_1 = \frac{dc(P_1)}{dt} = k_1 \cdot c(A) \cdot c(B)$$

$$A + B \rightarrow P_2 \quad \text{rate constant}: k_2; \quad \text{rate}: v_2 = \frac{dc(P_2)}{dt} = k_2 \cdot c(A) \cdot c(B)$$

The relative proportion of products formed is given by the ratio of the two rates, and thus the two rate constants:

$$\frac{v_1}{v_2} = \frac{k_1}{k_2} = \frac{c(P_1)}{c(P_2)} \tag{6.49}$$

The ration given by Eq. 6.49 is due to the kinetic control over the proportion of products formed. Notably, this refers to the concentration of products produced at any stage during the reaction, i.e. before the reaction reached equilibrium. Kinetic control is also the reason for the occurrence of electrochemical overpotentials; cf. Sect. 4.2.1). If a reaction is allowed to reach equilibrium, then the proportion of products is determined by the thermodynamic considerations. The ratio is then under thermodynamic control.

6.6.7   The Lindemann–Hinshelwood Mechanism

Examples for unimolecular reactions such as

$$A \rightarrow P$$

are homogeneous gas phase reactions that appear to follow 1st-order kinetics, such as e.g. the isomerisation of cyclo-propane:

For a 1st-order kinetics, the following rate law can be formulated:

$$v = k \cdot c(A)$$

However, in order to react in this gas phase process, an individual molecule needs to acquire enough energy through collision. Since the collision is a bimolecular process, the question arises how this may still adhere to a kinetic law of 1st-order.

A closer look reveals that there is an elementary unimolecular step in this process (hence 1st-order gas phase reactions are called unimolecular reactions), but there is also a bimolecular step in the overall reaction. Lindemann and Hinshelwood provided a plausible mechanism:

$A + A \rightarrow A^* + A$ A can become activated in a collision with another molecule.
$A + A^* \rightarrow A + A$ The activated molecule may lose its excess energy by collision.
$A^* \rightarrow P$ Alternatively, the activated molecule might react to products.

If the unimolecular step of product formation is slow enough to be rate-determining, then the overall reaction will have a 1st-order kinetics (as observed).

The reaction rates for the above set of elementary reactions can be derived as:

$$A + A \rightarrow A^* + A \qquad \frac{dc(A^*)}{dt} = k_1 \cdot c(A)^2$$

$$A + A^* \rightarrow A + A \qquad \frac{dc(A^*)}{dt} = -k_{-1} \cdot c(A) \cdot c(A^*)$$

$$A^* \rightarrow P \qquad \frac{dc(A^*)}{dt} = -k_2 \cdot c(A^*)$$

From these individual rates, the net rate of formation of A^* is thus:

$$\frac{dc(A^*)}{dt} = k_1 \cdot c(A)^2 - k_{-1} \cdot c(A) \cdot c(A^*) - k_2 \cdot c(A^*)$$

Applying the steady-state approximation, $\frac{dc(A^*)}{dt} \approx 0$, yields:

$$\frac{dc(A^*)}{dt} = k_1 \cdot c(A)^2 - k_{-1} \cdot c(A) \cdot c(A^*) - k_2 \cdot c(A^*) \approx 0$$

This can be re-arranged to:

$$k_1 \cdot c(A)^2 = k_{-1} \cdot c(A) \cdot c(A^*) + k_2 \cdot c(A^*)$$

$$c(A^*) = \frac{k_1 \cdot c(A)^2}{k_{-1} \cdot c(A) + k_2}$$

which yields for the rate of product formation from this rigorous consideration:

$$\frac{dc(P)}{dt} = k_2 \cdot c(A^*) = \frac{k_1 \cdot k_2 \cdot c(A)^2}{k_{-1} \cdot c(A) + k_2} \qquad (6.50)$$

However, this is not a 1st-order rate law!

Two special cases can be singled out. If the overall pressure in the reaction vessel is sufficiently high, the rate of de-activation by collisions between A^* and A is greater than the rate of the product formation, then

$$\left| \frac{dc(A \rightarrow A^*)}{dt} \right| = |-k_{-1} \cdot c(A) \cdot c(A^*)| \gg \left| \frac{dc(A \rightarrow P)}{dt} \right| = |-k_2 \cdot c(A^*)|$$

$$k_{-1} \cdot c(A) \cdot c(A^*) \gg k_2 \cdot c(A^*)$$

$$k_{-1} \cdot c(A) \gg k_2$$

and we can neglect k_2 in the denominator of above product rate and obtain

$$\frac{dc(P)}{dt} = \frac{k_1 \cdot k_2 \cdot c(A)^2}{k_{-1} \cdot c(A)} = \frac{k_1 \cdot k_2}{k_{-1}} \cdot c(A) \qquad (6.51)$$

which describes a rate law for a 1st-order reaction.

What happens if the overall pressure in this gas phase reaction is lowered, for example by decreasing the concentration of gas A?

If the number of A molecules is decreased, the activation and de-activation reactions are affected. In particular, it becomes less likely that A^* molecules are de-activated, since less A molecules are available for a collision. In relation to the de-activation, the product formation step will be more likely. Therefore, at low concentration of A:

$$k_{-1} \cdot c(A) \ll k_2$$

This means we can neglect $k_{-1} \cdot c(A)$ in the denominator of Eq. 6.50, and thus obtain:

$$\frac{dc(P)}{dt} = \frac{k_1 \cdot k_2 \cdot c(A)^2}{k_2} = k_1 \cdot c(A)^2 \qquad (6.52)$$

This means that if we reduce the pressure of A (i.e. the concentration), then the reaction should switch to a 2nd-order rate law, since the activation step becomes rate-determining.

The Lindemann–Hinshelwood mechanism can be experimentally tested, because the 1st-order rate law (Eq. 6.51) was obtained with the assumption that the rate of de-activation by collisions between A^* and A is greater than the rate of the product formation. This will only be the case, if there are enough molecules A available for frequent collisions. For the second case of low concentration of A (or, more generally, low overall pressure), a 2nd-order rate law is obtained (Eq. 6.52).

6.7 Polymerisation Kinetics

There are two main types of polymerisation processes that differ in their underlying reaction mechanism and the variation of the average molecular mass of the products over time (Table 6.2).

6.7.1 Stepwise Polymerisation

For illustration of stepwise polymerisation processes, we will consider the polymerisation of polyamides (see Fig. 6.17). Examples of naturally occurring polyamides of technological importance are proteins which constitute wool and silk. Synthetic polyamides are typically made by stepwise polymerisation to produce materials such as Nylons and Aramids. Due to their high durability and strength, most polyamides are used in textiles, automotive applications, carpets and sportswear.

Here, two different monomers are joined together, and the resulting product is thus called a copolymer. In regular copolymers such as polyamides, polyesters and polyurethanes, the repeating unit consists of one of each monomer, so that they alternate in the chain. In the example shown in Fig. 6.17, each monomer has the same reactive group on both ends; the direction of the amide bond therefore alternates between each monomeric unit. This is different in natural polyamides (proteins), where the direction of the amide bond stays the same throughout.

The reaction rate in a stepwise polymerisation reaction depends on the concentration of $—NH_2$ (amine) and $—COOH$ (acid) groups:

Table 6.2 Characteristics of the two main types of polymerisation reactions

Stepwise polymerisation	Chain polymerisation
• Any two monomers in the mixture can link together at any time.	• An activated monomer attacks another monomer and links to it.
• Growth of the polymer is not confined to chains already growing.	• This unit then attacks another monomer and links to it.
• Monomers are consumed early in the reaction.	• Long polymers are formed rapidly.
• Average molecular mass of the product grows with time.	• The yield, but not the molecular mass of the product, increases with longer reaction times.

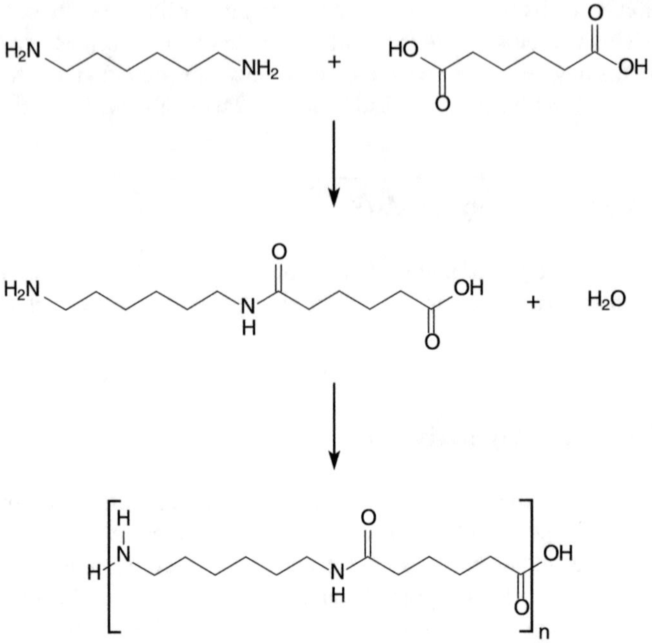

Fig. 6.17 Schematic illustration of the formation of Nylon 6-6 from hexane-1,6-diamine and hexanedioic acid (adipic acid). Instead of adipic acid, adipoyl chloride may be used

$$\frac{dc(\text{acid})}{dt} = -k \cdot c(\text{amine}) \cdot c(\text{acid})$$

Because there is one amine group for each acid group (assuming that enough amine reactants are provided), it follows:

$$\frac{dc(\text{acid})}{dt} = -k \cdot c(\text{acid})^2 \tag{6.53}$$

which constitutes a differential 2nd-order rate law.

If we assume that the rate constant for the reaction k remains constant with growing chain length, Eq. 6.53 can be integrated and yields (cf. Table 1.11):

$$c(\text{acid}) = \frac{c_0(\text{acid})}{1 + k \cdot t \cdot c_0(\text{acid})} \tag{6.54}$$

which is the integrated 2nd-order rate law. It provides a means to calculate the concentration of 'free' (not yet condensed) acid groups at any time during the reaction.

The fraction of condensed acid groups, p, can then be calculated as the concentration of condensed acid groups, divided by the initial concentration of acid groups. By using the expression for the concentration of acid groups that are still available (Eq. 6.54), one obtains:

$$p = \frac{c_0(\text{acid}) - c(\text{acid})}{c_0(\text{acid})} = 1 - \frac{c(\text{acid})}{c_0(\text{acid})} = 1 - \frac{1}{1 + k \cdot t \cdot c_0(\text{acid})} \tag{6.55}$$

This equation can be arithmetically re-arranged to yield:

$$p = \frac{1 + k \cdot t \cdot c_0(\text{acid})}{1 + k \cdot t \cdot c_0(\text{acid})} - \frac{1}{1 + k \cdot t \cdot c_0(\text{acid})} = \frac{k \cdot t \cdot c_0(\text{acid})}{1 + k \cdot t \cdot c_0(\text{acid})} \tag{6.56}$$

As a practical useful parameter, one defines the degree of polymerisation, $\langle N \rangle$, as the average number of monomer residues per polymer molecule. $\langle N \rangle$ therefore indicates the average length of the polymer. It is calculated as the ratio between the initial concentration of acid, $c_0(\text{acid})$, and the concentration of acid end groups remaining non-condensed, $c(\text{acid})$. By using the relationship from Eq. 6.56, this yields Carothers' equation (Carothers 1932) for stepwise polymerisation of linear polymers:

$$\langle N \rangle = \frac{c_0(\text{acid})}{c(\text{acid})} = \frac{1}{1 - p} = 1 + k \cdot t \cdot c_0(\text{acid}) \tag{6.57}$$

One can thus conclude that the average length $\langle N \rangle$ obtained from a stepwise polymerisation reaction grows linearly with time. The longer a stepwise polymerisation proceeds, the higher the average molecular mass of the product.

6.7.2 Chain Polymerisation

Many gas-phase reactions (for example the synthesis of hydrogen bromide from its elements, Sect. 6.6.1), but also liquid-phase polymerisations are chain reactions. In a chain reaction, an intermediate produced in one step generates another intermediate in the subsequent step. These intermediates are called chain carriers. In a radical chain reaction, the chain carriers are radicals, i.e. the intermediates possess unpaired electrons. Chain polymerisation occurs by addition of monomers to a rapidly growing polymer. The manufacture of frequently used plastics such as polyethylene (PE), polypropylene (PP), and polyvinyl chloride (PVC) is based on chain polymerisation (albeit these are mostly conducted catalytically, e.g. by using Ziegler–Natta catalysts). Mechanistically, the free radical mechanism consists of three stages: chain initiation, chain propagation, and chain termination (see Fig. 6.18).

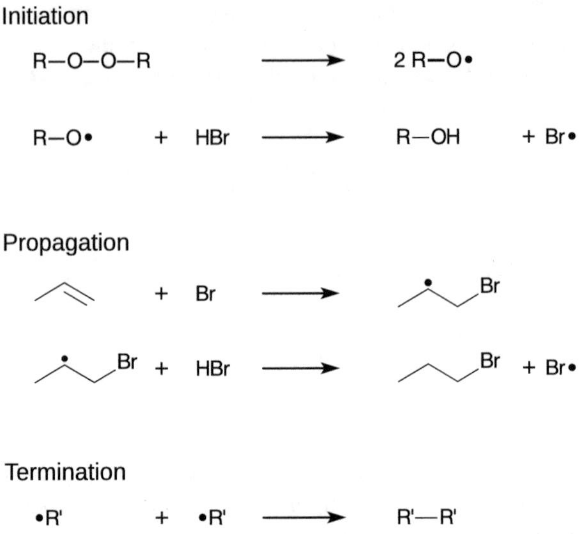

Fig. 6.18 Illustration of the three stages of a radical chain reaction that will result in polypropene. The radical starter in this example is a peroxide (R—O—O—R)

Initiation

R—O—O—R $\longrightarrow$ 2 R—O•

R—O•   +   HBr $\longrightarrow$ R—OH   + Br•

Propagation

+ Br $\longrightarrow$ Br

Br + HBr $\longrightarrow$ Br + Br•

Termination

•R'   +   •R' $\longrightarrow$ R'—R'

The central feature of the kinetic treatment of chain polymerisation reactions is that the rate is proportional to the square root of the initiator concentration:

$$v = k \cdot \sqrt{c(\text{initiator})} \cdot c(\text{monomer}) \qquad (6.58)$$

In chain polymerisation processes, the kinetic chain length v is defined as the number of monomer units consumed per number of activated centres produced in the initiation step. The kinetic chain length can be expressed in terms of the rate expressions: it is the rate of propagation of the chains (i.e. monomers are consumed at the rate with which chains propagate) divided by the rate of production of radicals. Without rigorously deriving this relationship, we appreciate that this yields the following expression for the kinetic chain length:

$$v = \frac{k_{\text{prop}} \cdot c(\text{monomer})}{2 \cdot k_{\text{term}} \cdot c(\text{radicale})} = \frac{k \cdot c(\text{monomer})}{\sqrt{c(\text{initiator})}} \qquad (6.59)$$

The final polymer produced by the chain mechanism may arise from mutual termination. In this case, the average number of monomers in the final polymer, $<N>$, is the sum of the numbers in the two combining chains. Since the average number of monomers in each chain is v, we obtain:

$$\langle N \rangle = 2 \cdot v = \frac{2 \cdot k \cdot c(\text{monomer})}{\sqrt{c(\text{initiator})}} \qquad (6.60)$$

It follows that the *slower the initiation of the chain* (i.e. the lower the concentration of initiator and the slower the rate of initiation), the *greater the kinetic chain length*, and therefore the higher the average molecular mass of the polymer.

6.8 Collision Theory

After having discussed the overall dynamics of chemical reactions in the preceding chapters, we now want to have a closer look at what happens to molecules at the climax of reactions. We can certainly expect that extensive structural changes take place and energies are re-distributed among bonds.

The calculation of rates of such processes from first principles is very difficult, but a quantitative account of reaction rates can be given in terms of

- collision theory for gas-phase reactions
- diffusion theory for diffusion-controlled reactions in solution
- transition state theory for activation-controlled reactions in solution

6.8.1 Collision Theory for Bimolecular Gas-phase Reactions

In Sect. 6.5.2, we introduced the concept that products are only formed if the collision between two reactants is of sufficiently high energy; this required activation energy E_a is described by the Arrhenius equation. An encounter of two reactants with less energy than E_a will lead to separation without product formation.

This concept of collision can well be used to describe gas-phase reactions. For solution reactions, one can imagine the two reactants approaching by diffusion and then acquiring energy from their immediate surroundings while they are in contact.

We consider the following bimolecular elementary reaction

$$A + B \rightarrow P \quad \text{with the rate law} \quad v = k \cdot c(A) \cdot c(B). \tag{6.61}$$

The rate of the reaction will be proportional to the rate of collisions, and thus to the mean speed of molecules, $\bar{c} \sim \sqrt{\dfrac{T}{M}}$ (cf. Sect. 5.1.1, Eq. 5.5). Also, the size of the reactants is an important parameter for successful collision; this is described by the cross-sectional area σ. (also called collisional cross-section). We thus obtain:

$$v \sim \sigma \cdot \bar{c} \cdot c(A) \cdot c(B) \quad \text{or} \quad v \sim \sigma \cdot \sqrt{\frac{T}{M}} \cdot c(A) \cdot c(B)$$

We have already established that not every collision will be reactive. Only those collisional encounters where the kinetic energy of the reactants exceeds the activation energy E_a will lead to product formation. This is expressed by the Boltzmann factor: $e^{-\frac{E_a}{R \cdot T}}$, leading to:

$$v \sim \sigma \cdot \sqrt{\frac{T}{M}} \cdot e^{-\frac{E_a}{R \cdot T}} \cdot c(A) \cdot c(B)$$

Still, despite considering the cross-sectional area σ, not every collision may lead to a reaction, since the reactants may have to collide in a particular orientation. This additional steric requirement is encoded in the factor P, so we obtain:

$$v \sim P \cdot \sigma \cdot \sqrt{\frac{T}{M}} \cdot e^{-\frac{E_a}{R \cdot T}} \cdot c(A) \cdot c(B)$$

and from comparison with the rate law in Eq. 6.61, we derive the following relationship for the rate constant:

$$k \sim P \cdot \sigma \cdot \sqrt{\frac{T}{M}} \cdot e^{-\frac{E_a}{R \cdot T}} \qquad (6.62)$$

In words, Eq. 6.62 can be expressed as:

$$\text{rate constant} \sim \text{steric requirement} \qquad P \cdot \sigma$$

$$\times \text{ encounter rate minimum} \qquad \sqrt{\frac{T}{M}}$$

$$\times \text{ energy requirement} \qquad e^{-\frac{E_a}{R \cdot T}}$$

In the following, we will see that this concept describes all aspects of a successful collision and can be rigorously derived by the collision theory.

6.8.2 Collision Rates in Gases

In Sect. 5.1.1, we have briefly introduced the kinetic molecular theory of gases, and derived an expression for the speed of the molecules in a homogeneous system filled with gas. Extending this consideration to the more general case of two different gases A and B, the relative mean speed of two species can be derived from that theory:

$$\bar{c}_{rel} = \sqrt{\frac{8 \cdot k_B \cdot T}{\pi \cdot \mu}} \qquad (6.63)$$

with $\mu = \frac{m_A \cdot m_B}{m_A + m_B}$ being the reduced mass.

The collision frequency Z_{AB} between molecules of gas A and gas B can be derived as:

$$Z_{AB} = \sigma \cdot \bar{c}_{rel} \cdot \frac{N(A)}{V} \cdot \frac{N(B)}{V} = \sigma \cdot \bar{c}_{rel} \cdot \frac{n(A) \cdot N_A}{V} \cdot \frac{n(B) \cdot N_A}{V}$$

which yields

$$Z_{AB} = \sigma \cdot \sqrt{\frac{8 \cdot k_B \cdot T}{\pi \cdot \mu}} \cdot N_A^2 \cdot c(A) \cdot c(B) \tag{6.64}$$

The collision cross-section σ is the area within which a projectile (A) must enter around the target (B) in order for a collision to occur. Figure 6.19 illustrates that, under the assumption of spherical particles, this requires a distance of the centres of gravity of the two particles of $(r_A + r_B)$.

The cross-section area is thus the area of a circle with radius $(r_A + r_B)$:

$$\sigma = (r_A + r_B)^2 \cdot \pi \tag{6.65}$$

The values of collision cross-sections of some select gases are given in 6.3. Using Eq. 6.64 and tabulated values of the collision cross-section, the collision frequency Z can be calculated. As an example, N_2 under normal conditions ($T_{normal} = 298.15$ K, $c = 1$ M) has a collision frequency of $Z = 5 \cdot 10^{34}$ m^{-3} s^{-1}.

Fig. 6.19 Illustration of the collision cross-section of two spherical particles

Table 6.3 Collision cross-section of select gases

Gas	σ in nm^2	Gas	σ in nm^2
H_2	0.27	Ar	0.36
He	0.21	CH_4	0.46
N_2	0.43	CO_2	0.52
O_2	0.40	C_2H_4	0.64
Ne	0.24	SO_2	0.58

6.8.3 Energy Requirement for Successful Collision

We have previously established that for a successful collision to occur, the reactants
need to have sufficient kinetic energy. The ratio of particles that possess energies ε_i
and ε_j is given by the Boltzmann distribution (see Sect. 5.1.2):

$$\frac{N_i}{N_j} = e^{-\frac{\varepsilon_i - \varepsilon_j}{k_B \cdot T}} \tag{5.7}$$

with Boltzmann's constant $k_B = 1.381 \ 10^{-23} \ J \ K^{-1}$

Note that the product of Boltzmann's and Avogadro's constant yields the gas
constant $k_B \cdot N_A = R = 8.3144 \ J \ K^{-1} \ mol^{-1}$.

Here, ε denotes extensive energies measured in J, and E molar energies (intensive) measured in $J \ mol^{-1}$.

From this equation it follows that the number of molecules that possess an energy
$E_i \geq E_{min}$ is:

$$N_i(E_i \geq E_{min}) = N_{total} \cdot e^{-\frac{e_{min}}{k_B \cdot T}} = N_{total} \cdot e^{-\frac{E_{min}}{R \cdot T}}$$

We can now combine the encounter (collision frequency Z_{AB}) and the energy
requirement (Boltzmann factor) to obtain a preliminary reaction rate:

$$\text{rate} \sim -Z_{AB} \cdot e^{-\frac{E_{min}}{R \cdot T}} = -\sigma \cdot \sqrt{\frac{8 \cdot k_B \cdot T}{\pi \cdot \mu}} \cdot e^{-\frac{E_{min}}{R \cdot T}} \cdot N_A^2 \cdot c(A) \cdot c(B) \tag{6.66}$$

6.8.4 Steric Requirements

When comparing experimental data from gas phase reactions with the theory
developed so far, it becomes obvious that there is agreement in some, but not all
cases. Insights come from experiments using molecular beams, where it was
observed that there can be three kinds of collisions:

- elastic (molecules separate again with total kinetic energy conserved)
- in-elastic (molecules separate again, but there is partial conversion of kinetic to
  potential energy)
- reactive (products are produced)

It thus seems appropriate to introduce a scaling factor (P) for the collision cross-
section, that can be used to define the reactive cross-section σ^*:

$$\sigma^* = P \cdot \sigma \tag{6.67}$$

It is immediately obvious, that the steric factor P is indeed a scaling factor that describes the fraction of the cross-sectional area that delivers successful encounters:

$$\frac{\sigma^*}{\sigma} = P$$

Intuitively, one would expect that the steric factor P assumes a maximum number of 1, and less than 1 for most instances. This is indeed the case, although exceptions are known. The reaction between K and Br_2, for example, has a steric factor of $P = 4.8$, which indicates that this reaction proceeds by an unusual mechanism.

6.8.5   Combining the Collision Parameters

After having derived the individual components of molecular collision in the previous sections, we can now combine all of them into a theoretical reaction rate:

$$\text{rate} = -P \cdot Z_{AB} \cdot e^{-\frac{E_{min}}{R \cdot T}} = -P \cdot \sigma \cdot \sqrt{\frac{8 \cdot k_B \cdot T}{\pi \cdot \mu}} \cdot e^{-\frac{E_{min}}{R \cdot T}} \cdot N_A^2 \cdot c(A) \cdot c(B) \tag{6.68}$$

This agrees with the formal obtained for a bi-molecular reaction:

$$\text{rate} = \frac{dc(A)}{dt} = -k \cdot c(A) \cdot c(B)$$

with the rate constant being a rate constant for the collision process

$$k_{\text{collision}} = \sigma \cdot P \cdot \sqrt{\frac{8 \cdot k_B \cdot T}{\pi \cdot \mu}} \cdot e^{-\frac{E_{min}}{R \cdot T}} \cdot N_A^2$$

However, when we compare the units of the rate constant obtained from a bi-molecular reaction with the units for the collision rate constant, we find:

$$[k_{\text{reaction}}] = 1 m^3 mol^{-1} s^{-1}$$

$$[k_{\text{collision}}] = \left[\sigma \cdot P \cdot \sqrt{\frac{8 \cdot k_B \cdot T}{\pi \cdot \mu}} \cdot e^{-\frac{E_{min}}{R \cdot T}} \cdot N_A^2 \right]$$

$$[k_{\text{collision}}] = 1\text{m}^2 \cdot 1 \cdot \sqrt{\frac{1\text{J} \cdot 1\text{K}}{1\text{K} \cdot 1\text{kg}}} \cdot 1 \cdot \frac{1}{\text{mol}^2}$$

$$= 1\text{m}^2 \cdot \sqrt{\frac{1\text{kg m}^2 \cdot 1\text{K}}{1\text{K} \cdot 1\text{kg} \cdot \text{s}^2}} \cdot \frac{1}{\text{mol}^2} = 1\text{m}^3\text{mol}^{-2}\text{s}^{-1}$$

This shows that the units of the two rate constants are not the same, but they should! Both rate constants k_{reaction} and $k_{\text{collision}}$ contain the units of 'volume per molar amount and time', but due to the definition of the collision reaction in terms of molecules (rather than molar amounts), we now have to adjust for one additional molar amount unit. This is done by multiplying in Avogadro's constant:

$$k_{\text{reaction}} \cdot N_A = k_{\text{collision}}$$

And we finally obtain the rate constant for a gas-phase bimolecular reaction (and substitute 'k_{reaction}' with 'k'):

$$k = P \cdot \sigma \cdot \sqrt{\frac{8 \cdot k_B \cdot T}{\pi \cdot \mu}} \cdot N_A \cdot e^{-\frac{E_{\text{min}}}{R \cdot T}} = P \cdot A \cdot e^{-\frac{E_{\text{min}}}{R \cdot T}} \qquad (6.69)$$

with the pre-exponential factor A being

$$A = \sigma \cdot \sqrt{\frac{8 \cdot k_B \cdot T}{\pi \cdot \mu}} \cdot N_A. \qquad (6.70)$$

A describes the theoretical encounter rate minimum and the energy requirement for a successful collision of the two gas molecules based on collision theory. It may need to be adjusted for the reactive cross-section (σ^*), which is done by multiplication with the steric factor P.

Estimation of the steric factor for a reaction
Consider the following gas-phase reaction

$$H_2 + C_2H_4 \rightarrow C_2H_6$$

and determine the steric factor P at $T = 628$ K, given that the experimental pre-exponential factor is $A_{\text{exp}} = 1.24 \cdot 10^6$ dm^3 mol^{-1} s^{-1}.
    In order to calculate the steric factor, we need to calculate the theoretical pre-exponential factor for the reaction (A), and compare it to the experimental value (A_{exp}). The ratio between the two will be P, since $A_{\text{exp}} = P \cdot A$.

(continued)

$$A = \sigma \cdot \sqrt{\frac{8 \cdot k_B \cdot T}{\pi \cdot \mu}} \cdot N_A \qquad (6.70)$$

Collision cross-sections (see Table 6.3):

$$\sigma(H_2) = 0.27\,nm \quad \sigma(C_2H_4) = 0.64\,nm$$

We approximate the cross-section as the average of the two:

$$\sigma = 0.46\,nm$$

For the mean speed of gas particles, we need the reduced mass:

$$\mu = \frac{2.016\,Da \cdot 28.05\,Da}{2.016\,Da + 28.05\,Da} = 1.881 Da = 1.881 \cdot 1.661 \cdot 10^{-27}\,kg$$
$$= 3.124 \cdot 10^{-27}\,kg$$

Therefore:

$$A = 0.46 nm^2 \cdot \sqrt{\frac{8 \cdot 1.381 \cdot 10^{-23} \cdot 628 \cdot J \cdot K}{3.142 \cdot 3.124 \cdot 10^{-27} \cdot kg \cdot K}} \cdot 6.022 \cdot 10^{23}\,mol^{-1}$$

$$A = 0.46 \cdot \left(10^{-9}\right)^2 \cdot m^2 \cdot \sqrt{7068481 \frac{kg \cdot m^2}{kg \cdot s^2}} \cdot 6.022 \cdot 10^{23}\,mol^{-1}$$

$$A = 0.46 \cdot 10^{-18} \cdot m^2 \cdot 2659 \cdot \frac{m}{s} \cdot 6.022 \cdot 10^{23} \cdot mol^{-1}$$

$$A = 0.46 \cdot 2659 \cdot 6.022 \cdot 10^{-18+23} \frac{m^3}{mol \cdot s}$$

$$A = 7.37 \cdot 10^8 \frac{m^3}{mol \cdot s} = 7.37 \cdot 10^8 \frac{10^3 dm^3}{mol \cdot s}$$

$$A = 7.37 \cdot 10^{11} \frac{dm^3}{mol \cdot s}$$

It follows:

(continued)

$$P = \frac{A_{\text{exp}}}{A} = \frac{1.24 \cdot 10^6}{7.37 \cdot 10^{11}} = 1.7 \cdot 10^{-6}$$

This low value for P is one reason, why catalysts are required to achieve hydration of ethane at reasonable rates.

6.9   Reactions in Solution

In the previous chapter, we considered reactions in the gas phase, which proceed by collisional encounter of molecules in space. Reactants in solution are encountering each other in a different way than in the gas-phase.

First, in solution, reactants have to find their way through the solvent (i.e. a network of other molecules), so we can expect the frequency of reactive encounters to be considerably less than in the gas-phase. Second, after an encounter, the reactants are leaving their current positions slower than in the gas phase, since they are held in place by the surrounding solvent. This is called the cage effect.

Conceptually, it is, of course, still required that the reactants require the minimum energy (activation energy), in order for the reaction to proceed. However, an encounter pair may accumulate enough energy to react in due course, even though it may not have had sufficient energy initially. This gives rise to two classes of reactions, namely those with

- diffusion control
- activation control.

The complicated overall process can be divided into two stages, the encounter step and the post-encounter; both stages can subsequently be combined into the overall rate law.

The Encounter Step
We suppose, the encounter of two reactants was of 1st-order with respect to each of the reactants which need to diffuse towards each other in order to meet up:

$$A + B \rightarrow AB \quad v_d = k_d \cdot c(A) \cdot c(B)$$

The rate of this step is determined by the constant k_d that specifies the diffusional characteristics of A and B.

Post-Encounter
The encounter AB can either break up into its individual partners, or react to product P. If we assume pseudo 1st-order for both processes, we obtain:

$$\text{AB} \rightarrow \text{A} + \text{B} \quad v'_d = k'_d \cdot c(\text{AB})$$

$$\text{AB} \rightarrow \text{P} \qquad v_a = k_a \cdot c(\text{AB})$$

The rate constant for the break-up is denoted k'_d, as this step is the reverse reaction of the encounter step.

Combining Both Steps to Yield the Overall Rate Law
We can thus formulate an expression for the change of concentration of AB over time:

$$v = v_d - v'_d - v_a = \frac{dc(\text{AB})}{dt}$$

$$\frac{dc(\text{AB})}{dt} = k_d \cdot c(\text{A}) \cdot c(\text{B}) - k'_d \cdot c(\text{AB}) - k_a c(\text{AB}) \qquad (6.71)$$

Using the steady-state approximation for intermediates, we can claim:

$$\frac{dc(\text{AB})}{dt} \approx 0 \qquad (6.45)$$

and thus obtain:

$$c(\text{AB}) = \frac{k_d \cdot c(\text{A}) \cdot c(\text{B})}{k_a + k'_d}$$

The rate of product formation is thus:

$$\frac{dc(\text{P})}{dt} = k_a \cdot c(\text{AB}) = k_a \cdot \frac{k_d \cdot c(\text{A}) \cdot c(\text{B})}{k_a + k'_d} \qquad (6.72)$$

which may be denoted simpler by combining all individual rate constants into an overall rate constant k:

$$\frac{dc(\text{P})}{dt} = k \cdot c(\text{A}) \cdot c(\text{B}) \text{ with } k = \frac{k_a \cdot k_d}{k_a + k'_d} \qquad (6.73)$$

Considering the reaction scheme and the expression of the rate constant in 6.73, we can now distinguish two different cases:

Diffusion Control

The reaction to product happens faster than the separation of AB. This means, the break-up reaction (characterised by k'_d) is much slower than the reaction to product (characterised by k_a).

$$k'_d \ll k_a \Rightarrow k \approx k_d$$

If k'_d is much smaller than k_a, k'_d has a negligible contribution to the denominator in Eq. 6.73, and can thus be eliminated. This means that the overall rate constant k is dominated by contributions of k_d, which is the encounter rate constant. The reaction is then said to be under diffusion control.

Activation Control

In the reverse case, the break-up of AB (characterised by k'_d) happens faster then the reaction to product (characterised by k_a); k'_d is much larger than k_a:

$$k_a \ll k'_d \Rightarrow k \approx \frac{k_d}{k'_d} \cdot k_a$$

If k_a is much smaller than k'_d, k_a has a negligible contribution to the denominator in Eq. 6.73, and can thus be eliminated. This results in an overall rate constant k that is characterised by the encounter equilibrium, but also has contributions from the rate constant that characterises the actual reaction to product. Such reactions are said to be under activation control.

6.10   Diffusion Control

For diffusion-controlled reactions, we will need to consider the rate at which the reactants diffuse to each other. In Sect. 5.2.7, we learned that the diffusion coefficient D of a solute is given by the Stokes-Einstein equation:

$$D = \frac{k_B \cdot T}{6 \cdot \pi \cdot \eta \cdot r} \tag{5.48}$$

with η being the viscosity of the solution, and r the radius of the solute. Without rigorously deriving the following relationship, we appreciate that the rate constant of diffusion-controlled reactions can be calculated as follows:

$$k_d = 4 \cdot \pi \cdot r^* \cdot N_A \cdot \sum_{i=1}^{N} D_i \tag{6.74}$$

Here, the parameter r^* describes the maximum distance the reactants may adopt such that a reaction can occur. The factor $\sum_{i=1}^{N} D_i$ is the sum of the diffusion coefficients of all reactants.

If we consider a bi-molecular reaction and substitute with the diffusion coefficients as calculated by the Stokes-Einstein relationship, we obtain:

$$D_A = \frac{k_B \cdot T}{6 \cdot \pi \cdot \eta \cdot r_A} \quad \text{and} \quad D_B = \frac{k_B \cdot T}{6 \cdot \pi \cdot \eta \cdot r_B}$$

As an approximation, we assume that $r_A \approx r_B \approx \frac{1}{2} \cdot r^*$, so we can build the sum of the diffusion coefficients

$$D = D_A + D_B = 2 \cdot D_A = \frac{k_B \cdot T}{6 \cdot \pi \cdot \eta \cdot \frac{1}{2} \cdot r^*} + \frac{k_B \cdot T}{6 \cdot \pi \cdot \eta \cdot \frac{1}{2} \cdot r^*} = \frac{2 \cdot k_B \cdot T}{3 \cdot \pi \cdot \eta \cdot r^*}$$

and then obtain for the rate constant of diffusion-controlled reactions (according to Eq. 6.74):

$$k_d = 4 \cdot \pi \cdot r^* \cdot N_A \cdot \frac{2 \cdot k_B \cdot T}{3 \cdot \pi \cdot \eta \cdot r^*} = \frac{8 \cdot k_B \cdot N_A \cdot T}{3 \cdot \eta}$$

We remember that the product of Boltzmann's and Avogadro's constant equals the gas constant, $k_B \cdot N_A = R$, and therefore arrive at the following expression for the rate constant of a bi-molecular diffusion-controlled reaction:

$$k_d = \frac{8 \cdot R \cdot T}{3 \cdot \eta}. \tag{6.75}$$

A notable observation from this approximation is the fact that the rate constant is independent of the identity of the reactants. Therefore, the rate constant for the diffusion-controlled reaction only depends on the temperature and the viscosity of the solvent!

6.11 Transition State Theory

In Sect. 6.5.2, we discussed molecular aspects of the Arrhenius equation and introduced the concept, that an activated complex needs to be formed that possesses the minimum energy (activation energy E_a) required for the successful reaction.

With regards to terminology, the 'activated complex' and the 'transition state' are intimately linked, but not the same thing. The term 'activated complex' describes a chemical assembly, i.e. a cluster of atoms. In contrast, the 'transition state' describes

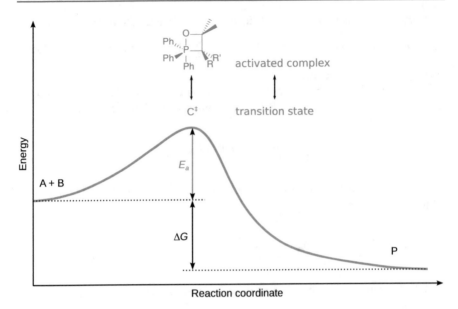

Fig. 6.20 Illustration of the transition state in an energy profile of a reaction. The 'activated complex' describes the atomic assembly; the 'transition state' comprises the entire configuration of that state, including nuclear and electronic configuration, and the internal motions. The energy profile is the same as seen earlier in Fig. 6.14

the nuclear and electronic configurations of the reactants which assume a form that transforms them into the products (see Fig. 6.20).

6.11.1 Formal Kinetics of the Activated Complex

The appearance of an activated species in the course of a reaction can be subjected to a formal kinetics treatment. Formation of the transition state $C^{\ddagger}$ in the course of a reaction can be formulated as a rapid pre-equilibrium

$$A + B \rightleftharpoons C^{\ddagger}$$

If we consider, for illustration, a gas phase reaction, then we can describe the equilibrium constant as follows:

$$K^{\ddagger} = \frac{p_{C^{\ddagger}} \cdot p^{\varnothing}}{p_A \cdot p_B} \tag{6.76}$$

with the partial pressures $p_i = R \cdot T \cdot c_i$ used instead of the concentrations c_i. Because the equilibrium constant $K^{\ddagger}$ has no unit, we need to multiply with the standard pressure $p^{\varnothing}$. This yields for the concentration of the activated complex:

$$\frac{1}{p^{\varnothing}} \cdot K^{\ddagger} \cdot p_A \cdot p_B = p_{C^{\ddagger}}$$

$$\frac{1}{p^{\varnothing}} \cdot K^{\ddagger} \cdot [R \cdot T \cdot c(A)] \cdot [R \cdot T \cdot c(B)] = R \cdot T \cdot c(C^{\ddagger})$$

$$\frac{1}{p^{\varnothing}} \cdot K^{\ddagger} \cdot R \cdot T \cdot c(A) \cdot c(B) = c(C^{\ddagger})$$

$$c(C^{\ddagger}) = \frac{R \cdot T}{p^{\varnothing}} \cdot K^{\ddagger} \cdot c(A) \cdot c(B) \qquad (6.77)$$

The overall reaction

$$A + B \rightarrow P \quad v = \frac{dc(P)}{dt} = k \cdot c(A) \cdot c(B)$$

is a bi-molecular reaction and thus follows a second order rate law. Since the activated complex can decay to the products in a unimolecular reaction

$$C^{\ddagger} \rightarrow P \quad v = \frac{dc(P)}{dt} = k^{\ddagger} \cdot c(C^{\ddagger})$$

we can also formulate the rate law with respect to product formation based on the decay of the activated complex. For the rate of product formation we can therefore conclude:

$$v = \frac{dc(P)}{dt} = k \cdot c(A) \cdot c(B) = k^{\ddagger} \cdot c(C^{\ddagger}) = k^{\ddagger} \cdot \frac{R \cdot T}{p^{\varnothing}} \cdot K^{\ddagger} \cdot c(A) \cdot c(B) \quad (6.78)$$

The above equation yields an expression of the macroscopic rate constant in terms of parameters that characterise the activated complex:

$$k = k^{\ddagger} \cdot K^{\ddagger} \cdot \frac{R \cdot T}{p^{\varnothing}} \qquad (6.79)$$

6.11.2 Internal Coordinates of the Activated Complex

In a more rigorous treatment of the activated complex that considers the internal motions (vibrations, rotations), it becomes apparent that not all of these motions (degrees of freedom) send the complex into the way of the products. Since we now think of the activated complex (atomic arrangement) together with its internal motions, we will use the term 'transition state' for the following discussion.

Based on the idea that not all internal motions lead to product formation, a transmission coefficient κ is introduced. The transmission coefficient describes the successful passage of the reactants through the transition state, on to the products. In the absence of specific information about a reaction, one assumes that $\kappa = 1$.

Within this concept, a modified equilibrium constant, $\bar{K}_C^{\ddagger}$, is introduced; it describes the transition state $C^{\ddagger}$ with one vibrational mode being discarded. We have seen in the previous section, how the macroscopic 2nd-order rate constant is linked to the transition state:

$$k = k^{\ddagger} \cdot K^{\ddagger} \cdot \frac{R \cdot T}{p^{\varnothing}} \qquad (6.79)$$

By taking into account the internal motions and the associate modified equilibrium constant, Henry Eyring, and, independently, Meredith Evans and Michael Polanyi, have further developed the theory of the activated complex to arrive at a description of the transition state (Evans and Polanyi 1935; Eyring 1935).

The Eyring equation

$$A + B \rightleftharpoons C^{\ddagger} \rightarrow P \quad v = k \cdot c(A) \cdot c(B)$$

The 2nd-order rate constant can be developed as follows:

$$k = k^{\ddagger} \cdot K^{\ddagger} \cdot \frac{R \cdot T}{p^{\varnothing}} = \kappa \cdot \bar{K}_C^{\ddagger} \cdot \frac{k_B \cdot T}{h}$$

This expression is called the Eyring equation and links the macroscopic reaction kinetics and the molecular parameters of the reactants:

$$k = \kappa \cdot \bar{K}_C^{\ddagger} \cdot \frac{k_B \cdot T}{h} \qquad (6.80)$$

The factor $\left(\frac{k_B \cdot T}{h}\right)$ describes a general frequency with respect to individual particles where h is the Planck constant. Notwithstanding the importance of this relationship, the required knowledge of the partition function of the transition state $C^{\ddagger}$ (represented by $\bar{K}_C^{\ddagger}$) makes its use in applied settings challenging. Internal modes of the activated complexes are difficult to access, since their characterisation (e.g. by means of spectroscopy) is rather complex.

6.12 Exercises

1. Consider the gas-phase reaction

$$H_2 + I_2 \rightarrow 2\,HI$$

   (a) Assume that the reaction order is as suggested by the chemical equation. Calculate the rate constant at 681 K, assuming that from an initial pressure of iodine of 823 N m^{-2}, the rate of loss of iodine was 0.192 N m^{-2} s^{-1}. The initial pressure of hydrogen was 10,500 N m^{-2}.
   (b) What is the rate of the reaction if the iodine pressure was unchanged and the initial hydrogen pressure was 39,500 N m^{-2}?

2. The rate constant for the decomposition of a particular substance is $2.80 \cdot 10^{-3}$ dm^3 mol^{-1} s^{-1} at 30 °C, and $1.38 \cdot 10^{-2}$ dm^3 mol^{-1} s^{-1} at 50 °C. Evaluate the Arrhenius parameters of the reaction.

3. The reaction mechanism for the reaction of A_2 and B to product P involves the intermediate A:

$$A_2 \rightleftharpoons A + A \quad \text{(fast)}$$

$$A + B \rightarrow P \quad \text{(slow)}$$

   Deduce the rate law for the reaction assuming a pre-equilibrium.

4. The rate constant of a first-order reaction was measured as $1.11 \cdot 10^{-3}$ s^{-1}.
   (a) What is the half-life of the reaction?
   (b) What time is needed for the concentration of the reactant to fall to 1/8 of its initial value?
   (c) What time is needed for the concentration of the reactant to fall to 3/4 of its initial value?

Catalysis

7

Catalysts accelerate reactions but do not undergo a net chemical change themselves. The process of catalysis is achieved by lowering the activation energy of a reaction, and providing an alternative path to circumvent the slow, rate-determining step of the non-catalysed reaction (see Fig. 7.1).

While the catalyst does not undergo a net chemical reaction, it can take part in the reaction, but will be regenerated in due course. Two types of catalysts can be distinguished:

- Homogeneous catalyst: a catalyst that exists in the same phase as the reaction mixture
- Heterogeneous catalyst: exists in a different phase than the reaction mixture.

Examples for reactions with homogeneous catalysis include hydrogen peroxide decomposition by iodide and any reactions by biological enzymes; here, the iodide and the enzymes are present in solution, as are the reactants. In contrast, heterogeneous catalysts do not exist in the same phase as the reactants. Examples for such processes include the olefin hydrogenation by palladium, platinum or nickel catalysis.

7.1 Homogeneous Catalysis

In order to appreciate the kinetics of a homogeneously catalysed reaction, we will have a closer look at the decomposition of hydrogen peroxide, a process which can be catalysed by iodide ions. The reactant (H_2O_2) and the catalyst (I^-) are both present in solution, hence this constitutes a case of homogeneous catalysis.

© Springer International Publishing AG, part of Springer Nature 2018
A. Hofmann, *Physical Chemistry Essentials*,
https://doi.org/10.1007/978-3-319-74167-3_7

Fig. 7.1 Comparison of the energy profile of a reaction in the absence and presence of a catalyst. The free energy stabilisation of products over reactants is not altered; the catalyst only lowers the activation energy for the reaction

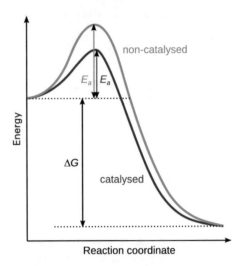

The decomposition of H_2O_2 is a disproportionation reaction, i.e. a reaction where an element changes its oxidation to a higher and a lower state simultaneously. It follows the net reaction:

$$2\,H_2O_{2(aq)} \rightarrow 2\,H_2O_{(l)} + O_{2(g)}$$

The reaction is believed to happen via the following pre-equilibrium that leads to generation of the hydroxyoxidanium ion $H_3O_2^+$:

$$H_3O^+ + H_2O_2 \rightleftharpoons H_3O_2^+ + H_2O$$

I^- can act as a catalyst by reacting with the hydroxyoxidanium ion to hypoiodous acid, HOI:

$$H_3O_2^+ + I^- \rightarrow HOI + H_2O$$

$$HOI + H_2O_2 \rightarrow H_3O^+ + O_2 + I^-$$

The equilibrium reaction is characterised by the equilibrium constant K. Treating the reaction of iodide with the hydroxyoxidanium ion as an elementary reaction, we can formulate a rate law of second order.

$$H_3O^+ + H_2O_2 \rightleftharpoons H_3O_2^+ + H_2O \quad K = \frac{c(H_3O_2^+)}{c(H_2O_2) \cdot c(H_3O^+)} \qquad (7.1)$$

$$H_3O_2^+ + I^- \rightarrow HOI + H_2O \quad v = k_a \cdot c(H_3O_2^+) \cdot c(I^-) \qquad (7.2)$$

$$HOI + H_2O_2 \rightarrow H_3O^+ + O_2 + I^- \quad (\text{fast})$$

Since the reaction of hypoiodous acid with peroxide is an extremely fast reaction, the contribution of this step to the overall rate of peroxide decomposition is negligible. The reaction rate for the catalysed reaction is thus:

$$v = \frac{dc(O_2)}{dt} = \frac{dc(HOI)}{dt}$$

From Eq. 7.2, we obtain for the rate

$$v = \frac{dc(HOI)}{dt} = k_a \cdot c(H_3O_2^+) \cdot c(I^-)$$

For the concentration of the hydroxyoxidanium ion, we know from the equilibrium 7.1 that

$$c(H_3O_2^+) = K \cdot c(H_2O_2) \cdot c(H_3O^+)$$

This yields for the overall rate:

$$v = \frac{dc(O_2)}{dt} = k_a \cdot c(H_3O^{2+}) \cdot c(I^-) = k_a \cdot K \cdot c(H_2O_2) \cdot c(H_3O^+) \cdot c(I^-) \quad (7.3)$$

The notable observation from Eq. 7.3 is that an experimentally observed rate constant k is indeed the product of the rate constant of the reaction in 7.2 and the equilibrium constant of reaction 7.1 ($k_a \cdot K$). The derived rate law also shows that the catalysed reaction is dependent on the concentration of the catalyst, the iodide anion.

7.1.1   Acid and Base Catalysis

Acid and base catalysis are of particular importance as they are frequently observed features of organic reactions. We will only look briefly at catalysis involving Brönstedt acids and bases, but Lewis acids/bases can also take part as catalysts in reactions. The crucial step in acid catalysis involving Brönstedt acids is the transfer of a proton to the substrate. Examples include acid-catalysed ester hydrolysis and the keto-enol tautomerism illustrated in Fig. 7.2.

In base catalysis involving Brönstedt bases, abstraction of a proton from the substrate is the crucial step. Examples for such reactions include the isomerisation

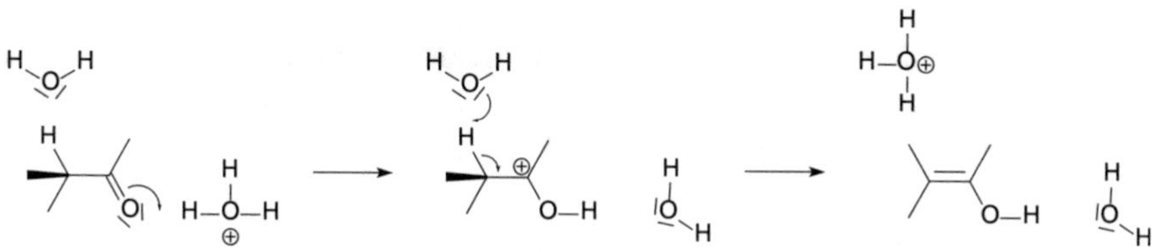

Fig. 7.2 The mechanism of acid catalysis in the transformation of a ketone to an enol (keto-enol tautomerism)

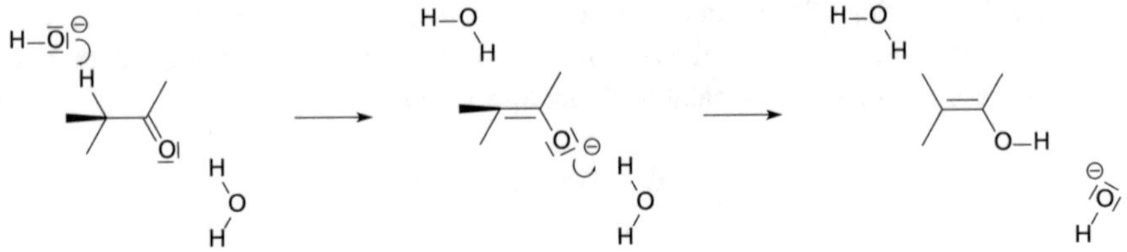

Fig. 7.3 The mechanism of base catalysis in the transformation of a ketone to an enol (keto-enol tautomerism)

and halogenation of organic compounds, Claisen condensation, aldol condensation, keto-enol tautomerism (see Fig. 7.3).

7.1.2 Enzymatic Catalysis

Of particular importance for many biochemical reactions is the homogeneous catalysis carried out by enzymes. Enzymatic functions are observed with particular proteins (enzymes) or nucleic acid molecules (ribozymes) that possess active sites which bind the substrates and process them to products. Enzyme-catalysed reactions are prone to inhibition by molecules that interfere with product formation. Many drugs for treatment of diseases work by inhibiting particular enzymes.

Experimental studies of enzyme kinetics are typically conducted by monitoring the initial rate of product formation in solution, with low concentrations of enzymes present. The general features for enzyme-catalysed reactions are:

- For a given initial substrate concentration $c_0(S)$, the initial rate of product formation v_0 is proportional to the total concentration of enzyme $c_0(E)$:

$$v_0 \sim c_0(E)$$

- For a given total concentration of enzyme $c_0(E)$ and at low concentrations of substrate $c_0(S)$, the rate of product formation is proportional to $c_0(S)$:

$$v_0 \sim c_0(S) \quad \text{for small } c_0(S)$$

- For a given total concentration of enzyme $c_0(E)$ and large substrate concentrations $c_0(S)$, the rate of product formation becomes independent of the substrate concentrations, reaching a maximum rate v_{max}:

$$v_0 = v_{max} \quad \text{for large } c_0(S)$$

7.1.3 The Michaelis-Menten Mechanism

The general features of enzyme-catalysed reactions (see previous section) are accounted for in the Michaelis-Menten mechanism, which assumes an enzyme-substrate complex as intermediate in the enzymatically catalysed product formation:

$$E + S \underset{k_{-1}}{\overset{k_1}{\rightleftharpoons}} [E\text{-}S] \overset{k_2}{\rightarrow} P + E$$

Investigating the kinetics of the hydrolysis of sucrose into glucose and fructose by the enzyme invertase, Leonor Michaelis and Maud Menten developed the above mechanism and derived the following expression for the rate of an enzymatic reaction, known as the Michaelis-Menten equation (Michaelis and Menten 1913):

$$v = \frac{k_2 \cdot c_0(E)}{1 + \frac{K_m}{c_0(S)}} \quad \text{or} \quad v = \frac{k_2 \cdot c_0(E) \cdot c_0(S)}{c_0(S) + K_m} \tag{7.4}$$

with the Michaelis-Menten constant

$$K_m = \frac{k_{-1} + k_2}{k_1} \tag{7.5}$$

Equation 7.4 describes the shape of the v_0–$c_0(S)$ curve as shown in Fig. 7.4, and thus explains the three observations introduced in Sect. 7.1.2 that characterise enzymatic catalysis.

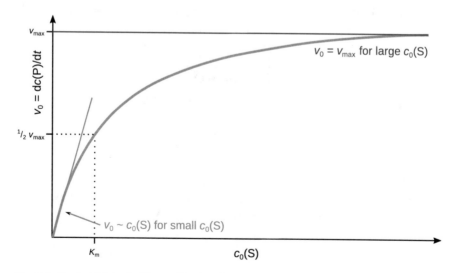

Fig. 7.4 Typical Michaelis-Menten kinetics of an enzyme in the v_0–$c_0(S)$ plot

Observation	Michaelis-Menten
$v_0 \sim c_0(E)$	$v = \frac{k_2 \cdot c_0(E) \cdot c_0(S)}{c_0(S) + K_m}$
$v_0 \sim c_0(S)$ for small $c_0(S)$	for $c_0(S) << K_m$ follows: $v = \frac{k_2}{K_M} \cdot c_0(E) \cdot c_0(S)$
$v_0 = v_{max}$ for large $c_0(S)$	for $c_0(S) >> K_m$ follows: $v = v_{max} = k_2 \cdot c_0(E)$

Table 7.1 The Michaelis-Menten kinetics describes the three general observations of enzymatic reactions

As apparent from Table 7.1, these considerations also deliver a definition of the maximum rate as per $v_{max} = k_2 \cdot c_0(E)$, which we can substitute into the initial Michaelis-Menten Eq. 7.4:

$$v = \frac{k_2 \cdot c_0(E) \cdot c_0(S)}{c_0(S) + K_M} \qquad (7.4)$$

and thus obtain

$$v = \frac{v_{max} \cdot c_0(S)}{c_0(S) + K_M} \qquad (7.6)$$

This form of the Michaelis-Menten equation is of practical importance, as it contains all parameters accessible in experimental measurements of enzymatic reactions. The equation can be transformed into a linear relationship, which has been of huge importance for enzymology studies. This analysis technique has been of great importance before the availability of computer-aided non-linear fitting software; it is still applied today in particular applications. By inverting both side of Eq. 7.6 one obtains:

$$\frac{1}{v} = \frac{c_0(S) + K_M}{v_{max} \cdot c_0(S)}$$

where the right side can be separated into two terms:

$$\frac{1}{v} = \frac{1}{v_{max}} + \frac{K_M}{v_{max}} \cdot \frac{1}{c_0(S)} \qquad (7.7)$$

By plotting $\frac{1}{v_0}$ on the ordinate (y-axis) and $\frac{1}{c_0(S)}$ on the abscissa (x-axis), a linear relationship with the positive slope $\frac{K_M}{v_{max}}$ and positive y-intercept $\frac{1}{v_{max}}$ is obtained. This plot is known as the Lineweaver-Burk plot (see Fig. 7.5).

Fig. 7.5 Illustration of the Lineweaver-Burk plot

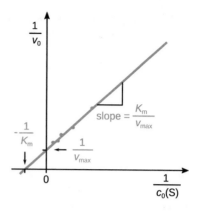

7.1.4 Enzyme Efficiency and Enzyme Inhibition

The turnover number of an enzyme, k_{cat}, is the number of catalytic cycles performed in a given time interval, divided by that time interval. k_{cat} is a first-order rate constant and equivalent to the rate constant for release of product from the enzyme-substrate complex, k_2:

$$E + S \;\; \underset{k_{-1}}{\overset{k_1}{\rightleftharpoons}} \;\; [\text{E-S}] \;\; \overset{k_2}{\rightarrow} \;\; P + E$$

$$k_{cat} = k_2 = \frac{v_{max}}{c_0(\text{E})} \tag{7.8}$$

Equation 7.8 is a direct consequence of one of the three general observations for enzymatic reactions (see Table 7.1).

The efficiency of an enzyme is dependent on

- How much substrate it requires
- How many turnover cycles it performs in a given time period.

The substrate concentration required to achieve 50% of the maximum rate of an enzymatic reaction is described by the Michaelis-Menten constant K_M. Therefore, a lower K_M will result in higher efficiency. Likewise, the efficiency will be higher, if the turnover number (k_{cat}) is high. The catalytic efficiency η of an enzyme is thus defined as:

$$\eta = \frac{k_{cat}}{K_m} \tag{7.9}$$

and, by using the relationships from Eqs. 7.5 and 7.8, can also be expressed in terms of the rate constants k_1, k_{-1} and k_2:

$$\eta = \frac{k_{cat}}{K_m} = \frac{k_1 \cdot k_2}{k_{-1} + k_2}$$

7.1.5   Enzyme Inhibition

Molecules that affect enzyme activity are called effectors and can either be activators or inhibitors. The discovery of enzyme inhibitors (and thus the study of enzyme inhibition) is an important biomedical discipline, since enzyme inhibitors may be used as therapeutics. Drug discovery therefore relies heavily on enzyme inhibition studies.

There are three different types of enzyme inhibition mechanisms:

- Competitive inhibition
- Un-competitive inhibition
- Non-competitive inhibition

In competitive inhibition, either the substrate or the inhibitor can bind to the enzyme, but not both of them simultaneously. This can come about by way of two different scenarios: The inhibitor may directly compete with the substrate for binding in the active site (one binding site); alternatively, the inhibitor may bind to a site different from the active site (two binding sites), but its binding will exclude binding of the substrate in the active site. This latter case is called allosteric inhibition. When comparing the enzyme kinetics data at different concentrations of effector in the Lineweaver-Burk plot, competitive inhibition shows the behaviour illustrated in Fig. 7.6, left panel.

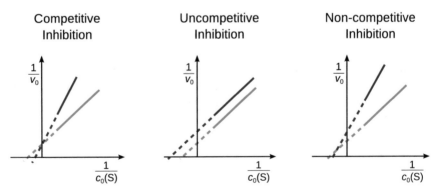

Fig. 7.6 Different types of inhibition show different changes in the Lineweaver-Burk plot. In this schematic illustration, the turquoise curves show enzyme activity data in the absence, and the magenta curves in the presence of an inhibitor

The mechanism of uncompetitive inhibition requires two different binding events, the substrate binding to the active site of the enzyme, and the inhibitor binding to a different site. Binding of the inhibitor decreases the activity of the enzyme. However, in contrast to the above allosteric competitive inhibition, in the case of uncompetitive inhibition the effector binds only when the substrate is present. Therefore, binding of the inhibitor is indeed to the enzyme-substrate complex [E-S]. In the Lineweaver-Burk plot, uncompetitive inhibition results in parallel lines with different x- and y-intercepts (Fig. 7.6, centre panel).

Non-competitive inhibition of an enzymatic reaction arises when the inhibitor reduces the binding of the substrate to the free enzyme as well as the enzyme-substrate complex [E-S]. This mechanism can be distinguished from the two other in the Lineweaver-Burk plot as illustrated in Fig. 7.6, right panel.

7.2   Heterogeneous Catalysis

We have established earlier that catalysts accelerate reactions, but do not undergo a net chemical change themselves. Catalysis is thus achieved by lowering the activation energy of a reaction, and by providing an alternative path to circumvent the slow, rate-determining step of the non-catalysed reaction.

A heterogeneous catalyst, exists in a different phase than the reaction mixture; its catalytic activity builds on the capability to bring the reactants into very close spatial vicinity. A typical example is the olefin hydrogenation reaction catalysed by metals such as palladium, platinum or nickel.

If we consider a heterogeneous catalytic process involving solids, we need to conceptualise the attachment of particles to the surface of the solid (adsorption), as well as the reverse process (desorption). The substance that attaches to the surface is called the adsorbate; the underlying material is called the adsorbent or substrate.

7.2.1   Solid Surfaces

While the surface of solids appears flat from a macroscopic perspective, this perception is a matter of resolution. Indeed, at sub-microscopic level, surfaces of solids are not flat but show various features and defects. At atomic detail, the flat surface looks like a layer of oranges in the grocery store. Importantly, there may be higher order in solids whereby the constituting atoms, ions or molecules are arranged in particular repeating structures. Such structures are called a crystalline lattice and the solid is then called a crystal.

The terrace step kink (also known as terrace ledge kink) model describes the surface formation and transformations of crystals, whereby the energy of a particular atom on a surface depends on the number of bonds that atom has with neighbouring atoms (Stranski 1928). As the term 'terrace step kink model' implies, the typical features on surfaces are terraces, steps and kinks (see Fig. 7.7).

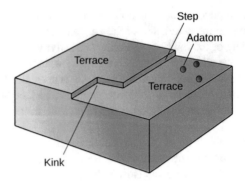

Fig. 7.7 Schematic illustration of the terrace step kink model of solid surfaces

Fig. 7.8 Different faces of a crystal grow at different speeds, indicated by the length of the turquoise arrows. Evidently, the slowest growing faces dominate the overall shape and macroscopic appearance of a crystal

When an atom arrives at a terrace, it is called an adatom and may bounces across the surface depending on the inter-molecular potential—this process is called accommodation (see Physisorption in the following section). If it comes to lie in a kink (kink atom) or at a step (step atom), it can interact with more than one surface atom, and the interaction may be strong enough to trap it. When ions deposit from solution, the loss of solvation energy is offset by strong Coulomb forces (i.e. the electrostatic forces) between the arriving ions and ions on the surface.

Macroscopically, many crystals are recognised by their shape which is constituted by flat faces with sharp angles. A distinct shape is not a necessary criterion for a solid to be a crystal, but it is a frequently observed property. How quickly a surface (or face) growths on a crystal, depends on the particular plane. As illustrated in Fig. 7.8, the slowest growing faces dominate the overall shape of the crystal.

7.2.2 Adsorption

When particles adsorb to surfaces, a fundamental observable is the fractional coverage of the available surface area. As occupancy of a surface area comes about by particles occupying individual sites, the fractional coverage Γ can be defined as per

$$\Gamma = \frac{N_{\text{occupied}}}{N_{\text{total}}} \tag{7.10}$$

Here, N_{occupied} is the number of occupied adsorption sites and N_{total} is the number of total adsorption sites on the surface. Another frequently used definition expresses the fractional coverage Γ in terms of the volume of adsorbate (V) and the volume of adsorbate required for total coverage (V_∞):

$$\Gamma = \frac{V}{V_\infty} \tag{7.11}$$

The rate of adsorption is the rate of change of surface coverage (dΓ) over time (dt) and can be determined using the observables of Eqs. 7.10 or 7.11:

$$v_{\text{ads}} = \frac{d\Gamma}{dt} \tag{7.12}$$

Two types of adsorption are typically distinguished, based on the types of forces involved in the interaction between adsorbate and adsorbent: physisorption and chemisorption. The problem of distinguishing between the two types is akin to that of distinguishing between chemical and physical interactions in general.

Physisorption

If molecules or atoms attach to a surface exclusively due to van der Waals interactions, this adsorption process is called physisorption. This type of adsorption is for example observed in dispersions and arises due to dipolar interactions. van der Waals interactions are long-range interactions of rather low energy. The rather small enthalpies of physisorption (~ -20 kJ mol^{-1}) are not sufficient to break bonds, so the adsorbate retains its identity on the surface; geometric distortions may be possible, though. Despite being rather weak interactions, there are examples of notable roles in natural process. For example, the ability of geckos to climb walls and ceilings rests on the van der Waals attraction between surfaces and their foot-hairs. The energy released upon physisorption can be absorbed as vibrations of the substrate lattice (increased thermal motion) and thus lead to an approaching particle bouncing across the surface, until it has lost its energy. This process is called accommodation. The enthalpy of physisorption can be measured by monitoring the temperature of a sample of known heat capacity during the adsorption process.

Chemisorption

Atoms or molecules that stick to the surface by forming a covalent bond perform chemisorption. The enthalpy of chemisorption is much higher than in the case of physisorption (~ -200 kJ mol^{-1}) and may thus lead to bond breakage in the adsorbate, which allows the resulting species (molecular fragments) to maximise their coordination number with the substrate. The existence of molecular fragments on the substrate is one of the reasons why solid surfaces act as catalysts.

Except for special cases, chemisorption is exothermic ($\Delta H < 0$). An example of an endothermic process is the adsorption of H_2 on glass. Since spontaneous processes require a change in the free energy of $\Delta G < 0$, the positive enthalpy change during H_2 adsorption needs to be overcome. In the case of H_2 on glass, there is a significant increase in entropy as H_2 molecules dissociate into H atoms, thereby outnumbering the enthalpy change and thus leading to a negative ΔG:

$$\Delta G = \Delta H - T \cdot \Delta S < 0 \qquad (2.42)$$

7.2.3 Adsorption Isotherms

During the adsorption process, free and adsorbed particles are in dynamic equilibrium. If we consider a gas as the adsorbing species, then the fractional surface coverage Γ depends on the pressure of the gas overlying the substrate. The variation of fractional coverage Γ with pressure p at constant temperature is called the adsorption isotherm.

If we make the following assumptions:

- Adsorption cannot proceed beyond monolayer coverage
- All sites are equivalent and the surface is uniform (at microscopic scale)
- The ability of a particle to adsorb is independent of the occupancy of neighbouring sites (i.e. there is no interactions between adsorbed particles)
- Every particle provides a single species that will be adsorbed (no dissociation)

then the rate of adsorption can then be expressed as:

$$v_{ads} = \frac{d\Gamma}{dt} = k_{ads} \cdot p \cdot N \cdot (1 - \Gamma) \qquad (7.13)$$

This equation is reminiscent of a second-order rate law. Here, the rate constant is k_{ads}, the concentration of one reactant is given by the pressure p, and the concentration of the second reactant is the number of free sites on the substrate, $[N \cdot (1 - \Gamma)]$.

Similarly, the rate of desorption is given by:

$$v_{des} = \frac{d\Gamma}{dt} = -k_{des} \cdot N \cdot \Gamma \qquad (7.14)$$

This is a first-order rate law, with the rate constant k_{des} and the concentration of 'product' given by the number of adsorbate molecules bound to substrate sites, i.e. the number of occupied sites, $(N \cdot \Gamma)$.

At equilibrium, there is no net change, so the rate of adsorption equals the rate of desorption, and by substituting expressions from Eqs. 7.13 and 7.14 one obtains:

$$| v_{\text{ads}} | = | v_{\text{des}} | \quad \Rightarrow \quad | k_{\text{ads}} \cdot p \cdot N \cdot (1 - \Gamma) | = | -k_{\text{des}} \cdot N \cdot \Gamma |$$

$$k_{\text{ads}} \cdot p \cdot N - k_{\text{ads}} \cdot p \cdot N \cdot \Gamma = k_{\text{des}} \cdot N \cdot \Gamma$$

$$k_{\text{ads}} \cdot p \cdot N = k_{\text{ads}} \cdot p \cdot N \cdot \Gamma + k_{\text{des}} \cdot N \cdot \Gamma$$

$$k_{\text{ads}} \cdot p \cdot N = (k_{\text{ads}} \cdot p \cdot N + k_{\text{des}} \cdot N) \cdot \Gamma$$

$$\Gamma = \frac{k_{\text{ads}} \cdot p \cdot N}{k_{\text{ads}} \cdot p \cdot N + k_{\text{des}} \cdot N} = \frac{k_{\text{ads}} \cdot p}{k_{\text{ads}} \cdot p + k_{\text{des}}}$$

$$\Gamma = \frac{\frac{k_{\text{ads}}}{k_{\text{des}}} \cdot p}{\frac{k_{\text{ads}}}{k_{\text{des}}} \cdot p + 1}$$

This yields an expression for the Langmuir isotherm (see Fig. 7.9), which relates the fractional coverage to the pressure of adsorbate:

$$\Gamma = \frac{K \cdot p}{K \cdot p + 1} \tag{7.15}$$

whereby the Langmuir constant K is given by the ratio between the rate constants of the adsorption and desorption processes:

$$K = \frac{k_{\text{ads}}}{k_{\text{des}}} \tag{7.16}$$

Fig. 7.9 The Langmuir adsorption isotherm describes the saturation of binding sites on the catalyst (substrate) in dependence of the pressure of gas molecules (adsorbate)

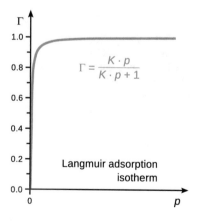

7.2.4 Adsorption Isotherms with Dissociation

When deriving the concept of the Langmuir isotherm in the previous section, we assumed that the particles arriving at the catalyst's surface do not dissociate. In many cases (for example the synthetically important hydrogenation reactions using H_2) this can not be assumed. If the adsorbate dissociates into two species upon surface attachment, then the rate of adsorption is proportional to the pressure and the probability that both fragments find sites, i.e. the square number of vacant sites. Instead of Eq. 7.13, one thus obtains:

$$v_{ads} = \frac{d\Gamma}{dt} = k_{ads} \cdot p \cdot [N \cdot (1 - \Gamma)]^2 \tag{7.17}$$

Similarly, for the rate of desorption, we need to take into account that two dissociated fragments re-combine and thus two occupied sites are being vacated:

$$v_{des} = \frac{d\Gamma}{dt} = -k_{des} \cdot (N \cdot \Gamma)^2 \tag{7.18}$$

The combination of Eqs. 7.17 and 7.18 yields the Langmuir isotherm for adsorption with dissociation:

$$\Gamma = \frac{\sqrt{K \cdot p}}{\sqrt{K \cdot p} + 1} \tag{7.19}$$

as illustrated in Fig. 7.10.

Fig. 7.10 Comparison of Langmuir isotherms for non-dissociating and dissociating ($N = 2$) adsorbates

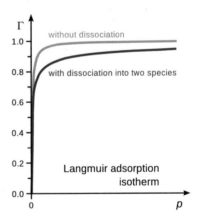

7.2.5 The Isosteric Enthalpy of Adsorption

As per the definition, the isotherms describe the adsorption behaviour at constant temperature. Therefore, if adsorption experiments are carried out at different temperatures, different isotherms are obtained. The variation between those isotherms is captured by different values of $K = \frac{k_{ads}}{k_{des}}$. Since K is an equilibrium constant, we can use the van't Hoff equation to obtain an enthalpy:

$$\left(\frac{\delta(\ln K)}{\delta T}\right)_{\Gamma} = \frac{\Delta H^{\varnothing}_{ads}}{R \cdot T^2} \tag{2.60}$$

$\Delta H^{\varnothing}_{ads}$ is called the isosteric enthalpy of adsorption and is the standard enthalpy of adsorption at a fixed surface coverage.

7.2.6 Mechanisms of Heterogeneous Catalysis

The previous sections have been concerned with the adsorption of particles at the surface of a solid catalyst. Since catalysts are used to facilitate particular reactions, we now want to look at the rates of these reactions. Two different mechanisms for bimolecular surface-catalysed reactions have been proposed. The Langmuir-Hinshelwood mechanism assumes that the two molecules adsorb to the surface and then undergo the reaction. In contrast, in the Eley-Rideal mechanism, only one of the molecules adsorbs and the other one reacts with it directly from the gas phase without adsorbing. In the following, we will see that the kinetics for the two mechanisms differ. However, the rate constant k of the catalysed reaction will always be much larger than that of the non-catalysed reaction, since the reaction on the surface has a much lower activation energy.

Langmuir-Hinshelwood Mechanism
In the Langmuir-Hinshelwood mechanism, a reaction takes place between two reactants A and B adsorbed on the surface of the catalyst. The reaction rate thus depends on the extent of surface coverage by the two species, Γ_A and Γ_B:

$$A_{(ads)} + B_{(ads)} \rightarrow P \quad \text{with} \quad v = k \cdot \Gamma_A \cdot \Gamma_B \tag{7.20}$$

If we use the partial pressures, p_A and p_B, for the two reactants, we can express the partial surface coverages Γ_A and Γ_B by Langmuir isotherms (assuming no dissociation):

$$\Gamma_A = \frac{K_A \cdot p_A}{1 + K_A \cdot p_A + K_B \cdot p_B} \quad \text{and} \quad \Gamma_B = \frac{K_B \cdot p_B}{1 + K_A \cdot p_A + K_B \cdot p_B}$$

Substituting these expressions in Eq. 7.20 yields the reaction rate for the Langmuir-Hinshelwood mechanism as:

Catalyst	Reaction
Pt	$2\,CO + O_2 \rightarrow 2\,CO_2$
Pt	$N_2O + H_2 \rightarrow N_2 + H_2O$
Pd	$2\,C_2H_4 + O_2 \rightarrow 2\,H_3C - CHO$
Cu	$C_2H_4 + H_2 \rightarrow C_2H_6$
ZnO	$CO + 2\,H_2 \rightarrow H_3C - OH$

Table 7.2 Examples of surface-catalysed reactions that follow the Langmuir-Hinshelwood mechanism

Catalyst	Reaction
Pt	$4\,NH_3 + 3\,O_{2(ads)} \rightarrow 2\,N_2 + 6\,H_2O$
Ni, Fe	$C_2H_2 + H_{2(ads)} \rightarrow C_2H_4$
Ag	$2\,C_2H_4 + O_{2(ads)} \rightarrow 2\,H_2COCH_2$

Table 7.3 Examples of surface-catalysed reactions that follow the Eley-Rideal mechanism

$$v = \frac{k \cdot K_A \cdot K_B \cdot p_A \cdot p_B}{\left(1 + K_A \cdot p_A + K_B \cdot p_B\right)^2} \tag{7.21}$$

Examples for reactions following the Langmuir-Hinshelwood mechanism include (Table 7.2):

Eley-Rideal Mechanism

Examples for reactions following the Eley-Rideal mechanism include (Table 7.3):

In the Eley-Rideal mechanism, a reactant B in the gas phase collides with a reactant A already adsorbed on the surface. The rate of reaction is thus dependent on the partial pressure of B and the extent of surface coverage of A:

$$A_{(ads)} + B_{(g)} \rightarrow P \quad \text{with } v = k \cdot \Gamma_A \cdot p_B \tag{7.22}$$

If the adsorption of A follows a Langmuir isotherm without dissociation, one obtains the rate of the Eley-Rideal mechanism as:

$$v = \frac{k \cdot K \cdot p_A \cdot p_B}{1 + K \cdot p_A} \tag{7.23}$$

7.3 Methods to Investigate Surfaces and Surface Processes

In heterogeneous catalysis, catalysis happens on the surface of the catalyst (also called adsorbent or substrate). Therefore, in order to investigate processes at surfaces, or indeed their make-up, experimental techniques are required. Two types of methods of surface investigation can be distinguished: spectroscopic methods and imaging methods.

The choice of a particular spectroscopic method depends on what exactly is to be investigated. From Fig. 7.11, which shows the electromagnetic spectrum, it is obvious that different energy regions of the spectrum need to be employed,

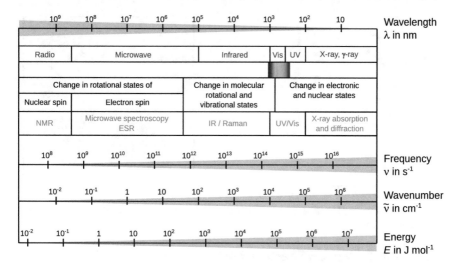

Fig. 7.11 The electromagnetic spectrum

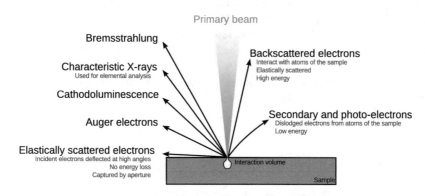

Fig. 7.12 Interaction of a beam of high energy electromagnetic radiation with a solid specimen

depending on whether bonds between adsorbate and substrate molecules are to be investigated, or atomic information is required. A frequent problem in practical applications is, for example, the chemical composition of a surface. This can be determined by using ionising electromagnetic radiation such as X-rays (see also Sect. 13.5) or hard UV. The interaction of a beam of high energy electromagnetic radiation leads to a variety of effects in a solid sample (Fig. 7.12) which can be used to characterise the specimen.

Whereas spectroscopic methods result in 'abstract' information about a sample, imaging methods provide pictures as to the distribution of features or physical structures of samples and typically yield 'images' as an immediate result of the experiment. In the following, a few methods used in surface characterisation will be briefly introduced.

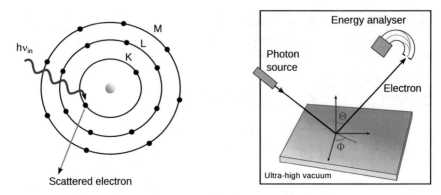

Fig. 7.13 Schematic illustration of the photoelectric effect (left) and its application in photoelectron spectroscopy

7.3.1 Photoelectron Spectroscopy

This method is based on the photoelectric effect which was originally observed by James Franck and Gustav Hertz, and later explained by Albert Einstein. Light of sufficiently high energy can be used to ionise a sample, giving rise to photo electrons from species present. Depending on whether X-rays or hard UV radiation is used, the method is called XPS or UPS (see also Sect. 13.5.1). Since the kinetic energy of electrons ejected from an atom depends on the internal electronic structure of that atom, information about the electronic structure and chemical composition is obtained (Fig. 7.13).

7.3.2 Auger Electron Spectroscopy

The Auger effect is the emission of an electron after incident high energy caused loss of an electron. The hole left by the first departing electron is filled by a second electron from a higher shell that takes the place of the first electron. This transition releases energy which can either be emitted as a photon (X-ray fluorescence; see Sect. 7.3.3) or lead to ejection of a third electron, called the Auger electron (see Fig. 7.14). The energies of the Auger electrons are characteristic of the species present which makes Auger electron spectroscopy a useful tool for determination of the chemical identities in samples. The incident energy may be provided by X-ray, hard UV or electron beams. The Auger process is discussed in more detail in Sect. 13.6.3.

7.3.3 X-ray Fluorescence

Arising from the same physical process as the Auger electrons, photons may be emitted instead of secondary electrons (Fig. 7.15, left panel). In this case, the process

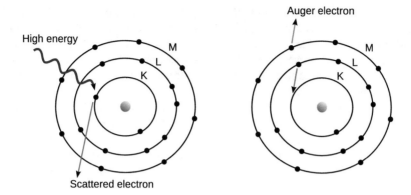

Fig. 7.14 Schematic illustration of the Auger effect

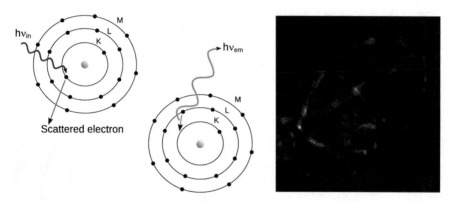

Fig. 7.15 Left: The emission of a photon instead of an Auger electron (after electron loss due to high energy impact) is called X-ray fluorescence. Right: An X-ray fluorescence micrograph mapping iron in the egg of the parasitic worm *Schistosoma japonicum*. Figure adapted from Hofmann et al. (2014)

is called X-ray fluorescence (see also Sect. 13.6.2). Again, spectral analysis of the emitted energies allows conclusions as to the species present. By rastering a sample through the incoming beam of high energy photons and determining the energy of the emitted X-rays, a spatial mapping can be performed (Fig. 7.15, right panel). In doing so, the spectroscopic method of X-ray fluorescence becomes an imaging method and is thus called X-ray fluorescence microscopy.

7.3.4 LEED

The arrangement of atoms close to the surface can be determined by low energy electron diffraction (LEED). Here, the wave character of electrons is used. An incident electron beam of relatively low energy (typically in the range 20–200 eV) is diffracted by the atomic lattice formed by atoms on or close to the surface. The diffracted electrons can be observed by a fluorescent screen, and the resulting diffraction pattern is then visible from behind the sample (modern instruments use a position sensitive detector instead of the fluorescent screen, thus enabling digital data acquisition). The diffraction pattern represents the two-dimensional reciprocal lattice of the specimen's surface.

As introduced in Sect. 7.2.1, hardly any surface is an ideal surface (i.e. like a cut through the bulk material). Instead, there are many terraces, steps and kinks in surfaces of solids which can be identified with LEED. Besides characterisation of the basic lattice of a solid sample, further information can be gained from particular features caused by super lattices, step lattices or domain formation (Fig. 7.16).

Like photo-electron and Auger electron spectroscopy, LEED requires the sample to be brought into an chamber of ultra high vacuum. The correct alignment of the crystalline sample can be achieved with the help of X-ray diffraction. After mounting, the specimen is cleaned chemically by oxidation/reduction cycles and the surface flattened using exposure to high temperature. This process (called annealing) may lead to re-surfacing of impurities. Therefore, Auger electron spectroscopy is often used in conjunction with LEED to monitor the purity of the sample.

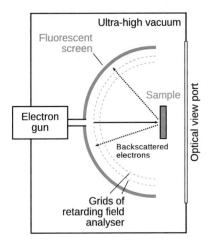 

Fig. 7.16 Left: Schematics of a LEED instrument. The concentric grids are used for filtering inelastically scattered electrons. Instead of optical observation of the LEED pattern, modern instruments have a position-sensitive digital detector (called delay-line detector) instead of the fluorescent screen. Right: LEED pattern of an Ir(100)−1 × 1 surface recorded at an energy of 195 eV. The image was kindly provided by L Hammer (Friedrich-Alexander-Universität Erlangen-Nürnberg, Germany) and is reproduced with permission

7.3.5   Scanning Probe Methods

Most of the surface methods mentioned in the previous sections deliver information about the surface of materials from which information about topological features is available by means of reconstruction. However, we have also seen that X-ray fluorescence can be used in conjunction with raster scanning of the sample, thus directly resulting in an image.

Scanning probe methods are based on a similar principle; here, a probe is raster scanning over a surface, and the interaction between the probe and the surface is measured. The most scanning probe methods are scanning tunneling microscopy and atomic force microscopy (AFM). In both cases, the physical elevation of surfaces (z-direction) is probed.

Scanning Tunneling Microscopy (STM)

Gerd Binnig and Heinrich Röhrer (Nobel prize in 1986) developed STM in the early 1980s. A sharp metal tip is brought into a distance of 2–5 Å of a conducting surface, and a potential of about 2 V is applied between the tip and the sample. In this configuration, electrons can tunnel across the gap between the tip and the sample, and the current depends on the distance between the tip and surface in an exponential fashion. This strong dependence is the reason for the high resolving power of STM which allows imaging an true atomic resolution.

Atomic Force Microscopy (AFM)

This surface-sensitive method enables imaging of surfaces or detection of interactions between molecules; it was conceived by Binnig and colleagues in the mid-1980s as a further development of STM. In AFM, the surface is visualised in three dimensions by touching it with a tiny probe called the tip. The physical phenomenon probing surface elevation is the interaction force between the tip and the sample surface.

Attractive or repulsive forces between tip and surface are observed through measurement of the reflection of a laser beam by the cantilever. The reflected laser beam is detected by a photodiode detector which converts the optical into an electrical signal. The microscope uses a feedback loop which triggers a z-movement of the piezoelectric scanner upon bending of the cantilever. The parameters of the feedback loop are optimised such as to minimise and maintain a constant value of the force between the tip and the sample.

The surface area (x- and y-direction) is scanned in a raster-like fashion with a tip brought very close to the sample. The tip is attached to the free end of a cantilever (Fig. 7.17) which will be bent due to interactions between the tip and the sample surface. It provides an image of the surface topography of the specimen at high resolution.

The lateral resolution depends on the diameter and geometry of the tip apex. With conventional instruments, high resolution images at a lateral resolution of 0.5–1 nm ($= 5$–10 Å) can be obtained. The vertical resolution is higher, but limited by mechanical vibrations and thermal fluctuations of the cantilever. Typically, vertical resolutions of about 0.1–0.2 nm ($= 1$–2 Å) can be achieved. Further developments of

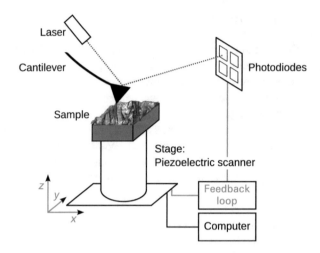

Fig. 7.17 Schematic illustration of an atomic force microscope. Figure kindly provided by A Simon (Université Lyon 1, France) and adapted from Hofmann et al. (2014)

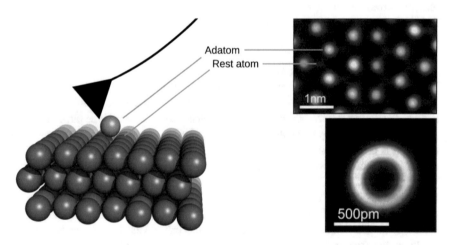

Fig. 7.18 AFM at sub-nanometer scale can visualise individual atoms. Right top: AFM image of the Si(111)-(7 × 7) reconstruction using a CO-terminated metal tip. The rest atoms and adatoms are clearly visible. Right bottom: AFM image of a Cu adatom on Cu(111), showing a ringlike symmetry that is caused by a toroidal charge density of the adatoms. Experimental AFM images are reproduced from Emmrich et al. (2015) with permission (License No 3691111422111)

this technique even allows imaging and characterisation of surfaces at subatomic resolution (Sugimoto et al. 2007; Emmrich et al. 2015) whereby single adatoms appear as toroidal structures and multi-atom clusters as connected structures, showing each individual atom as a torus (Fig. 7.18).

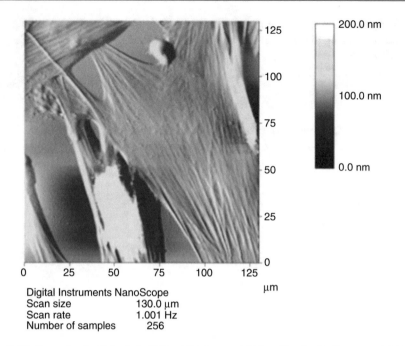

Digital Instruments NanoScope
Scan size   130.0 μm
Scan rate   1.001 Hz
Number of samples   256

Fig. 7.19 Example of a biological AFM application: an AFM profile of cells (Simon et al. 2003) with a colour gradient indicating the vertical height. Figure kindly provided by A Simon (Université Lyon 1, France)

AFM is a technique with a wide variety of applicability. It can operate in vacuum, air or liquid and be used to probe conductor or non-conductor surfaces. It can also be used with biological samples (see Fig. 7.19) and has become very popular in many fields, including materials science, polymer science, physics, life sciences and nanobiotechnology.

7.4 Exercises

1. The initial rates of the myosin-catalysed hydrolysis of ATP were measured in the presence of varying starting concentrations of ATP:

c_0(ATP)	mmol dm^{-3}	0.005	0.010	0.020	0.030	0.050	0.100	0.200	0.300
v_0	μmol dm^{-3} s^{-1}	0.051	0.083	0.118	0.138	0.158	0.178	0.190	0.194

Assume that the enzymatic reaction follows a Michaelis-Menten mechanism and determine the maximum rate and the Michaelis constant.

2. The adsorption behaviour of 1 g activated carbon at 0 °C has been quantitatively assessed using different pressures of N_2. The following molar amounts of N_2 are adsorbed:

$p(N_2)$	kPa	0.524	1.73	3.06	4.13	7.50	10.3
$n_{ads}(N_2)$	10^{-4} mol	0.440	1.35	2.27	3.14	4.60	5.82

Assuming Langmuir adsorption behaviour, determine (a) the maximum amount of N_2 that can be adsorbed by 1 g activated carbon at 0 °C and (b) the Langmuir constant.

The Fabric of Atoms

8

8.1 Properties of Light and Electrons

Several simple experiments indicate that neutral substances, i.e. such that possess no overall charge, are composed of charged particles whose individual charges balance each other. For example, if a salt is dissolved in water, the conductivity of the resulting solution is larger than that of pure water. Historically, the first findings in this context (Elster and Geitel 1882; Hertz 1887) were those showing that a metal can emit charged particles when either heated (thermionic emission) or exposed to ultraviolet light (photoelectric effect).

When the emitted charged particles in those experiments are accelerated by an electric field, they can even penetrate thin metal layers without being diverted from their initial direction of propagation (Lenard 1903). The emitted particles were thus termed cathode rays. Lenard therefore concluded that if atoms make up matter, then they cannot be solid particles that fill all the space occupied by matter. From those scattering experiments, the atomic radius was estimated to be approx 10^{-8} cm. The part of the atom that carries most of the mass (the nucleus) was estimated to possess a radius of approx 10^{-12} cm.

8.1.1 Charge and Mass of the Electron

Indeed, the charged particles emitted from metals upon exposure to high-energy light or heating are electrons, and the fact that they can be emitted by matter indicates that they are a constituting component of matter.

The charge of the electron was first estimated in a landmark experiment by Millikan who injected oil droplets into the slit between two horizontally arranged capacitor plates (Millikan 1913). The tiny charged droplets become charged and suspended against gravity. The electric field E provided by the capacitor plates, gives rise to an accelerating electrical force onto the charge Q which is given by

© Springer International Publishing AG, part of Springer Nature 2018
A. Hofmann, *Physical Chemistry Essentials*,
https://doi.org/10.1007/978-3-319-74167-3_8

$$F_{el} = Q \cdot E.$$

This accelerating force is either increased or lessened by the gravity, depending on whether the particles move upwards or downwards. The force on the particles due to gravity is calculated as per:

$$F_{gravity} = m \cdot g.$$

For a spherical particle, one obtains the following relationship between mass and radius:

$$m = V \cdot \rho = \frac{4}{3} \cdot \pi \cdot r^3 \cdot \rho$$

and thus

$$F_{gravity} = \frac{4}{3} \cdot \pi \cdot r^3 \cdot \rho \cdot g.$$

The total accelerating force on the charged particle in this experiment is thus

$$F_{acc} = F_{el} \pm F_{gravity} = Q \cdot E \pm \frac{4}{3} \cdot \pi \cdot r^3 \cdot \rho \cdot g.$$

This is counteracted by the frictional force, as the particles are moving in air atmosphere. The frictional force is calculated according to Stokes for spherical particles that move with velocity v through medium of viscosity η:

$$F_{friction} = 6 \cdot \pi \cdot \eta \cdot r \cdot v.$$

A steady movement is observed when the two forces balance each other:

$$F_{friction} = F_{acc}.$$

From this, the velocity of the moving particles can be calculated for the two cases, one where the particles move downwards (along the gravitational field) and one where they move upwards (against the gravitational field):

$$v_{down} = \frac{Q \cdot E + \frac{4}{3} \cdot \pi \cdot r^3 \cdot \rho \cdot g}{6 \cdot \pi \cdot \eta \cdot r}$$

$$v_{up} = \frac{Q \cdot E - \frac{4}{3} \cdot \pi \cdot r^3 \cdot \rho \cdot g}{6 \cdot \pi \cdot \eta \cdot r}.$$

This constitutes two equations with two unknowns, Q and r; one of the unknown parameters can thus be calculated by measuring the velocities v_{up} and v_{down} for a sufficiently large number of particles. Since the oil droplets are macroscopic particles and might be multiply charged, the experimental values of Q are integer multiples of the elementary charge $Q(e^-) = e$. This constant charge of the electron has been determined to be

$$e = 1.60210 \cdot 10^{-19} \, C. \tag{8.1}$$

In experiments with the cathode rays, it became apparent that the flight path of electrons emitted from the metal cathode can be altered by exposing the electron beam to an external magnetic field. This phenomenon was used by Thomson to quantitatively characterise the electron (Thomson 1897), allowing the calculation of its mass.

A charged particle with velocity v that is exposed to a magnetic field B experiences an acceleration by the Lorentz force, giving rise to a circular flight path. The Lorentz force acts as a centripetal force (directed towards the centre of the circle), and is balanced by the centrifugal force:

$$F_{\text{Lorentz}} = Q \cdot v \cdot B = \frac{m \cdot v^2}{r} = F_{\text{centrifugal}}.$$

This yields an expression for the ratio between mass and charge of the particle, and in particular for the electron:

$$\frac{m}{Q} = \frac{m_e}{e} = \frac{v}{B \cdot r} \Rightarrow \frac{m_e^2}{e^2} = \frac{v^2}{B^2 \cdot r^2}. \tag{8.2}$$

If the electrons in the cathode ray tube are accelerated by applying an external potential U, then we know that the electrons entering the magnetic field have a kinetic energy of

$$E_{\text{kin}} = \frac{1}{2} \cdot m_e \cdot v^2 = e \cdot U$$

from which it follows that

$$v^2 = \frac{2 \cdot e \cdot U}{m_e}.$$

The above expression can be substituted into Eq. 8.2 which then yields:

$$\frac{m_e^2}{e^2} = \frac{v^2}{B^2 \cdot r^2} = \frac{2 \cdot e \cdot U}{m_e \cdot B^2 \cdot r^2}$$

and thus

$$\frac{m_e}{e} = \frac{2 \cdot U}{B^2 \cdot r^2}. \tag{8.3}$$

The mass-to-charge ratio of the electron can thus be obtained by knowledge of the external magnetic field strength B, the applied acceleration voltage U and the radius of the circular flight path of the electron. Importantly, as this relationship applies to any charged particle, its mass-to-charge ration m/Q can be obtained from the parameters B, U and r, when forcing the particle onto a circular flight path—a concept used for detectors in mass spectrometers.

With the known elementary charge e (Eq. 8.1), the mass of the electron has been determined to

$$m_e = 9.1091 \cdot 10^{-31} \text{ kg.} \tag{8.4}$$

8.1.2   The Wave Properties of Light

The corpuscular theory of light was put forward at the end of the sixteenth century by several philosophers and taken up by Isaac Newton (1704), explaining the optical phenomena known at the time. The Dutch mathematician and scientist Christiaan Huygens proposed in 1677 that light was a wave phenomenon (Huygens 1690); however, that theory only gained wider acceptance when the observation of interference could not be explained by the corpuscular theory.

The phenomenon of interference was demonstrated by experiments attributed toThomas Young (Young 1802), where light from a coherent source passes two parallel slits and then observed behind the slits on a screen (Fig. 8.1). With just one open slit, a projection of the slit geometry is produced on the screen by a smooth distribution of light, as would be expected for a stream of particles. However, when both slits are open, a fringe-like pattern appears on the screen. This varying intensity of light after it passed the slits is called interference and is a phenomenon that cannot be explained by light behaving as particles, but rather waves.

In 1871, James Clerk Maxwell formulated, based on ideas of Faraday and experiments by Hertz, the concept of light as an electromagnetic phenomenon, and light was then seen as a sole wave phenomenon.

Yet once again, these ideas were challenged by results obtained from black-body radiation. A black-body is a perfectly insulated enclosure that absorbs all incident electromagnetic radiation and only emits radiation through a hole made in its wall.

Fig. 8.1 Double-slit experiment to demonstrate interference

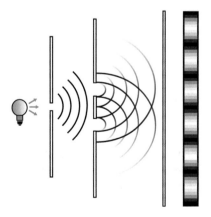

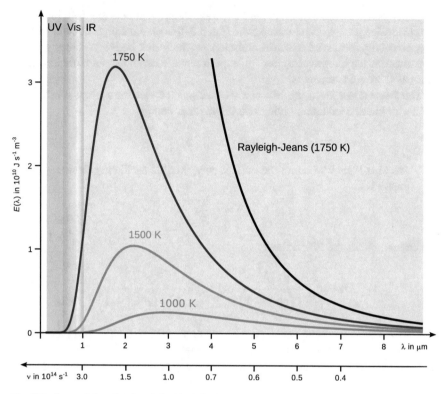

Fig. 8.2 Spectral flux density of the black-body

The emitted radiation is independent of the chemical nature of the body and only depends on its temperature; it is hence called thermal radiation. The higher the temperature of the black-body, the more radiation it emits at every wavelength (see Fig. 8.2). For example, at room temperature, it emits radiation that is mostly in the infrared and invisible region of the spectrum. At higher temperatures, the emission of infrared light increases; it can be felt as heat and the body glows visibly red. At even higher temperatures (e.g. stars like the sun with a temperature of 5780 K at the surface), the body appears bright yellow or blue-white and emits significant amounts of short wavelength radiation, including UV and even X-rays.

Rayleigh and Jeans proposed in 1900 that electrons at the surface of the black-body oscillate with a frequency ν, thus giving rise to electromagnetic radiation according to Maxwell's theory of light for which the spectral flux density $E(\nu)$ can be calculated:

$$E(\nu) = \frac{\nu^2}{c^2} \cdot k_B \cdot T \quad \text{or} \quad E(\lambda) = \frac{c}{\lambda^4} \cdot k_B \cdot T. \tag{8.5}$$

These expressions (known as the Rayleigh-Jeans law) describe the intensity of emitted radiation at a particular frequency ν or wavelength λ. This law predicts that the flux density increases with decreasing wavelength, and becomes infinite at very small wavelengths. As can be seen from Fig. 8.2, this predication is at odds with the experimentally observed spectral emission of the black-body. Whereas there is agreement at longer wavelengths, the experimental data shows an emission maximum at shorter wavelengths.

The issue arises from the assumption that surface electrons adopt a behaviour similar to linear oscillators, giving rise to the inner energy

$$U(e^-) = k_B \cdot T.$$

It was Max Planck who realised that the energy of the oscillating electrons should be expressed as

$$U(e^-) = \frac{h \cdot \nu}{e^{\frac{h \cdot \nu}{k_B \cdot T}} - 1}$$

which then yields the Planck law:

$$E(\nu) = \frac{h \cdot \nu^3}{c^2 \cdot \left(e^{\frac{h \cdot \nu}{k_B \cdot T}} - 1\right)} \text{ or } E(\lambda) = \frac{h \cdot c^2}{\lambda^5 \cdot \left(e^{\frac{h \cdot c}{\lambda \cdot k_B \cdot T}} - 1\right)}. \tag{8.6}$$

Planck's law correctly predicts the radiation emitted by the black-body and introduces a constant $h = 6.626176 \cdot 10^{-34}$ J s which is called the Planck constant and possesses a finite value. The Planck constant is at the core of quantum theory:

▶   Any action (= energy·time) of a natural event is an integer multiple of the quantum of action h. The smallest possible action is h.

This concept is part of a set of a distribution statistics for thermal equilibria. According to Planck, the energy is distributed among the atoms according to the temperature of the object. A few atoms have low energy, many have medium energy and a few have high energy. The large amount of atoms possessing medium energy increases as the temperature increases. Each atom can emit electromagnetic radiation. For very high frequencies ν, the energy needed to emit one quantum of energy is very large, and only a few atoms in the black-body have that much energy available, so only a few high-frequency quanta are radiated. It is much easier for atoms to emit low-energy (low frequency) radiation. In between the two extremes, however, there are many atoms that have enough energy to emit radiation of moderate energy. These add up to produce the peak in the emission curve of the black-body. This peak shifts to higher frequencies for bodies at higher temperature, since there are more individual atoms that possess higher energy.

8.1.3 The Photoelectric Effect

When light of sufficient energy is hitting onto a metal plate, electrons (now called photoelectrons) are released from the metal that possess a certain kinetic energy $\left(E_{kin} = \frac{1}{2} \cdot m \cdot v^2\right)$. This energy can be measured by clamping the metal plate against another electrode and apply an electric field that opposes the stream of photoelectrons. By varying the electric potential such that the stream (= current I) of photoelectrons reduces to zero (U_0), one can measure the maximum kinetic energy of the photoelectrons:

$$E_{max} = \frac{1}{2} \cdot m \cdot v_{max}^2 = e \cdot U_0.$$

Since the photoelectrons are released as a result of the incident light, the wave theory of light predicts that the kinetic energy of the photoelectrons should vary with the intensity of the incoming light wave whose impulse is being transferred onto the electrons. However, whereas more intense light causes a larger current of photoelectrons, the maximum kinetic energy of those electrons, measured via the potential U_0, remains constant.

If the energy (h·ν) of the incident light is increased by shining light of higher frequency ν (shorter wavelength λ) onto the metal electrode, the kinetic energy of the photoelectrons increases in a linear fashion (Fig. 8.3). If the metal electrode is changed to a different material, the same type of relationship is observed, but the line is shifted parallel along the x-axis. This relationship can be mathematically formulate as:

$$U_0 = \text{const.} \cdot (\nu - \nu_{cath})$$

and multiplied with the elementary charge e to obtain the same relationship in terms of energies:

$$e \cdot U_0 = (e \cdot \text{const.}) \cdot (\nu - \nu_{cath}). \tag{8.7}$$

The first factor on the right hand side of Eq. 8.7 is again a constant and turns out to be the Planck constant h. Re-arrangement of this equation then yields Einstein's frequency law (Einstein 1905) which earned him the Nobel prize in 1921:

$$h \cdot \nu = e \cdot U_0 + h \cdot \nu_{cath}. \tag{8.8}$$

These results from the photoelectron experiments do not agree with the wave theory of light. Instead, Eq. 8.8 suggests that light propagates in the form individual quanta that travel with the speed of light c. These light quanta are called photons which have an energy

$$E = h \cdot \nu \tag{8.9}$$

and can thus be likened to corpuscular radiation. The term h·ν_{cath} on the right hand side of Eq. 8.8 is dependent on the material of the photoelectrode and constitutes the work required to remove an electron from the solid matter.

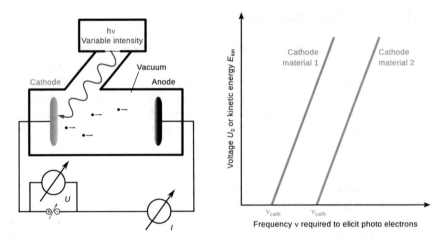

Fig. 8.3 The photoelectric effect. Left: Schematics of an apparatus to measure the kinetic energy of photoelectrons. Right: The higher the energy of the incident photons (i.e. the higher the frequency ν), the higher the kinetic energy (measured via the potential U_0 required to reduce the current to zero) of the photoelectrons emitted by the cathode. Different cathode materials show the same relationship, but the energy of the incoming photons required to elicit photoelectrons is different

8.1.4   The Dual Nature of Light

The photoeffect is proof that light, besides its wave behaviour, also possesses corpuscular characteristics. Emission of photoelectrons is only possible if the incoming light possess an energy larger than the work required to release an electron from the solid. The kinetic energy of photoelectrons cannot be changed by the intensity of the incoming light, but only by changing its energy. The intensity only affects the current (i.e. number) of released photoelectrons.

The interference experiments described in Sect. 8.1.3 (Fig. 8.1), can be modified such as to measure photons as they pass through the slits. These measurements show that each photon passes through one slit as it would be expected for a particle. Furthermore, the photons detected on the screen behind the slits are always detected at discrete points, demonstrating again behaviour of a single particle. Only the varying density of these hits (i.e. a distribution counted over many individual events) gives rise to the interference pattern (see also the discussion of electron diffraction in the next section).

8.1.5   The Dual Nature of the Electron

In Sect. 8.1.1, we discussed the corpuscular characteristics of the electron. If a thin layer of solid polycrystalline material is subjected to an incident electron beam (Fig. 8.4), then the corpuscular theory would predict a single point of incident on

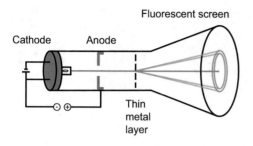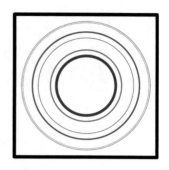

Fig. 8.4 Electrons accelerated in a cathode ray tube onto a thin metal layer show electron diffraction on the fluorescent screen

the detector. However, diffraction rings are observed instead, similar to the observation made in X-ray diffraction experiments (see Sect. 13.5.2).

The diffraction of electrons cannot be explained by the corpuscular theory as it is an interference phenomenon that requires waves. Just as light, the electron therefore also possesses properties of a particle and a wave.

In the cathode ray tube, it is possible to accelerate the electrons by varying the potential difference that gives rise to the electric field experienced by the electrons. Experimentally, it can be shown that the wavelength of the electrons is indirectly proportional to their momentum:

$$\lambda_{electron} \sim \frac{1}{p_{electron}} = \frac{1}{m_{electron} \cdot v_{electron}} \Rightarrow \lambda = \frac{const.}{m_{electron} \cdot v_{electron}}.$$

This important relationship is known as the DeBroglie relationship and links the corpuscular property (momentum) of the electron with its wave property (wavelength). The proportionality constant is thus of pivotal importance as it is the very factor that enables this conversion. It will not come as a big surprise that this constant is the Planck constant h, which renders the DeBroglie relationship as

$$\lambda = \frac{h}{m_{electron} \cdot v_{electron}}. \qquad (8.10)$$

Optical resolution and electron microscopes
The wave properties of electrons are the basis for the important practical application in electron microscopes. The resolving power of optical systems defines the ability of an imaging device to separate points that are located at a small distance. The minimum resolvable distance is often called (angular) resolution. Physically, this is determined by the Rayleigh criterion which

(continued)

states that two point-like light sources are just resolved, when the main diffraction maximum of one image coincides with the first minimum of the other. Any greater distance between the two point sources means they are well resolved; any smaller distance makes the two points non-resolvable. In terms of a circular aperture, this leads to the equation

$$\sin\theta = \frac{1.22 \cdot \lambda}{D}$$

where θ is the angular resolution, λ the wavelength of light used to image and D the diameter of the aperture of the lens system; the factor of 1.22 is derived from the Bessel function of first order first kind, divided by π ($J_1/\pi = 1.22$). It is obvious from above equation that the resolution (θ) will assume smaller values (i.e. "increase"), as the wavelength of the light used becomes smaller.

The human eyes are sensitive to wavelengths of light in the range of 750 nm to 400 nm; the maximum resolution is equivalent to an object size at the order of 0.1 mm. With the help of optics, the light microscope (operating in the same wavelength range) can achieve a resolution of about 200 nm. In electron microscopes, the wave behaviour of electrons yields radiation of a wavelength of about 0.1 Å and contemporary instruments achieve a resolution of about 1–2 Å.

The dualism raises the question of what individual objects, such as a single electron is at a particular point in time. Is it simultaneously a particle and wave? We therefore consider again an electron beam produced in an evacuated tube (Fig. 8.5).

If there is no solid material in the way of the electron beam, it can be focussed and appears as a single intense spot on the detector (Fig. 8.5, left panel). When the beam penetrates a thin sample of solid polycrystalline matter, a diffraction image appears on the detector that is reminiscent of those obtained with X-rays as mentioned above (Fig. 8.5 centre panel). If one now lowers the intensity of the electron beam, the intensity of the diffraction rings gradually decreases. However, when the intensity is lowered such that it is no longer a beam of electrons, but rather a stream of individual electrons hitting the solid sample, the diffraction rings disappear from the detector and only individual and independent spots of incidence are recorded (Fig. 8.5 right panel). When observed over a long time, the individual spots of incidence will compose the original diffraction image.

If the electrons indeed possessed wave properties throughout, the complete diffraction image in above experiment should have been visible at all times. These observations suggest, that in this type of experiment the wave properties only exist for statistics of many particles.

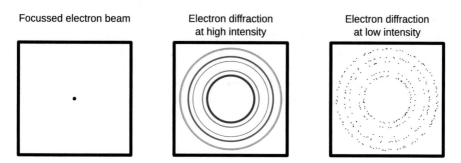

Fig. 8.5 Electron diffraction experiments. See main text for discussion

8.1.6   The Wave-Particle Dualism

In the previous sections, we came to appreciate that light as well as electrons possess characteristics of particles as well as waves. With the characteristic properties of waves being the wavelength λ, and the characteristic properties of moving particles being their momentum p, the pivotal relationship of the wave-particle dualism is the DeBroglie relationship:

$$\lambda = \frac{h}{p} = \frac{h}{m \cdot v}. \qquad (8.10)$$

For photons, the corpuscular energy is $E = m \cdot c^2$, so their momentum is $p = m \cdot c$. With Eq. 8.9 it then follows that

$$p = \frac{h \cdot v}{c}$$

and substitution in Eq. 8.10 then leads to the relationship between wavelength λ and frequency v of photons:

$$\lambda = \frac{h}{\frac{h \cdot v}{c}} = \frac{c}{v}. \qquad (8.11)$$

Whereas these discoveries were originally made with light and electrons, it has been shown by using atomic and molecular rays that the wave-particle dualism also exists for larger and thus heavier particles. From the DeBroglie relationship 8.10 it is obvious that the wavelengths of heavy particle rays have very small wavelengths (due to the high mass as compared to electrons or photons) and these small wavelengths are very difficult to detect. The largest entities for which the wave-particle dualism has been experimentally tested were molecules with some 800 atoms and a total mass of approx. 10,000 Da (Eibenberger et al. 2013).

In the previous sections, we discussed the interference of light and the diffraction of electrons. If we describe the beam of photons or electrons as a stream of particles, then a measure for the intensity of the beam is the number of particles passing a tiny volume slice per time. If we describe the beam as a wave characterised by an amplitude Ψ, then the intensity of the beam in a small volume element is proportional to the squared amplitude, Ψ^2.

In the case of the low intensity beam in above diffraction experiment (Fig. 8.5 right panel), we can not predict where an electron will be hitting the detector area. However, the probability will be high in those regions where a strong diffraction ring is located (known from the experiment that worked with the intense electron beam). Notably, the intensity of a diffraction ring is a function of the intensity (= squared amplitude) of the electron wave. The squared amplitude Ψ^2 therefore is a measure of the probability to find an electron at a particular location (in this case on the detector area).

In this particular experiment, we can determine with confidence the kinetic energy of the electrons forming the beam; it is given by the electric potential applied to accelerate the electrons in the tube. We thus know with great certainty what the momentum $p = m \cdot v$ and thus the wavelength λ of the electron beam is. However, we cannot be certain where exactly an electron will hit the detector, the exact three-dimensional coordinates (x, y, z) are unknown.

This observation illustrates Heisenberg's uncertainty relationship which states that

▶   The more precise the momentum of a particle is defined, the less
     certainty there is about the particle's location. And vice versa, the more
     precisely the location of a particle is known, the less certainty there is
     about its momentum.

Wave packets

For (heavy) particle waves, neither the wavelength λ, nor the frequency ν can be measured. The same is true for the phase velocity which is the rate at which the phase of the wave propagates in space (note: for light, the phase velocity is c and can be measured).

If waves of slightly differing wavelengths are superimposed, this yields a so-called wave packet (Fig. 8.6). In such packets, oscillations of the individual constituting waves amplify each other in some areas, and they cancel the amplitudes in other areas. The wave packet as whole travels through space with the group velocity.

The propagation of an electron beam—when considered as a wave—may thus be described as a travelling wave packet, where the areas of large amplitudes indicate a high probability of the location of an electron and the areas of low or zero amplitude indicate low or zero probability of finding an electron.

A wave packet that describes a moving particle requires a particular width Δx. If this width is very small, the location of the particle is well defined. However, in order to obtain a very thin wave packet, the superposition of many

(continued)

waves of varying wavelengths λ is required; the wavelength range $\Delta\lambda$ is large. In contrast, if only few waves with differing wavelengths are superimposed ($\Delta\lambda$ is small), the resulting wave packet is rather broad (Δx is large), therefore giving a considerable ambiguity as to the actual location of the particle. Earle Hesse Kennard then derived the inequality of the product of the two distributions Δx and $\Delta\lambda$:

$$\Delta x \cdot \frac{1}{\Delta\lambda} \geq \frac{1}{4\pi}$$

from which one derives with the DeBroglie relationship 8.10:

$$\Delta x \cdot \Delta p \geq \frac{h}{4\pi} \tag{8.12}$$

which summarises Heisenberg's uncertainty relationship in a numerical fashion.

Importantly, this quantum mechanical principle does not violate observations made with macroscopic objects. For heavy particles, which are the subject of classical mechanics, the mass m is orders of magnitudes larger than for (sub)atomic particles. If we replace the momentum p by the product of mass and velocity, we obtain:

$$\Delta x \cdot \Delta(m \cdot v) \geq \frac{h}{4\pi}$$

$$\Delta x \cdot \Delta v \geq \frac{h}{4\pi \cdot m}. \tag{8.13}$$

In formula 8.13, the right side of the inequality assumes values near zero in the case of heavy particles due to the large numerical value of their mass. This means that the uncertainty on the left hand side of may take very small values or indeed zero. In other words, there is virtually no uncertainty as to the location (Δx) and velocity (Δv) of the particle.

Since we have recognised that the wave-like behaviour of electrons and light can only be established through a distribution of many individual events, we can understand that it is not possible to track the path of an individual wave, for example behind the slits in the interference experiment in Fig. 8.1. Any screen or detector would only ever measure the absorption of a particle like-photon at a discrete point. The act of observation therefore affects the result obtained with the measurement. It is not possible to observe the particle and wave aspects of these beams simultaneously. This phenomenon is termed the complementarity principle.

Fig. 8.6 Illustration of a
wave packet

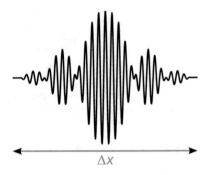

$$\Delta x$$

8.2   Properties of Atoms

8.2.1   Atomic Spectroscopy

Equipped with some knowledge about electrons and photons, we next want to
investigate their interactions with atoms in order to learn more about the inner makings
of atoms. Therefore, we again employ accelerated electrons, such as in a cathode ray
tube (Fig. 8.7), and probe their collisions with gas molecules. The cathode ray tube is
thus filled with mercury vapour such that the electrons can collide with Hg atoms in
the gas phase. A grid is placed between the cathode and the anode and a counter
voltage is applied between grid and anode to slow down the electrons.

With increasing acceleration voltage, an increased current (= stream of electrons)
is registered, as more and more electrons are being emitted by the cathode. The
collisions with Hg atoms thus need to be elastic, i.e. they appear without loss of
kinetic energy. However, once the acceleration voltage is larger than a threshold
voltage (here 4.9 V), the current drastically drops, indicating that a large number of
electrons no longer possesses the necessary energy to overcome the potential
difference between the grid and the anode. Apparently, the electrons lose their
energy in an inelastic collision with the Hg atoms. Once the voltage is increased
beyond the threshold voltage, the mercury vapour starts to emit light with a wavelength of 253.6 nm, which corresponds to an energy of 4.9 eV which is exactly the
energy lost by the electrons in their inelastic collisions. This energy appears to be
transferred onto the gas atoms, which then are in an excited state. They return from
this excited state in to the ground state by emitting this energy in form of photons.

This experiment by Franck and Hertz indicates that atoms only absorb and emit
energy in discrete portions, demonstrating the quantum nature of atoms (Franck and
Hertz 1914). Since the collisions between the accelerated electrons and atoms indeed
happen by collisions between the accelerated electrons and the electrons in the
atomic sphere, one can conclude that indeed the atomic electrons exists in discrete
energy states.

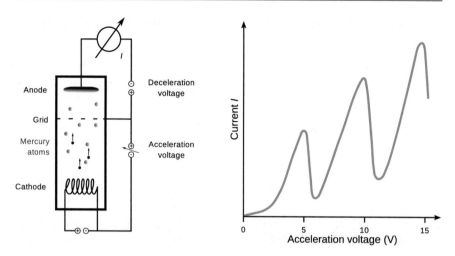

Fig. 8.7 Illustration of the experiment by Franck and Hertz. Left: Electrons emitted by the cathode are accelerated towards a grid by a varying acceleration potential and eventually collide with mercury atoms in the gas phase (magenta). After being slowed down, they hit the anode and a current is registered. Right: Electron current in dependence of the acceleration potential. The sharp drop of current on the right side of each peak indicates a loss of energy of the electrons. This energy is absorbed by the mercury atoms, which transit into an excited state

Atoms can not only be excited by collisions with electrons; if sufficient thermal energy is provided, this may also lead to formation of excited atomic states. A simple experiment in this context is the observation of excited sodium atoms in the flame of a Bunsen burner or a plasma torch (Fig. 8.8 left). Sodium-containing salts introduced into the flame of a Bunsen burner result in a yellow colour of the flame. As the excited sodium atoms return to the ground state, they emit yellow light which can visually be observed. More generally, it becomes apparent that atoms only emit light of discrete wavelengths which is why atomic spectra are called line spectra.

A similar experimental setup can be used to investigate absorption of light by atoms (Fig. 8.8 right). Using a light source that provides light in the entire visual spectral range, a continuous spectrum is observed when that light is refracted by a wavelength discriminator. If a Bunsen burner or plasma torch is placed between the light source and the wavelength discriminator, the thermal energy is used to vapourise individual atoms from salts introduced into the plasma or flame. Individual atoms in their ground states will absorb light of particular wavelengths coming from the light source and thus transition into excited states. These particular wavelengths are missing from the light that reaches the detector and appear as dark lines in the observed spectrum.

These concepts form the basis of atomic absorption (AAS) and atomic emission spectroscopy (AES). Both methodologies have been in use in analytical chemistry laboratories; however, atomic emission spectrometers are most frequently employed in contemporary laboratories, typically with an inductively coupled plasma as source of thermal energy (ICP-AES).

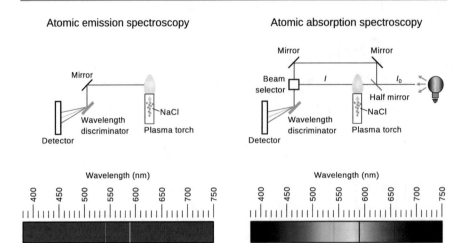

Fig. 8.8 Comparison of atomic emission (AES) and absorption spectroscopy (AAS). In AES, the emission of light from atoms in their excited states is measured and yields lines at discrete wavelengths observed against a dark background. In contrast, atoms may exist in their ground states and an external light source can be used to cause transition to an excited state. As this energy is then missing from the transmitted light, discrete wavelengths appear as absorption (black) lines against the continuous spectrum. The spectra shown illustrate the sodium line spectra

For illustration, we will consider the line spectrum of atomic hydrogen in more detail. Johann Balmer found empirically in 1885 that the spectral lines of atomic hydrogen can be classified into series which follow a mathematical relationship further developed by Johannes Rydberg and Walter Ritz:

$$\frac{1}{\lambda} = R_\infty \cdot \left(\frac{1}{n_0^2} - \frac{1}{n^2} \right). \tag{8.14}$$

Here n and n_0 are integer numbers and n is always larger than n_0. For a particular series of lines, n_0 is a constant. The constant R_∞ is known as the Rydberg constant and assumes a value of $R_\infty = 1.097 \cdot 10^7 \text{ m}^{-1}$ for hydrogen. The reciprocal wavelength is known as the wavenumber $\tilde{\nu}$, and from Eq. 8.11, the following relationship can be derived:

$$\frac{1}{\lambda} = \frac{\nu}{c} = \tilde{\nu}. \tag{8.15}$$

In Eq. 8.14, the expressions $T_n = -\dfrac{R_\infty}{n^2}$ and $T_{n_0} = -\dfrac{R_\infty}{n_0^2}$ are called atomic terms. For varying numbers n one thus obtains (Table 8.1):

Table 8.1 Spectroscopic terms

n	T_n
1	$T_1 = -R_\infty$
2	$T_2 = -\frac{R_\infty}{4}$
3	$T_3 = -\frac{R_\infty}{9}$
4	$T_4 = -\frac{R_\infty}{16}$
∞	$T_\infty = 0$

Table 8.2 Spectral series of atomic hydrogen

Lyman series	$n_0 = 1$	$\tilde{v} = T_n - T_1$	$n = 2, 3, \ldots$
Balmer series	$n_0 = 2$	$\tilde{v} = T_n - T_2$	$n = 3, 4, \ldots$
Paschen series	$n_0 = 3$	$\tilde{v} = T_n - T_3$	$n = 4, 5, \ldots$
Bracket series	$n_0 = 4$	$\tilde{v} = T_n - T_4$	$n = 5, 6, \ldots$
Pfund series	$n_0 = 5$	$\tilde{v} = T_n - T_5$	$n = 6, \ldots$

Equation 8.14 can thus be expressed as:

$$\tilde{v} = T_n - T_{n_0} \tag{8.16}$$

The different series of spectral lines observed for atomic hydrogen are then obtained from Eq. 8.16 and given particular names as summarised in Table 8.2. Importantly, the wavenumbers $\tilde{v}$ of the individual lines can now be obtained by the difference of two atomic terms. Graphically, this can be visualised in a so-called term scheme (Fig. 8.9) whereby every term is shown as a horizontal line. The individual terms are ordered vertically with the ascending running number n. In such a term scheme, a specific spectral line then appears as a vertical difference line between two particular terms.

The interpretation of the term scheme was a breakthrough by Niels Bohr in 1913. If Eq. 8.16 is multiplied with (h·c), one obtains:

$$h \cdot c \cdot \tilde{v} = h \cdot v = h \cdot c \cdot T_n - h \cdot c \cdot T_{n_0} = \frac{h \cdot c \cdot R_\infty}{n_0^2} - \frac{h \cdot c \cdot R_\infty}{n^2}.$$

Since the left side of this equation describes an energy, the same must be true for the right hand side:

$$h \cdot v = E_{\text{end}} - E_{\text{start}} = \frac{h \cdot c \cdot R_\infty}{n_0^2} - \frac{h \cdot c \cdot R_\infty}{n^2}$$

$$= h \cdot c \cdot R_\infty \cdot \left(\frac{1}{n_0^2} - \frac{1}{n^2} \right), \text{with } n > n_0. \tag{8.17}$$

The energies E_{end} and E_{start} are correlated with energy levels of the atom. The emitted light observed as the spectral line in the atomic emission spectrum—or

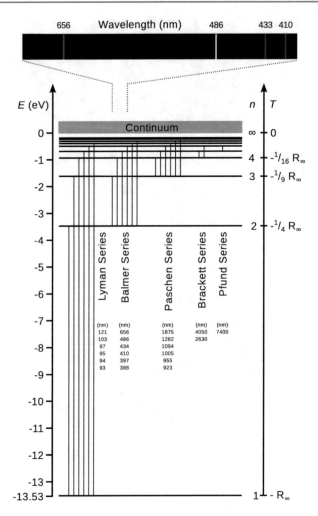

Fig. 8.9 The term scheme of atomic hydrogen

similarly, the absorbed light observed as line in the atomic absorption spectrum—is thus the difference of two particular energy levels. Absorption of light of a particular wavelength therefore causes transition of an atom from a lower to a higher energy state. Vice versa, atoms can return from a high to a low energy state by emission of light of a particular wavelength.

A closer look at the wavelengths of the spectral lines in Fig. 8.9 shows that within each of the series, their values are closer together the shorter the wavelengths (i.e. the larger the energy of the absorbed/emitted light) become. For each series, there is thus a shortest wavelength called the series limit. If the energy of light used in an absorption experiment is larger than the energy of the series limit, the incident photon carries so much energy that it ionises the atom when absorbed. Atomic spectroscopy can thus be used to determine the ionisation energy of atoms.

8.2.2 The Hydrogen Model by Bohr

According to Rutherford, atoms should consist of a positively charged nucleus around which electrons orbit in relatively large distances. This idea of atoms having a similar constellation like a solar system was further developed by Bohr. In contrast to planets orbiting a sun based on a gravitational attractive force, the orbiting electrons should be attracted to the nucleus by an electrostatic (i.e. the Coulomb) force. However, according to the macroscopic laws, an orbiting electron should thus be able to assume any energy and orbiting radius, depending on its velocity. A loss of energy would eventually lead to the electron collapsing onto the nucleus. Bohr therefore postulated that the quantisation discovered by Planck should also be applied to the electrons orbiting an atomic nucleus. By combining this quantum mechanical principle with the planet model, individual states of energies for the orbiting electrons are obtained:

$$E_n = -\frac{m_e \cdot e^4}{8 \cdot \varepsilon_0^2 \cdot n^2 \cdot h^2} \tag{8.18}$$

where $\varepsilon_0 = 8.8542 \cdot 10^{-12}$ A^2 s^4 m^{-3} kg^{-1} is the vacuum permittivity. These energy levels depend on particular states characterised by the principal quantum number n:

$$E_n = -\frac{m_e \cdot e^4}{8 \cdot \varepsilon_0^2 \cdot h^2} \cdot \frac{1}{n^2} = -E_A \cdot \frac{1}{n^2}. \tag{8.19}$$

The jump from a higher (n_2) to a lower (n_1) energy state therefore requires an energy change in the system which is given by:

$$\Delta E = E_2 - E_1 = -E_A \cdot \left(\frac{1}{n_2^2} - \frac{1}{n_1^2}\right) = E_A \cdot \left(\frac{1}{n_1^2} - \frac{1}{n_2^2}\right), \text{whereby } n_2 > n_1.$$

With $\Delta E = h \cdot \nu$, this yields:

$$h \cdot \nu = E_A \cdot \left(\frac{1}{n_1^2} - \frac{1}{n_2^2}\right). \tag{8.20}$$

Comparison with Eq. 8.17 shows that the above equation for atomic energy states based on the principal quantum number n is identical with the equation derived for the atomic line spectra. E_A is the lowest possible energy of the atom, i.e. the energy of the ground state. It corresponds to the energy of the electron which is orbiting closest to the nucleus ($n = 1$; see Eq. 8.19). By comparison with Eq. 8.17 it transpires that this energy is linked to the Rydberg constant:

$$R_\infty = \frac{E_A}{h \cdot c}.$$

In case of atomic hydrogen, E_A is the ionisation energy as it represents the energy that one has to provide to remove the electron from the atom.

Whereas the model developed by Bohr successfully describes spectroscopic observations for single electron species such as atomic hydrogen and similar ions (He^+, Li^{2+}), it fails to explain spectra of multi-electron species (He, Li^+, ...).

The orbiting radius of an electron in the atomic model by Bohr

The planet model by Bohr uses the macroscopic laws of mechanics and electrostatics. An object moving around a centre is subject to two forces, the centripetal force F_{cp} and the centrifugal force F_{cf}. The centripetal force in Bohr's model is provided by the electrostatic attraction between the positively charged nucleus and the negatively charged electron:

$F_{cp} = \frac{1}{4\pi\cdot\varepsilon_0} \cdot \frac{e^2}{r^2}$.

The centrifugal force is a function of the angular velocity ω:

$F_{cf} = p \cdot \omega = m_e \cdot \omega^2 \cdot r$, where the momentum is $p = m_e \cdot \omega \cdot r$.

For a stable circular orbit, this yields the requirement:

$$F_{cp} = F_{cf} = \frac{1}{4\pi \cdot \varepsilon_0} \cdot \frac{e^2}{r^2} = m_e \cdot \omega^2 \cdot r. \tag{8.21}$$

In order to achieve discrete energy values, Bohr postulated that the angular momentum L is only allowed discrete values, which would then need to be multiples of the quantum of action h:

$$L = p \cdot r = m_e \cdot \omega \cdot r^2 = n \cdot \frac{h}{2\pi}. \tag{8.22}$$

Combining Eqs. 8.21 and 8.22, this yields:

$$\frac{1}{4\pi \cdot \varepsilon_0} \cdot \frac{e^2}{r^2} = \left[n \cdot \frac{h}{2\pi}\right]^2 \cdot \frac{1}{m_e \cdot r^3}$$

$$\frac{1}{4\pi \cdot \varepsilon_0} \cdot e^2 \cdot r = n^2 \cdot \frac{h^2}{4\pi^2} \cdot \frac{1}{m_e}.$$

This delivers an expression for the radius of the orbiting electron:

(continued)

$$r = n^2 \cdot \frac{h^2}{4\pi^2} \cdot \frac{1}{m_e} \cdot \frac{4\pi \cdot \varepsilon_0}{e^2} = \frac{n^2 \cdot h^2 \cdot \varepsilon_0}{\pi \cdot m_e \cdot e^2}. \qquad (8.23)$$

For the electron in the atomic hydrogen, the orbiting radius would thus be:

$$r = \frac{n^2 \cdot h^2 \cdot \varepsilon_0}{\pi \cdot m_e \cdot e^2} = \frac{1 \cdot \left(6.626 \cdot 10^{-34}\right)^2 J^2 \, s^2 \cdot 8.854 \cdot 10^{-12} \, A^2 \, s^4 \, m^{-3} \, kg^{-1}}{\pi \cdot 9.110 \cdot 10^{-31} \, kg \cdot \left(1.602 \cdot 10^{-19}\right)^2 \, C^2}$$

$$r = \frac{1 \cdot 6.626^2 \cdot 8.854 \cdot 10^{-68} \cdot 10^{-12} \, kg^2 \, m^4 \, s^{-4} \, s^2 \, A^2 \, s^4 \, m^{-3} \, kg^{-1}}{\pi \cdot 9.110 \cdot 1.602^2 \cdot 10^{-31} \cdot 10^{-38} \, kg \cdot A^2 \, s^2}$$

$$r = \frac{1 \cdot 6.626^2 \cdot 8.854 \cdot 10^{-80} \, m}{\pi \cdot 9.110 \cdot 1.602^2 \cdot 10^{-69}} = \frac{1 \cdot 6.626^2 \cdot 8.854}{\pi \cdot 9.110 \cdot 1.602^2} \cdot 10^{-11} \, m$$

$$r = 5.295 \cdot 10^{-11} \, m = 0.53 \, \text{Å}$$

8.3 Introduction to Quantum Mechanics

8.3.1 The Schrödinger Equation

Whereas the planet-like model chosen by Bohr can easily be visualised and is thus a fairly accessible model, the postulates required to achieve agreement with experimental observations are not immediately accessible. The general problem with this approach is the direct application of macroscopic laws to processes at the atomic level.

A different concept was suggested by Erwin Schrödinger based on the dualism of wave and matter, giving rise to the so-called wave mechanics. In that context, we have previously introduced the wave function Ψ which describes matter as a wave and as such does not require an individual point in space and time for characterisation. It has also become clear that the squared amplitude, Ψ^2, is a measure of the probability to find a particle in a volume element of space. In order to learn about states in atoms, it will be sufficient to analyse the wave function Ψ for its properties in various locations, i.e. $\Psi(x, y, z)$. If we are dealing with processes such as radiation, however, the time time-dependence of the wave function will also need to be considered, i.e. $\Psi(x, y, z, t)$.

For instances independent of time, Schrödinger suggested the following equation for the wave function $\Psi(x, y, z)$ in three dimensions

Fig. 8.10 Illustration of a
one-dimensional
standing wave

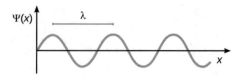

$$\Delta\Psi + \frac{8\pi^2 \cdot m}{h^2} \cdot \left(E - E_{pot}\right) \cdot \Psi = 0 \tag{8.24}$$

Notably, this equation cannot be proven from first principles. Very much like the laws of thermodynamics, the equation is a description of naturally occurring phenomena, whose 'proof' is the fact that it correctly describes these phenomena.

In Eq. 8.24, 'Δ' does not indicate a difference, but rather the Laplace operator which describes the second derivative:

$$\Delta = \frac{\delta^2}{\delta x^2} + \frac{\delta^2}{\delta y^2} + \frac{\delta^2}{\delta z^2}. \tag{8.25}$$

For simplicity, we will reduce further discussion of this problem to just one dimension (x). Visually, this suggests that we are dealing with a standing wave (Fig. 8.10). The mathematical expression for such a wave (cf. harmonic oscillator) is:

$$\Psi(x) = e^{\frac{2\pi \cdot i \cdot x}{\lambda}} = \left(\cos \frac{2\pi \cdot x}{\lambda} + i \cdot \sin \frac{2\pi \cdot x}{\lambda} \right), \tag{8.26}$$

where x is the coordinate in space (one dimension), λ is the wavelength and i is defined as per $i^2 = -1$. Calculation of the first and second derivative of the expression for the one-dimensional wave $\Psi(x)$ yields:

$$\frac{d\Psi(x)}{dx} = \frac{2\pi \cdot i}{\lambda} \cdot e^{\frac{2\pi \cdot i \cdot x}{\lambda}} \text{ and}$$

$$\frac{d^2\Psi(x)}{dx^2} = -\left(\frac{2\pi}{\lambda}\right)^2 \cdot e^{\frac{2\pi \cdot i \cdot x}{\lambda}}. \tag{8.27}$$

The expression for the second derivative can be re-arranged to read:

$$\frac{d^2\Psi(x)}{dx^2} + \left(\frac{2\pi}{\lambda}\right)^2 \cdot e^{\frac{2\pi \cdot i \cdot x}{\lambda}} = 0$$

$$\frac{d^2\Psi(x)}{dx^2} + \left(\frac{2\pi}{\lambda}\right)^2 \cdot \Psi(x) = 0. \tag{8.28}$$

From the DeBroglie relationship (Eq. 8.10), we can derive that

$$v = \frac{h}{m \cdot \lambda}$$

and then can obtain an expression for the kinetic energy E_{kin} that can be resolved for λ^2:

$$E_{kin} = \frac{1}{2} \cdot m \cdot v^2 = \frac{1}{2} \cdot m \cdot \left(\frac{h}{m \cdot \lambda}\right)^2 \Rightarrow \lambda^2 = \frac{h^2}{2 \cdot m \cdot E_{kin}},$$

and the above expression can be substituted in the wave Eq. 8.28:

$$\frac{d^2\Psi(x)}{dx^2} + \frac{8\pi^2 \cdot m}{h^2} \cdot E_{kin} \cdot \Psi(x) = 0.$$

Since the total energy of a particle is the sum of its kinetic and potential energy ($E = E_{kin} + E_{pot}$), we obtain:

$$\frac{d^2\Psi(x)}{dx^2} + \frac{8\pi^2 \cdot m}{h^2} \cdot (E - E_{pot}) \cdot \Psi(x) = 0, \tag{8.29}$$

and find that Eq. 8.29 is the one-dimensional form of the Schrödinger Eq. 8.24. Importantly, this agreement itself is not proof that the Schrödinger equation is correct.

Conceptually, the use of this equation for quantum mechanical problems assumes a model whereby particles with a mass m can be described as waves of matter. This raises the question of what exactly is propagating through space in a wave of matter. This question was answered by Max Born in 1928 who interpreted waves of matter as probability waves. Intriguingly, quantum mechanics therefore is intrinsically probabilistic. Whereas in classical mechanics, one can assign precise coordinates to a particle (limited only by the instrumentation used for measurement), quantum mechanics only allows assignment of probabilities to find a particle in one volume element or another. The Schrödinger equation acts as the link between both of these 'worlds'. The particles possess a potential energy E_{pot} and a mass m both of which can be determined using physical laws of the macroscopic world. Based on these numerical values the Schrödinger equation delivers the wave function Ψ and the total energy E which describe the quantum mechanical behaviour of the particle.

8.3.2   Basic Properties of Wave Functions

We will find later that wave functions can be determined with exception of a constant factor (Sect. 9.1.1, Eq. 9.3). However, since we concluded that the squared amplitude of the wave function is a measure of the probability to find the particle in a volume element dV, then the total probability to find the particle somewhere has to equal one. Mathematically, the total probability corresponds to an integration of the function $|\Psi|^2$ over the entire volume V:

$$\int_0^V |\Psi|^2 dV = 1. \tag{8.30}$$

The constant factor will thus have to be set such, that Eq. 8.30 is adhered to. In that case, the wave function Ψ is called a normalised wave function. A close look at the above equation shows that $|\Psi|^2$ carries the units of probability per volume and therefore constitutes a probability density.

The reason we have now introduced $|\Psi|^2$ instead of Ψ^2 for the probability density is that wave functions will involve complex numbers (as opposed to real numbers), in which the case the square amplitude is computed as the product of the amplitude and its complex conjugated:

$$\Psi \cdot \Psi^* = |\Psi|^2 \tag{8.31}$$

Complex numbers consist of a real ($\mathfrak{R}$) and an imaginary ($\mathfrak{J}$) part: $C = \mathfrak{R} + i \cdot \mathfrak{J}$ The complex conjugate of C is defined as $C^* = \mathfrak{R} - i \cdot \mathfrak{J}$. It is thus obvious that the product of C and C^* yields the sum of the squared real and imaginary parts:

$$C \cdot C^* = |C|^2 = (\mathfrak{R} + i \cdot \mathfrak{J}) \cdot (\mathfrak{R} - i \cdot \mathfrak{J}) = \mathfrak{R}^2 + i \cdot \mathfrak{J} \cdot \mathfrak{R} - i \cdot \mathfrak{J} \cdot \mathfrak{R} - i^2 \mathfrak{J}^2$$
$$= \mathfrak{R}^2 + \mathfrak{J}^2$$

If the imaginary part is zero ($\mathfrak{J} = 0$) then C is a real number and the square operation reduces to the case well-known for real numbers:

$$C \cdot C^* = \mathfrak{R}^2, \text{if } \mathfrak{J} = 0.$$

Above, we came to appreciate that in order for $|\Psi|^2$ to assume a physical meaning, we require the wave function Ψ to adhere to Eq. 8.30. Therefore, the function Ψ has to fulfil the following pre-requisites:

- Ψ needs to be a continuous function and we need to be able to determine its derivative. This requires that the first and second derivative of Ψ are also continuous functions. In other words, Ψ must not have any kinks are jumps.
- Ψ has to be unambiguous; for a particular set of values of the independent variables (x, y, z, t), there has to be only one value of Ψ.
- Ψ needs to assume finite values throughout and approach values of zero when the spatial variables x, y and z become infinite.

It is frequently necessary to calculate properties of a particle or system that are not immediately obtained by solving the Schrödinger equation. Obvious examples are the potential energy E_{pot} or the momentum p of a particle that are not directly available from the Schrödinger equation which only allows direct calculation of the total energy E.

In order to outline the general procedure for such cases, we will, for simplicity, again consider a one-dimensional wave function $\Psi(x)$. The probability density function $\rho(x)$ is given based on Eq. 8.31:

$$\rho(x) = |\Psi(x)|^2 = \Psi^*(x) \cdot \Psi(x), \text{assuming that } \Psi(x) \text{ is normalised.} \tag{8.32}$$

The probability to find the particle in the interval x and $x + dx$ is then available via the integral $\int_x^{x+dx} \rho(x)dx$. Therefore, if we are interested in the average value of the potential energy E_{pot}, which itself is a function of the location x, the function $E_{pot}(x)$ needs to be multiplied with the probability $\rho(x)$ to find the particle at each location x, and then integrate over all location values x. In this context (of probability theory), the average value of a quantity (e.g. $\bar{E}_{pot}$) is called the expected value of this quantity, denoted as $\langle E_{pot} \rangle$:

$$\langle E_{pot} \rangle = \int_{-\infty}^{+\infty} E_{pot}(x) \cdot \rho(x)dx = \int_{-\infty}^{+\infty} \Psi^*(x) \cdot E_{pot}(x) \cdot \Psi(x)dx. \tag{8.33}$$

8.4 Exercises

1. Calculate the wavelength of an electron that is accelerated by a potential difference of 10.0 kV.

2. The atomic model suggested by Niels Bohr in 1913 depicts atoms as systems very much like a solar system, where electrons travel in circular orbits around the positively charged nucleus. If one wanted to determine the location of an electron at a particular point in time with a certainty of ± 0.05 Å, what is the uncertainty with respect to the speed of the electron?

3. The wavelength of macroscopic objects: what is the wavelength of a person of 65 kg walking at a speed of 0.8 m s^{-1}?

4. Two consecutive lines in the atomic spectrum of hydrogen have the wavenumbers $\tilde{v}_i = 2.057 \cdot 10^6$ m^{-1} and $\tilde{v}_{i+1} = 2.304 \cdot 10^6$ m^{-1}. Calculate to which series these two transitions belong and which transitions they describe.

5. Assuming that (a) the sun ($T = 5780$ K) and (b) the earth ($T = 298$ K) behave as black-bodies, calculate and plot the spectral flux densities for the sun and the earth. What are the similarities and differences between both radiation curves? Use Planck's law to calculate $E(\lambda)$ for the wavelength range 100 nm–8 µm. Calculation and plotting might be best done with a spreadsheet software.

Quantum Mechanics of Simple Systems

9

9.1 The Particle in a Box

9.1.1 The Free Particle

As a first application of the Schrödinger equation, we will characterise a free particle that exists in space without any potential ($E_{pot} = 0$). For convenience, we will also consider just one dimension. Given these conditions, the Schrödinger equation 8.28 becomes

$$\frac{d^2\Psi(x)}{dx^2} + \frac{8\pi^2 \cdot m}{h^2} \cdot E \cdot \Psi(x) = 0$$

For simplicity, we introduced the parameter k to substitute the factor containing the mass of the particle (m), its energy (E) and the constants:

$$k = \frac{2\pi}{h} \cdot \sqrt{2 \cdot m \cdot E}, \tag{9.1}$$

which yields

$$\frac{d^2\Psi(x)}{dx^2} + k^2 \cdot \Psi(x) = 0. \tag{9.2}$$

If we assume that the wave function is $\Psi(x) = e^{b \cdot x}$, then the second derivative of $\Psi(x)$ is $d^2\Psi(x) = b^2 \cdot e^{b \cdot x}$. Substitution of those values in Eq. 9.2 yields:

$$b^2 \cdot e^{b \cdot x} + k^2 \cdot e^{b \cdot x} = 0$$

Since the function $e^{b \cdot x}$ never assumes the value of zero, the above equation is only true if:

© Springer International Publishing AG, part of Springer Nature 2018
A. Hofmann, *Physical Chemistry Essentials*,
https://doi.org/10.1007/978-3-319-74167-3_9

$$b^2 + k^2 = 0.$$

For this quadratic equation, two solutions are possible. We also remember that b might be a complex number, and therefore obtain:

$$b = \pm i \cdot k.$$

This results in two solutions for the wave function:

$$\Psi_1(x) = e^{i \cdot k \cdot x} = \cos(k \cdot x) + i \cdot \sin(k \cdot x) \text{ and}$$
$$\Psi_2(x) = e^{-i \cdot k \cdot x} = \cos(k \cdot x) - i \cdot \sin(k \cdot x)$$

Using these two partial solutions, we one can solve the differential Eq. 9.2. Without performing this mathematical exercise in a rigorous fashion, we appreciate that the general solution for a wave function fulfilling Eq. 9.2 is:

$$\begin{aligned}
&\Psi(x) = A \cdot \Psi_1(x) + B \cdot \Psi_2(x), \text{which is equivalent to} \\
&\Psi(x) = A \cdot e^{i \cdot k \cdot x} + B \cdot e^{-i \cdot k \cdot x}, \text{or} \\
&\Psi(x) = A \cdot \sin(k \cdot x) + B \cdot \cos(k \cdot x), \text{or} \\
&\Psi(x) = C \cdot \sin(k \cdot x + \delta)
\end{aligned} \tag{9.3}$$

where A, B, C and δ are constants resulting from the integration required when solving the differential Eq. 9.2.

It can be shown that for $B = 0$, the particle possesses a momentum that is positive with respect to the x-axis, i.e. it moves along the direction of the positive x-axis branch. In contrast, for $A = 0$, the momentum is negative, indicating that the particle moves in the direction of the negative x-axis branch. Inspection of the general solution of the wave function, e.g. in the form of

$$\Psi(x) = A \cdot \sin(k \cdot x) + B \cdot \cos(k \cdot x) \tag{9.3}$$

shows that Ψ is the superposition of two sine-like waves. The two waves are running in opposite directions and, by interference, result in a standing wave.

In the above discussion, it has not been necessary to demand any particular values for the parameter k. Upon re-arrangement of Eq. 9.1, we obtain the energy E of the particle:

$$E = \frac{h^2 \cdot k^2}{8\pi^2 \cdot m} = E_{kin}$$

which constitutes the kinetic energy of this particle, since we set out with the requirement that the potential energy was zero. Since k can assume any value, the kinetic energy of the free particle can assume any value—it exists in a continuum.

9.1.2 The Particle in a One-dimensional Box with Infinite Potential Walls

In the previous section, we introduced the general solution of the Schrödinger equation and derived a wave function that describes a freely moving particle of the kinetic energy E_{kin} and no potential energy ($E_{pot} = 0$).

We now want to consider the case that the particle is enclosed in a box of width a in which it can move around freely (no potential energy), but from which it cannot escape. This can be realised by imposing a steeply rising potential at the walls of the box (Fig. 9.1). The potential at the walls of the boxes will be set infinitely high. Therefore, three regions need to be considered:

$$\begin{array}{lll} \text{Region 1}: & x \leq 0 & E_{pot} = \infty \\ \text{Region 2}: & 0 \leq x \leq a & E_{pot} = 0 \\ \text{Region 3}: & x \geq a & E_{pot} = \infty \end{array}$$

For regions 1 and 3, the Schrödinger equation is:

$$\frac{d^2 \Psi(x)}{dx^2} + \frac{8\pi^2 \cdot m}{h^2} \cdot (E - \infty) \cdot \Psi(x) = 0$$

This equation can only be adhered to, if the wave function is $\Psi(x) = 0$. Notably, this results in a probability density of $\rho(x) = 0$, which means that the particle does not exist in those regions:

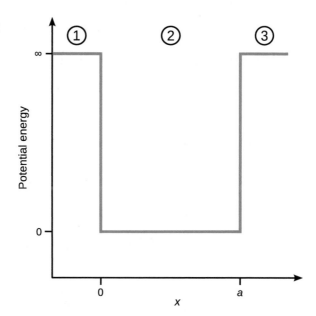

Fig. 9.1 The potential well for the particle in a box

$$\rho(x) = |\Psi(x)|^2 = 0.$$

In region II, the particle can move without restriction, since $E_{pot} = 0$. This is the same case we have discussed in the previous section. The Schrödinger equation is thus:

$$\frac{d^2\Psi(x)}{dx^2} + k^2 \cdot \Psi(x) = 0, \text{with } k = \frac{2\pi}{h} \cdot \sqrt{2 \cdot m \cdot E}, \qquad (9.2)$$

and we obtain the general solution

$$\Psi(x) = A \cdot \sin(k \cdot x) + B \cdot \cos(k \cdot x). \qquad (9.3)$$

It is important to remember now that a general prerequisite for any wave function Ψ is that it is a continuous function (Sect. 8.3.2). Due to the high potential at the walls of the box, which made us conclude for $\Psi(x) = 0$ in regions I and III, this means that:

$$\Psi(0) = 0 \text{ and } \Psi(a) = 0$$

The condition of $\Psi(x) = 0$ can only be met, if in Eq. 9.3 the cosine term disappears, i.e. $B = 0$. The second condition, $\Psi(a) = 0$, requires then that

$$\Psi(a) = A \cdot \sin(k \cdot a) = 0.$$

Since a sine function periodically assumes values of zero when the argument is an integer multiple of π (Fig. 9.2), the product $k \cdot a$ needs to fulfil the requirement:

$$k \cdot a = n \cdot \pi, \text{with } n = 0, 1, 2, 3, \ldots$$

which means that k can only take particular values:

$$k = \frac{n \cdot \pi}{a}, \text{with } n = 0, 1, 2, 3, \ldots \qquad (9.4)$$

The allowed solutions for the Schrödinger Eq. 9.2 thus are:

Fig. 9.2 The function sin x periodically assumes a value of zero

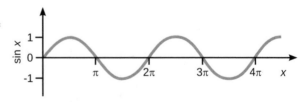

$$\Psi_n(x) = A \cdot \sin\left(\frac{n \cdot \pi}{a} \cdot x\right), \text{ with } n = 1, 2, 3, \ldots$$

Note that at this stage, only solutions with $n \neq 0$ remain possible, since a wave function of $\Psi_0(x) = 0$ would not be a sensible solution; the particle has to be in the box. The value of A is chosen such that the wave function is normalised, and this process yields:

$$\Psi_n(x) = \sqrt{\frac{2}{a}} \cdot \sin\left(\frac{n \cdot \pi}{a} \cdot x\right), \text{ with } n = 1, 2, 3, \ldots \tag{9.5}$$

Importantly, the energy values that the particle in the box can assume are no longer continuous, but subject to the condition 9.4:

$$E_n = \frac{h^2 \cdot k^2}{8\pi^2 \cdot m} = \frac{h^2 \cdot n^2 \cdot \pi^2}{8\pi^2 \cdot m \cdot a^2} = \frac{h^2}{8 \cdot m \cdot a^2} \cdot n^2, \text{ with } n = 1, 2, 3, \ldots \tag{9.6}$$

So whereas the freely moving particle can assume any energy value, the particle confined in a potential well (box) is only allowed to assume discrete energy values (Eq. 9.6) which are a function of the principal quantum number n. The allowed discrete energy values are called eigenvalue of the Schrödinger equation

Figure 9.3 summarises the allowed energy levels, the wave functions and probability densities for the first four quantum numbers n. It becomes obvious that the probability of finding the particle in various areas within the box is not the same at all locations x, and it also is a function of the quantum number n. The graphs of the wave functions in Fig. 9.3 illustrate that the number of knots (where the wave function assumes a value of zero) equals $(n-1)$; note that the knots at the walls are not considered.

Lastly, Eq. 9.6 shows, that any particle in a potential well has a minimum energy and cannot assume a state of zero energy. Since the lowest value for the quantum number is $n = 1$, the minimum energy is:

$$E_1 = \frac{h^2}{8 \cdot m \cdot a^2}.$$

9.1.3 The Particle in a One-dimensional Box with Finite Potential Walls

In the previous section, the potential well was constructed with infinitely high potential such that it was impossible for the particle to leave, regardless of how much energy would be provided to the system. here, we will consider the more realistic case of potential walls with finite heights. Within the well, the particle of mass m shall not be exposed to any potential ($E_{pot} = 0$), outside the well, the particle

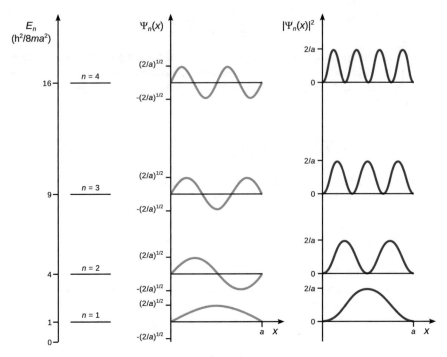

Fig. 9.3 Allowed energy levels E_n, wave functions Ψ_n and probability densities $|\Psi_n|^2$ for the particle in a linear potential well

shall be exposed to the potential energy V_0. As before, we need to distinguish three regions:

$$
\begin{array}{lll}
\text{Region I}: & x \le 0 & E_{\text{pot}} = V_0 \\
\text{Region II}: & 0 \le x \le a & E_{\text{pot}} = 0 \\
\text{Region III}: & x \ge a & E_{\text{pot}} = V_0
\end{array}
$$

but also two different cases with respect to the total energy E of the particle:

$E < V_0$: in this case, the energy of the particle is not sufficient to leave the well. This is the case of the bound particle.

$E > V_0$: here, the energy of the particle is sufficient to leave the well. However, when outside the well, it experiences the potential V_0, in contrast to the free particle (Sect. 9.1.1).

When solving the Schrödinger equation for the different cases, it is found in agreement with the previous simpler models, that the allowed energy of the particle inside the well are discrete (Fig. 9.4 left). In addition, one observes that the number of possible energy states increases with the potential energy value V_0 outside the well, and also with the width of the box a. And lastly, the values of the allowed energy levels (eigenvalue) increase with increasing potential V_0.

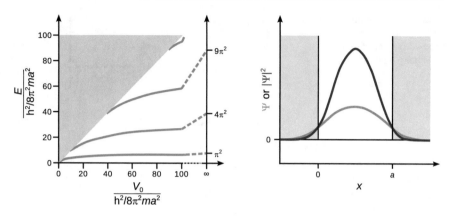

Fig. 9.4 Energy levels of a particle in a well with potential V_0. If the total energy of the particle E is less than V_0 (lower right half of the diagram), the particle is bound and can only assume discrete energy levels (eigenvalue). The number of the different possible levels as well as their numerical value depends on the value of V_0 and the width of the well a. If the total energy of the particle E is larger than V_0 (upper left half of the diagram), the particle is in an energy continuum (it may assume any energy) and no longer bound

A further notable observation of the box with finite potential walls is made when the wave functions Ψ and the probability densities $|\Psi|^2$ are analysed (Fig. 9.4 right). The solution of the Schrödinger equation in this case shows that both functions assume non-zero values outside the potential well, in contrast to the box with infinite walls (Fig. 9.3). The fact that the bound particle may also be found outside the box shows that it can exist in regions where the total energy E is less than the potential energy V_0; this is only possible if it possesses negative kinetic energy!

9.2 The Rigid Rotor

9.2.1 The Rigid Rotor with Space-fixed Axis

Following the introductory applications of the Schrödinger equation to individual particles, we now consider the circular movement of two masses m_1 and m_2 that are connected by a rigid connection—a very simple model for a two-atomic molecule. For rotational movement of a mass m, we need to consider the momentum of inertia I which is given by

$$I = m \cdot r^2 \tag{9.7}$$

where r is the distance of the mass to the axis of rotation. For convenience, the velocity of the mass m is given as an angular velocity ω which is related to the rotational frequency ν_{rot}:

Fig. 9.5 The reduced mass μ
allows substitution of a
two-body rigid rotor with a
one-body rigid rotor

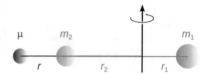

$$\omega = 2\pi \cdot \nu_{rot}. \tag{9.8}$$

If two connected masses m_1 and m_2 rotate around their centre of gravity (at distance r_1 from mass m_1 and at distance r_2 from mass m_2), the system can be substituted by one where a reduced mass μ (sometimes also called effective mass) rotates around a centre at distance r (Fig. 9.5):

$$r = r_1 + r_2. \tag{9.9}$$

The position of the centre of gravity is given by:

$$m_1 \cdot r_1 = m_2 \cdot r_2. \tag{9.10}$$

Combining Eqs. 9.9 and 9.10 yields the following expressions for the distances r_1 and r_2:

$$r_1 = \frac{m_2}{m_1 + m_2} \cdot r \quad \text{and} \quad r_2 = \frac{m_1}{m_1 + m_2} \cdot r. \tag{9.11}$$

The momentum of inertia for the system consisting of two connected masses is then:

$$I = m_1 \cdot r_1^2 + m_2 \cdot r_2^2$$

With Eq. 9.11 this yields:

$$I = \frac{m_1 \cdot m_2}{m_1 + m_2} \cdot r^2 = \mu \cdot r^2 \tag{9.12}$$

and thus defines the reduced mass μ.

In the simplest case, a rigid rotor ($r = $ const.) possesses a rotation axis which itself is fixed in space. The circular movement of a point around the rotation axis at distance r is then best described by monitoring the angle φ (measured against the positive branch of the x-axis in a Cartesian coordinate system). The location parameter describing the movement is thus chosen as $(r \cdot \phi)$. If we assume that the rotor is not subject to any outside potential ($E_{pot} = 0$), the Schrödinger equation 8.28 becomes:

$$\frac{d^2\Psi(r \cdot \phi)}{d(r \cdot \phi)^2} + \frac{8\pi^2 \cdot \mu}{h^2} \cdot E \cdot \Psi(r \cdot \phi) = 0$$

where the radius r can be isolated from the differential since the only variable in the location parameter in a given rotor is the angle ϕ:

$$\frac{d^2\Psi}{r^2 \cdot d\phi^2} + \frac{8\pi^2 \cdot \mu}{h^2} \cdot E \cdot \Psi = 0. \tag{9.13}$$

For further convenience it seems useful to introduce a rotational constant B which is defined as

$$B = \frac{h}{8\pi^2 \cdot c \cdot \mu \cdot r^2} = \frac{h}{8\pi^2 \cdot c \cdot I}. \tag{9.14}$$

Eq. 9.13 then becomes

$$\frac{d^2\Psi}{d\phi^2} + \frac{E}{h \cdot c \cdot B} \cdot \Psi = 0. \tag{9.15}$$

If we define a further constant k as

$$k^2 = \frac{E}{h \cdot c \cdot B}$$

Eq. 9.15 takes the form of

$$\frac{d^2\Psi}{d\phi^2} + k^2 \cdot \Psi = 0 \tag{9.16}$$

which we can solve immediately, recalling the general solution of the Schrödinger equation for a free particle that is not subjected to an external potential:

$$\Psi = A' \cdot e^{i \cdot k \cdot \phi} + B' \cdot e^{-i \cdot k \cdot \phi} \text{ or, equivalently,}$$
$$\Psi = C' \cdot \sin(k \cdot \phi + \delta). \tag{9.17}$$

Sensible solutions for Eq. 9.17 are obtained when the wave function Ψ assumes the same value after a full rotation, therefore

$$\Psi(\phi) = \Psi(\phi + 2\pi).$$

This requirement, however, is only met, if the constant k in the Schrödinger Eq. 9.16 assumes only integer values:

Fig. 9.6 Allowed energy
levels for a rigid rotor with
fixed rotation axis

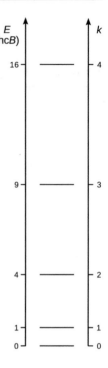

$$k = \pm\sqrt{\frac{E}{h \cdot c \cdot B}} = 0, \pm 1, \pm 2, \dots,$$

which means that the energy of the rigid rotor cannot assume any, but only discrete
values (Fig. 9.6):

$$E_k = h \cdot c \cdot B \cdot k^2, \text{with } k = 0, \pm 1, \pm 2, \dots$$

9.2.2 The Rigid Rotor with Space-free Axis

The rigid rotor with space-fixed axis only rotates in a plane perpendicular the
rotation axis. If we think about this as a model for an electron rotating around an
atomic nucleus, in the general—and more realistic—case, the rotation plane can
form any angle with the rotation axis. In addition to the two parameters r (distance of
the mass from the rotation axis) and ϕ (rotational angle of the mass with respect to
the positive x-axis), a third parameter needs to be considered that describes the angle
of the rotational plane with the positive z-axis; this angle is called θ.

Since these three parameters uniquely describe the position of a point in the
three dimensional space they can be used instead of the Cartesian coordinates x,

y and z. This new coordinate system is called the spherical coordinate system consisting of

- r radial coordinate
- ϕ azimuth angle
- θ polar (or inclination) angle.

The transformation of spherical to Cartesian coordinates is possible by:

$$
\begin{aligned}
x &= r \cdot \cos\phi \cdot \sin\theta \\
y &= r \cdot \sin\phi \cdot \sin\theta \\
z &= r \cdot \cos\theta
\end{aligned}
\tag{9.18}
$$

and graphically illustrated in Fig. 9.7.

When solving the Schrödinger equation, we now need to explicitly consider the wave function Ψ being a function of three coordinates, $\Psi(r, \phi, \theta)$. The mathematical treatment therefore becomes more advanced than in the previous cases where we considered just one dimension. without discussing a rigorous derivation of the solution of the Schrödinger equation for this case, we summarise the main points.

Since the rotor considered is a rigid rotor, the radial coordinate r is constant and therefore does not need to be considered in the differentiation. The wave function Ψ is therefore composed of two functions

- one that describes the behaviour of the azimuth angle ϕ: $\Phi(\phi)$, and
- one that describes the behaviour of the polar (inclination) angle θ: $\Theta(\theta)$

Therefore:

$$
\Psi(\phi, \theta) = \Phi(\phi) \cdot \Theta(\theta)
\tag{9.19}
$$

In the course of the solving the Schrödinger equation, it will become useful to substitute the function $\Theta(\theta)$ with a function that depends $\cos\theta$ instead:

Fig. 9.7 Polar coordinates and the Cartesian coordinate system

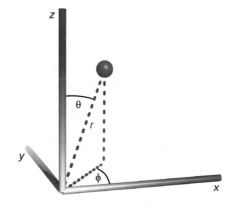

$$\Psi(\phi, \theta) = \Phi(\phi) \cdot P(\cos \theta) \tag{9.20}$$

For the azimuth function $\Phi(\phi)$, the following expression is obtained:

$$\frac{d^2\Phi(\phi)}{d\phi^2} + C \cdot \Phi(\phi) = 0 \tag{9.21}$$

and we recognise by comparison with Eq. 9.16, that this can be resolved to:

$$\Phi(\phi) = A' \cdot e^{i \cdot m \cdot \phi} + B' \cdot e^{-i \cdot m \cdot \phi}. \tag{9.22}$$

In analogy to the case of the rigid rotor with space-fixed axis, the system has to be in the same state after each full rotation, and therefore it can be concluded that

$$m = \sqrt{C}, \quad \text{with} \quad m = 0, \pm 1, \pm 2, \ldots \tag{9.23}$$

which introduces m as a quantum number.

Instead of the inclination function $\Theta(\theta)$ the function $P(\cos\theta)$ is used as a substitute and solving the Schrödinger equation yields the requirement

$$\frac{E}{h \cdot c \cdot B} = (m + s) \cdot (m + s + 1), \tag{9.24}$$

where m and s are two integer numbers that possess a particular relationship. If one introduces

$$l = m + s \tag{9.25}$$

then the following requirements arise from the relationship between m and s:

$$l = 0, 1, 2, 3, \ldots \text{ and } \quad l \geq |m|. \tag{9.26}$$

From these requirements, it is obvious that m can take the following values:

$$m = -l, \, -l + 1, \, \ldots, 0, 1, \, \ldots, l - 1, l. \tag{9.27}$$

The function $P(\cos\theta)$ turns out to be dependent on the two variables m and l and is called the associated Legendre function $P_l^m(\cos \theta)$.

When combining the solutions for the azimuth and the inclination function to yield an expression for the wave function Ψ, one obtains:

$$\Psi(\phi, \theta) = P_l^m(\cos \theta) \cdot e^{i \cdot m \cdot \phi}, \tag{9.28}$$

with the two quantum numbers l and m. The wave functions obtained are known as spherical harmonics.

Fig. 9.8 Allowed energy
levels for a rigid rotor with
free rotation axis

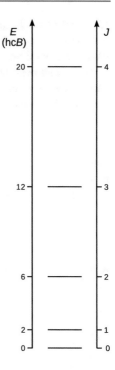

From Eq. 9.24, we can derive the allowed energy states of the rigid rotor with space-free axis (Fig. 9.8):

$$E = h \cdot c \cdot B \cdot (m + s) \cdot (m + s + 1) = h \cdot c \cdot B \cdot l \cdot (l + 1).$$

When this model is applied to a molecule, the quantum number J is typically used instead of l:

$$E_J = h \cdot c \cdot B \cdot J \cdot (J + 1), \text{with } J = 0, 1, 2, \ldots \qquad (9.29)$$

With setting $J = l$ and Eq. 9.27, it is obvious that for each quantum number l (J), there are $(2l + 1)$ different wavefunctions (one for each value of m) with the same energy. Levels with the same energy are said to be degenerate. For the rigid rotor, the degeneracy of each level is thus $(2J + 1)$.

9.3   The Harmonic Oscillator

In the previous models considering the rigid rotor, we did not require to pay particular attention to the potential energy, as the distance between the two masses m_1 and m_2 does not change. The potential energy of the rigid rotor is therefore zero.

If we consider a di-atomic molecule, however, the distance between the two atoms is not constant, due to the molecular vibration which causes the inter-atomic distance to fluctuate around an equilibrium distance.

From classical mechanics, we remember that the elongation of a mass m suspended by a spring leads to an oscillation behaviour (see Fig. 9.2) which is described by Hook's law:

$$F = -k \cdot x, \tag{9.30}$$

where the force of elongation F is proportional to the elongation distance x and related by the spring constant k.

Since the force is the first derivative of the potential energy

$$F = -\frac{\mathrm{d}E_{\mathrm{pot}}}{\mathrm{d}x}, \tag{9.31}$$

it becomes clear that

$$\frac{\mathrm{d}E_{\mathrm{pot}}}{\mathrm{d}x} = k \cdot x, \tag{9.32}$$

which can be integrated to allow calculation of a numerical value of the potential energy V and results in a parabolic dependence of the potential energy from the elongation distance x (Fig. 9.9):

$$E_{\mathrm{pot}} = \frac{1}{2} \cdot k \cdot x^2. \tag{9.33}$$

The frequency of vibrational oscillation of a di-atomic molecule can also be derived from the classical mechanics, if the reduced mass μ is used to describe the two connected masses m_1 and m_2:

$$\nu_0 = \frac{1}{2\pi} \sqrt{\frac{k}{\mu}}, \text{with} \, \mu = \frac{m_1 \cdot m_2}{m_1 + m_2}. \tag{9.34}$$

For the quantum mechanical treatment of the vibration of a di-atomic molecule, we need to use the potential energy (Eq. 9.33), as well as the reduced mass μ when compiling the Schrödinger equation:

$$\frac{\mathrm{d}^2\Psi(x)}{\mathrm{d}x^2} + \frac{8\pi^2 \cdot \mu}{\mathrm{h}^2} \cdot \left(E - \frac{1}{2} \cdot k \cdot x^2 \right) \cdot \Psi(x) = 0 \tag{9.35}$$

Due to some advanced mathematics required to solve this equation, we refrain from a rigorous discussion and appreciate that solutions for Eq. 9.35 can be found whereby the wave functions Ψ_n take the form of

Fig. 9.9 Potential energy of
the harmonic oscillator

$$\Psi_n(x) = N_v \cdot H_v(x) \cdot e^{-\frac{\beta}{2}x^2}, \text{ with } v = 0, 1, 2, 3, \ldots \tag{9.36}$$

introducing the new quantum number, v. The constant β is comprised of the mechanical properties of the di-atomic molecule (reduced mass μ and spring constant k), and N_v is a normalisation factor. Their definitions are:

$$\beta = \frac{2\pi}{h} \cdot \sqrt{\mu \cdot k} \text{ and } N_v = \frac{\beta^{\frac{1}{4}}}{\pi^{\frac{1}{4}} \cdot (2^v \cdot v!)^{\frac{1}{2}}} \tag{9.37}$$

The function $H_v(x)$ is a potential development series; the wave functions for the first five quantum numbers (v = 0, 1 ... , 4) are summarised in Table 9.1 and graphically illustrated in Fig. 9.10.

Table 9.1 Solutions to the Schrödinger Eq. 9.35 for the first five quantum numbers v

v = 0	$\Psi_0(x) = N_0 \cdot e^{-\frac{\beta}{2}x^2}$
v = 1	$\Psi_1(x) = N_1 \cdot \left(2 \cdot \beta^{\frac{1}{2}} \cdot x\right) \cdot e^{-\frac{\beta}{2}x^2}$
v = 2	$\Psi_2(x) = N_2 \cdot (4 \cdot \beta \cdot x^2 - 2) \cdot e^{-\frac{\beta}{2}x^2}$
v = 3	$\Psi_3(x) = N_3 \cdot \left(8 \cdot \beta^{\frac{3}{2}} \cdot x^3 - 12 \cdot \beta^{\frac{1}{2}} \cdot x\right) \cdot e^{-\frac{\beta}{2}x^2}$
v = 4	$\Psi_4(x) = N_4 \cdot \left(16 \cdot \beta^2 \cdot x^4 - 48 \cdot \beta \cdot x^2 + 12\right) \cdot e^{-\frac{\beta}{2}x^2}$

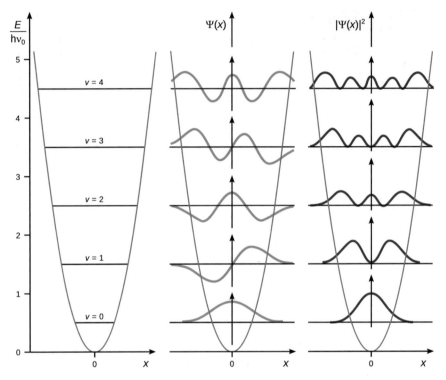

Fig. 9.10 Allowed energy levels (left), wave functions (centre) and probability density (right) of the harmonic oscillator

As observed in the previous systems, the harmonic oscillator assumes discrete energy levels (Fig. 9.10), which are a function of the quantum number v:

$$E_v = h \cdot \nu_0 \cdot \left(v + \frac{1}{2} \right), \tag{9.38}$$

where ν_0 is the frequency as defined in Eq. 9.34. Notably, the lowest energy state achieved at the quantum number v = 0 still comprises the zero-point energy of

$$E_0 = \frac{1}{2} \cdot h \cdot \nu_0. \tag{9.39}$$

Figure 9.10 (right) illustrates that the probability of finding the particle varies strongly with the location x, and this variation, in turn, differs with the quantum number v. Importantly, the quantum mechanical results also show that the particle can exist outside the region enclosed by the parabolic potential wall.

9.4   Tunnelling

In Sect. 9.1.3, we saw that the probability to find a particle outside a box with walls
of a finite potential is non-zero (Fig. 9.4), i.e. it may exist outside the box. A similar
observation has been made in the previous section when discussing the harmonic
oscillator (Fig. 9.10). Therefore, the question arises as to whether and how a particle
may be able to tunnel through a sufficiently thin potential barrier, despite its energy
E being lower than the potential energy $E_{pot} = V_0$.

The situation is illustrated in Fig. 9.11. Three regions need to be considered:

$$
\begin{aligned}
&\text{Region 1}: & x \leq 0 & \quad E_{pot} = 0 \\
&\text{Region 2}: & 0 \leq x \leq a & \quad E_{pot} = V_0 \\
&\text{Region 3}: & x \geq a & \quad E_{pot} = 0
\end{aligned}
$$

For regions 1 and 3, the Schrödinger equation is thus:

$$
\frac{d^2 \Psi(x)}{dx^2} + \frac{8\pi^2 \cdot m}{h^2} \cdot E \cdot \Psi(x) = 0,
$$

and for region 2, it reads:

$$
\frac{d^2 \Psi(x)}{dx^2} + \frac{8\pi^2 \cdot m}{h^2} \cdot (E - V_0) \cdot \Psi(x) = 0
$$

Fig. 9.11 Illustration of the
tunnelling through a
sufficiently thin potential wall

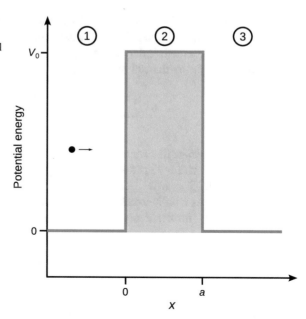

For both equations, we can define a constant k that comprises of the factor before $\Psi(x)$, i.e.:

$$k_1 = \frac{8\pi^2 \cdot m}{h^2} \cdot E \quad \text{and} \quad k_2 = \frac{8\pi^2 \cdot m}{h^2} \cdot (E - V_0),$$

which leads to the following solutions for the wave functions in the three regions 1–3:

$$\Psi_1 = A \cdot e^{i \cdot k_1 \cdot x} + B \cdot e^{-i \cdot k_1 \cdot x}$$
$$\Psi_2 = C \cdot e^{i \cdot k_2 \cdot x} + D \cdot e^{-i \cdot k_2 \cdot x}$$
$$\Psi_3 = F \cdot e^{i \cdot k_3 \cdot x} + G \cdot e^{-i \cdot k_3 \cdot x}$$

In all of the three above equations, the first term describes particles moving from the left to the right (see Fig. 9.11), and the second term describes particles moving from the right to the left. In regions 1 and 2 particles originally moving from the left to the right will be partially reflected on the potential walls, which leads to some particles moving from the right to left in those regions. Therefore, the coefficients B and D need to be non-zero. In region 3, if particles made it through to that region, they will keep on moving to the right and not return, so the coefficient G is zero.

Importantly though, the fraction of particles which made it from region 1 to region 3 depends on the ratio of the coefficients F and A. This ratio is called the transmission coefficient T:

$$T = \frac{|F|^2}{|A|^2}. \tag{9.40}$$

In order to determine the constants A, B, C, D and F, we need to consider that the overall wave function Ψ (and its derivatives) need to be continuous. Without discussing this mathematical derivation in detail, we appreciate that the following result is obtained for the transmission coefficient:

$$T = \frac{16 \cdot E \cdot (V_0 - E)}{V_0^2} \cdot e^{-2 \cdot \sqrt{\frac{8\pi^2 \cdot m \cdot a^2 \cdot (V_0 - E)}{h^2}}}. \tag{9.41}$$

This equation demonstrates that the tunnelling probability, given by the transmission coefficient T, mainly depends on the energy difference $(E-V_0)$ as well as the width a of the potential barrier. For a given energy E of the particle, the transmission coefficient decreases with increasing height of the potential barrier (V_0). The transmission coefficient further decreases, if the width a, i.e. the barrier thickness, increases.

9.5   Exercises

1. The radius of the first orbit in the Bohr model is $r_1 = 5.3 \cdot 10^{-9}$ cm. As a rough estimate, calculate the energy of the electron in the hydrogen atom, assuming it was a particle in a cubic box of the same volume as that of a sphere with radius r_1. Compare the result with the energy that is predicted by the Bohr model.

2. Calculate the zero-point energy of $^1H^{35}Cl$ (a) for one molecule, and (b) for 1 mol, assuming a force constant of 480.6 Nm^{-1}.

3. What is the value of the transmission coefficient for an electron with an energy of 1 eV that moves against a potential barrier of 5 eV and 2 nm thickness?

4. Calculate the probability of locating a particle in a potential-free one-dimensional box of length a between $1/4\ a$ and $3/4\ a$, assuming the particle being in its lowest energy state.

Quantum Theory of Atoms

10

10.1 The Hydrogen Atom

In the previous chapter, we familiarised ourselves with various applications of the Schrödinger equation, and can now discuss the hydrogen atom from a quantum mechanical view.

The hydrogen atom constitutes the simplest atomic constellation, comprising a proton and an electron. The potential energy of this system is of electrostatic nature and therefore given by the Coulomb attraction

$$E_{\text{pot}} = -\frac{e^2}{4\pi \cdot \varepsilon_0 \cdot r}.$$

The Schrödinger equation is thus

$$\Delta\Psi + \frac{8\pi^2 \cdot m_e}{h^2} \cdot \left(E + \frac{e^2}{4\pi \cdot \varepsilon_0 \cdot r}\right) \cdot \Psi = 0, \tag{10.1}$$

where Δ is the Laplace operator as introduced in Sect. 8.3.1. Since the electrostatic potential takes the form of a spherical potential, it will prove convenient to use the spherical coordinates, so the wave function Ψ depends on r, θ, ϕ: $\Psi(r,\theta,\phi)$. In order to solve the Schrödinger equation, the wave function is separated into three functions, each of which depends on one of the three spherical coordinates

$$\Psi(r, \theta, \phi) = R(r) \cdot \Theta(\theta) \cdot \Phi(\phi) = R(r) \cdot Y(\theta, \phi), \tag{10.2}$$

i.e. a radial function $R(r)$, an inclination function $\Theta(\theta)$ and an azimuth function $\Phi(\phi)$. The function $Y(\theta,\phi)$ combines the azimuth and inclination functions and resolves to the spherical harmonics we have already introduced when solving the Schrödinger equation for the rigid rotor with space-free axis:

© Springer International Publishing AG, part of Springer Nature 2018
A. Hofmann, *Physical Chemistry Essentials*,
https://doi.org/10.1007/978-3-319-74167-3_10

$$Y(\theta, \phi) = P_l^m(\cos\theta) \cdot e^{i \cdot m \cdot \phi}, \tag{9.28}$$

with the quantum numbers

$$\begin{aligned} m &= 0, \pm 1, \pm 2, \ldots \\ l &= 0, 1, 2, \ldots \\ l &\geq |m|. \end{aligned} \tag{10.3}$$

For the radial function $R(r)$, one finds the solution

$$R(r) = e^{-\sqrt{-2 \cdot \eta \cdot r}} \cdot P_{n,l}(r), \tag{10.4}$$

where $P_{n,l}(r)$ is a potential development series, and dependent on the quantum numbers

$$\begin{aligned} n &= 1, 2, 3, \ldots \\ l &= 0, 1, 2, \ldots \\ n &\geq |l + 1|. \end{aligned} \tag{10.5}$$

For convenience, the parameter η has been introduced in the exponential function of Eq. 10.4; it is defined as

$$\eta = \frac{4\pi^2 \cdot m_e \cdot E}{h^2}. \tag{10.6}$$

The wave functions that solve the Schrödinger equation for the hydrogen atom are thus functions

$$\Psi(r, \theta, \phi) = N \cdot e^{-\sqrt{-2 \cdot \eta \cdot r}} \cdot P_{n,l}(r) \cdot P_l^m(\cos\theta) \cdot e^{i \cdot m \cdot \phi} \tag{10.7}$$

that depend on the three quantum numbers n, l and m; N is the normalisation factor. From Eqs. 10.3 and 10.5 it becomes obvious, the quantum numbers n, l and m have particular relationships and therefore only select combinations are possible as illustrated in Table 10.1.

Since the wave functions Ψ that describe the atom are dependent on the three quantum numbers (Eq. 10.7), there will be a particular number of different wave functions (called linear independent wave functions). The number of possible wave functions is 1 for $n = 1$, 4 for $n = 2$, and 9 for $n = 3$. Due to the general hierarchy provided by the quantum number n, it is called the principal quantum number. Generally, the number of different (linear independent) wave functions is given by

Table 10.1 Possible combinations of quantum numbers for the first three principal quantum numbers ($n = 1, 2, 3$)

n	1	2				3								
l	0	0		1		0		1			2			
m	0	0	−1	0	1	0	−1	0	1	−2	−1	0	1	2

$$\sum_{l=0}^{n-1} (2 \cdot l + 1) = \frac{n}{2} \cdot (1 + 2 \cdot n - 1) = n^2. \tag{10.8}$$

Like in the discussion of the rigid rotor and the harmonic oscillator, the allowed energy levels of the hydrogen atom are not continuous but discrete, and are obtained by considering particular requirements when solving the Schrödinger equation such that the solutions are physically sensible. The allowed energy states (eigenvalues) of the hydrogen atom are obtained as

$$E_n = -\frac{m_e \cdot e^4}{8 \cdot \varepsilon_0^2 \cdot h^2 \cdot n^2}, \text{ with } n = 1, 2, 3, \ldots \tag{10.9}$$

As with previously discussed models, we find that the allowed energy levels of the hydrogen atom vary with a quantum number—specifically the principal quantum number n—and thus there are n different allowed energy levels. At the same time, we derived above (Eq. 10.8) that there are n^2 different wave functions, i.e. measurable different states. The energy levels are therefore said to be degenerate.

▶ Multiple states of a quantum mechanical system are degenerate if they possess the same energy value. Vice versa, an energy level is degenerate, if it corresponds to multiple different measurable states.

The wave functions solving the Schrödinger equation are called eigenfunctions. In the discussion above, we found that those wave functions are best separated into a function depending on the radial (r) and another function depending on the angular spherical coordinates (θ, ϕ). Since it is also necessary to normalise the wave function, a normalisation factor N became a third component; thus:

$$\Psi = N \cdot R_{n,l}(r) \cdot Y_{l,m}(\theta, \phi). \tag{10.10}$$

In the following sections, we will have a closer look at the radial eigenfunction as well as the spherical harmonics, and then compose the normalised eigenfunctions.

10.1.1 The Radial Eigenfunctions of the Hydrogen Atom

In the previous section, we introduced the solution for the radial function of the hydrogen atom as

$$R(r) = e^{-\sqrt{-2 \cdot \eta} \cdot r} \cdot P_{n,l}(r), \text{ with } \eta = \frac{4\pi^2 \cdot m_e \cdot E}{h^2} \tag{10.4}$$

An explicit calculation of the exponent in Eq. 10.4 thus yields:

$$\sqrt{-2 \cdot \eta} \cdot r = \sqrt{-2 \cdot \frac{4\pi^2 \cdot m_e \cdot E}{h^2}} \cdot r.$$

The allowed energy levels are given by Eq. 10.9; therefore:

$$\sqrt{-2 \cdot \eta} \cdot r = \sqrt{\frac{-2 \cdot 4\pi^2 \cdot m_e \cdot \left(-\frac{m_e \cdot e^4}{8 \cdot \varepsilon_0^2 \cdot h^2 \cdot n^2}\right)}{h^2}} \cdot r = \sqrt{\frac{8 \cdot \pi^2 \cdot m_e^2 \cdot e^4}{8 \cdot \varepsilon_0^2 \cdot h^4 \cdot n^2}} \cdot r$$

$$\sqrt{-2 \cdot \eta} \cdot r = \frac{\pi \cdot m_e \cdot e^2}{\varepsilon_0 \cdot h^2} \cdot \frac{1}{n} \cdot r$$

Comparison with Eq. 8.22 shows that the first factor in above equation is the reciprocal of the radius of the first orbit in the atomic model by Bohr:

$$r_{Bohr} = \frac{\varepsilon_0 \cdot h^2}{\pi \cdot m_e \cdot e^2} = 5.3 \cdot 10^{-11} \text{ m},$$

and thus allows us to measure the radius of the electron in multiples of the Bohr radius:

$\rho = \frac{r}{r_{Bohr}}$, yielding $\sqrt{-2 \cdot \eta} \cdot r = \frac{\rho}{n}$, and therefore:

$$R_{n,l}(r) = e^{-\frac{\rho}{n}} \cdot P_{n,l}(r). \tag{10.11}$$

Notably, the radial eigenfunction described in above Eq. 10.11 is not yet normalised. Table 10.2 summarises explicit normalised radial eigenfunctions for the hydrogen atom; Figure 10.1 illustrates these functions graphically.

Notably, the radial eigenfunction R only depends on the quantum numbers n and l, but not m. From Fig. 10.1, it is obvious that the number of zero-crossings of R increases with n, but decreases with l:

Table 10.2 Normalised radial eigenfunctions of the hydrogen atom for $n = 1, 2, 3$

n	l	$R_{n,l}$	No of radial nodes $n-l-1$
1	0	$R_{1,0} = 2 \cdot e^{-\rho}$	0
2	0	$R_{2,0} = \frac{1}{2\sqrt{2}} \cdot e^{-\frac{\rho}{2}} \cdot (2 - \rho)$	1
2	1	$R_{2,1} = \frac{1}{2\sqrt{6}} \cdot e^{-\frac{\rho}{2}} \cdot \rho$	0
3	0	$R_{3,0} = \frac{2}{81 \cdot \sqrt{3}} \cdot e^{-\frac{\rho}{3}} \cdot (27 - 18 \cdot \rho + 2 \cdot \rho^2)$	2
3	1	$R_{3,1} = \frac{4}{81 \cdot \sqrt{6}} \cdot e^{-\frac{\rho}{3}} \cdot (6 \cdot \rho - \rho^2)$	1
3	2	$R_{3,2} = \frac{4}{81 \cdot \sqrt{30}} \cdot e^{-\frac{\rho}{3}} \cdot \rho^2$	0

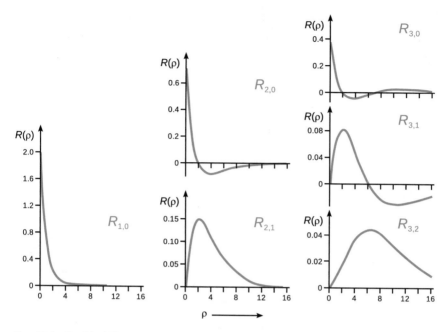

Fig. 10.1 Graphical illustration of normalised radial eigenfunctions of the hydrogen atom for $n = 1, 2, 3$

$$\text{No of zero-crossings} = \text{No of total nodes} = n - l - 1. \tag{10.12}$$

We have previously introduced for the wave function that the squared function is a probability density to find the particle in particular spatial locations. The same concept applies to the radial eigenfunction, where R^2 describes the probability density to find an electron along a beam originating in the nucleus.

When asking the question of the actual probability to find the electron at a particular distance from the nucleus, e.g. between ρ and $(\rho + d\rho)$, the function $R(\rho)^2$ needs to be integrated. For spherical symmetry, the volume of the spherical shell described by ρ and $(\rho + d\rho)$ is $(4\pi \cdot \rho^2 \cdot d\rho)$, which yields for the radial probability distribution

$$\text{radial probability density} = 4\pi \cdot \rho^2 \cdot R(\rho)^2 d\rho$$

The plotted probability distribution functions in Fig. 10.2 indicates that for the lowest possible principal quantum number $n = 1$, the highest probability to find the electron is at the distance of $\rho = 1$, i.e. $r = r_{Bohr}$. In this case, we see that the electron can be found in a spherical space around the nucleus, most likely at distance r_{Bohr}. However, the electron does not have a well-defined position and the space where the electron can be found certainly has no sharp boundary. The probability distribution decreases rapidly with increasing distance and anneals to zero at large distances.

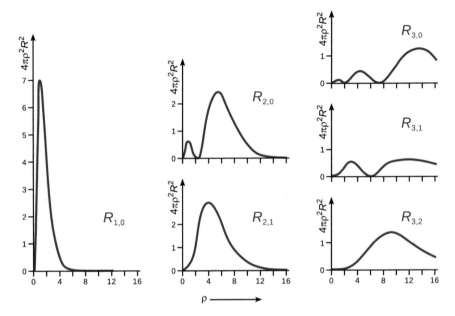

Fig. 10.2 Radial probability distribution functions of the hydrogen atom for $n = 1, 2, 3$

For the case $(n = 1, l = 0)$ in Fig. 10.2, we see that there are two maxima in the probability density function. Furthermore, there is a region where the probability to find the electron is zero. This region corresponds to the zero-crossing of the radial wave function $R(\rho)$ in Fig. 10.1 and is called a (radial) node. We can imagine this case as two concentric spheres with a node in between, which takes the shape of a spherical shell.

10.1.2 The Spherical Harmonics of the Hydrogen Atom

When solving the Schrödinger equation for the hydrogen atom, we obtained the spherical harmonics component as

$$Y_{l,m}(\theta, \phi) = P_l^m(\cos \theta) \cdot e^{i \cdot m \cdot \phi}. \tag{9.28}$$

Notably, these functions are not yet normalised and thus need to be multiplied with normalisation factors. The first nine normalised functions are summarised in Table 10.3.

In contrast to the radial eigenfunctions, which depend on only one variable (r or ρ), the spherical harmonics Y possesses two variables, the azimuth angle ϕ and the inclination angle θ. Illustration of the spherical harmonics therefore requires three dimensions. However, from Table 10.3, we learn that for ($l = 0, m = 0$), the spherical harmonics is independent of both ϕ and θ; it has a constant value of $\frac{1}{2 \cdot \sqrt{\pi}}$.

Table 10.3 The normalised spherical harmonics of the hydrogen atom for $l = 0, 1, 2$

l	m	$Y_{l,m}$	Number of angular nodes l
0	0	$Y_{0,0} = \frac{1}{2\cdot\sqrt{\pi}}$	0
1	−1	$Y_{1,-1} = \frac{1}{2}\cdot\sqrt{\frac{3}{2\pi}}\cdot\sin\theta\cdot e^{-i\phi}$	1
1	0	$Y_{1,0} = \frac{1}{2}\cdot\sqrt{\frac{3}{\pi}}\cdot\cos\theta$	1
1	1	$Y_{1,1} = \frac{1}{2}\cdot\sqrt{\frac{3}{2\pi}}\cdot\sin\theta\cdot e^{i\phi}$	1
2	−2	$Y_{2,-2} = \frac{1}{4}\cdot\sqrt{\frac{15}{2\pi}}\cdot\sin^2\theta\cdot e^{-2\cdot i\phi}$	2
2	−1	$Y_{2,-1} = \frac{1}{2}\cdot\sqrt{\frac{15}{2\pi}}\cdot\sin\theta\cdot\cos\theta\cdot e^{-i\phi}$	2
2	0	$Y_{2,0} = \frac{1}{4}\cdot\sqrt{\frac{5}{\pi}}\cdot(3\cdot\cos^2\theta - 1)$	2
2	1	$Y_{2,1} = \frac{1}{2}\cdot\sqrt{\frac{15}{2\pi}}\cdot\sin\theta\cdot\cos\theta\cdot e^{i\phi}$	2
2	2	$Y_{2,2} = \frac{1}{4}\cdot\sqrt{\frac{15}{2\pi}}\cdot\sin^2\theta\cdot e^{2\cdot i\phi}$	2

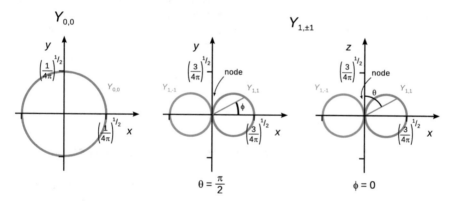

Fig. 10.3 The spherical harmonics functions (angular eigenfunctions) $Y_{0,0}$ (left) and $Y_{1,\pm1}$ (centre and right) of the hydrogen atom

Geometrically, this describes a sphere around the origin with a radius of $\frac{1}{2\cdot\sqrt{\pi}}$ (see Fig. 10.3 left panel).

The construction of the spherical harmonics for ($l = 1$, $m = \pm1$) is illustrated in Fig. 10.3 for a particular angle θ (centre panel) and a particular angle ϕ (right panel). The plots show the value of the function $Y_{1,1}$ and $Y_{1,-1}$ in dependence of the angles ϕ (centre panel) and q (right panel).

As before when discussing the general wave functions Ψ and the radial eigenfunctions R, we are next interested in the probability density which informs about the probability to find the particle in a particular location. For the spherical

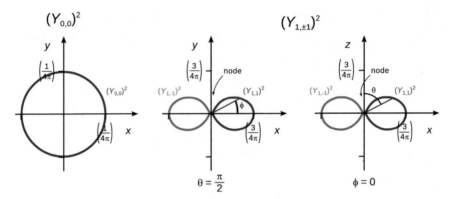

Fig. 10.4 The angular probability distribution functions $(Y_{0,0})^2$ (left) and $(Y_{1,\pm1})^2$ (centre and right) of the hydrogen atom

harmonics Y, the probability distribution is given (similar to Ψ and R) by the squared function, Y^2. (see Fig. 10.4).

Like with the radial eigenfunctions, we also find for the spherical harmonics that there are areas in which the function Y is zero. In the probability density Y^2, the values in these areas remain zero, thus giving rise to the (angular) nodes, i.e. areas in which the probability to find the electron is zero. From Table 10.3 it is obvious that the number of angular nodes is given by the quantum number l:

$$\text{No of angular nodes} = l. \tag{10.13}$$

10.1.3 The Normalised Eigenfunctions of the Hydrogen Atom

Solving the Schrödinger equation for the hydrogen atom led us to the wave functions given in Eq. 10.10, which represent the normalised eigenfunctions of the hydrogen atom.

In a process called separation of variables, we have found solutions for the radial (Sect. 10.1.1) as well as the angular component (Sect. 10.1.2) of those eigenfunctions. Now, we need to combine both of those components, R and Y, and determine the final normalising factor N to obtain the normalised eigenfunctions.

Additionally, since the final wave functions for $m \neq 0$ are complex (i.e. contain the number i), linear combinations of the corresponding $+m$ and $-m$ wave functions are generated, which leads to real eigenfunctions (i.e. not containing the number i). For example, the p_x and p_y orbitals are formed by linear combinations of p_{+1} and p_{-1}. In contrast, the p_z orbital is the same as p_0.

The results are summarised in Table 10.4 and show that the state of the electron is characterised by three quantum numbers:

Table 10.4 The real eigenfunctions of the hydrogen atom for the principal quantum number $n = 1, 2, 3$

Symbol	n	l	m	N	$R(r)$	$Y(\theta,\phi)$	Radial nodes	Angular nodes
$1s$	1	0	0	$\frac{1}{\sqrt{\pi}}$	$e^{-\rho}$	1	0	0
$2s$	2	0	0	$\frac{1}{4\cdot\sqrt{2\pi}}$	$(2-\rho)\cdot e^{-\frac{\rho}{2}}$	1	1	0
$2p_z$	2	1	0	$\frac{1}{4\cdot\sqrt{2\pi}}$	$\rho\cdot e^{-\frac{\rho}{2}}$	$\cos\theta$	0	1
$2p_x$	2	1	±1	$\frac{1}{4\cdot\sqrt{2\pi}}$	$\rho\cdot e^{-\frac{\rho}{2}}$	$\sin\theta\cdot\cos\phi$	0	1
$2p_y$	2	1	±1	$\frac{1}{4\cdot\sqrt{2\pi}}.$	$\rho\cdot e^{-\frac{\rho}{2}}$	$\sin\theta\cdot\sin\phi$	0	1
$3s$	3	0	0	$\frac{1}{81\cdot\sqrt{3\pi}}$	$(27-18\cdot\rho+2\cdot\rho^2)\cdot e^{-\frac{\rho}{3}}$	1	2	0
$3p_z$	3	1	0	$\frac{\sqrt{2}}{81\cdot\sqrt{\pi}}$	$(6-\rho)\cdot\rho\cdot e^{-\frac{\rho}{3}}$	$\cos\theta$	1	1
$3p_x$	3	1	±1	$\frac{\sqrt{2}}{81\cdot\sqrt{\pi}}$	$(6-\rho)\cdot\rho\cdot e^{-\frac{\rho}{3}}$	$\sin\theta\cdot\cos\phi$	1	1
$3p_y$	3	1	±1	$\frac{\sqrt{2}}{81\cdot\sqrt{\pi}}$	$(6-\rho)\cdot\rho\cdot e^{-\frac{\rho}{3}}$	$\sin\theta\cdot\sin\phi$	1	1
$3d_{z^2}$	3	2	0	$\frac{1}{81\cdot\sqrt{6\pi}}$	$\rho^2\cdot e^{-\frac{\rho}{3}}$	$3\cdot\cos^2\theta-1$	0	2
$3d_{xz}$	3	2	±1	$\frac{\sqrt{2}}{81\cdot\sqrt{\pi}}$	$\rho^2\cdot e^{-\frac{\rho}{3}}$	$\sin\theta\cdot\cos\theta\cdot\cos\phi$	0	2
$3d_{yz}$	3	2	±1	$\frac{\sqrt{2}}{81\cdot\sqrt{\pi}}$	$\rho^2\cdot e^{-\frac{\rho}{3}}$	$\sin\theta\cdot\cos\theta\cdot\sin\phi$	0	2
$3d_{x^2-y^2}$	3	2	±2	$\frac{1}{81\cdot\sqrt{2\pi}}$	$\rho^2\cdot e^{-\frac{\rho}{3}}$	$\sin^2\theta\cdot(\sin2\cdot\phi)$	0	2
$3d_{xy}$	3	2	±2	$\frac{1}{81\cdot\sqrt{2\pi}}$	$\rho^2\cdot e^{-\frac{\rho}{3}}$	$\sin^2\theta\cdot(\sin2\cdot\phi)$	0	2

- the principal quantum number $n = 1, 2, 3, \ldots$
- the quantum number $l = 0, 1, 2, \ldots$
- the quantum number $m = 0, \pm1, \pm2, \ldots$

In addition to the numerical description of electron states by the above quantum numbers, a further nomenclature system ascribes letters to the various different states. The principal energy level (quantum number n) is thought of as a shell in which the electrons orbit; the individual shells are often (especially when referring to X-ray processes) given the letters K, L, M, etc. The quantum number l describes the subshells, denoted by the letters s, p, d, etc (see Table 10.5). The subshells, in turn, comprise of the individual atomic orbitals. As shown in Table 10.4, particular subshells are most commonly denoted by combining the numerical value of the principal quantum number with the letter describing the subshell. The characteristic director of the individual atomic orbital is then added as a subscript. For example, $2p_x$ denotes the p orbital ($n = 2$, $l = 1$) which primarily extends along the x-axis.

In order to describe the final eigenfunctions, we now need to combine the radial and the angular components. It rapidly becomes clear that this is best done with three-dimensional shapes, since the spherical harmonics depend on two angles, ϕ and θ. The radial eigenfunction provides a measure for how far away from the atomic nucleus an orbital extends, but its nodes might also call for regions in which the

Table 10.5 Shell and atomic orbital nomenclature

Quantum number n	Shell	Quantum number l	Atomic orbital
1	K	0	s
2	L	1	p
3	M	2	d
4	N	3	f
5	O	4	g
6	P	5	h
7	Q	6	i
		7	k

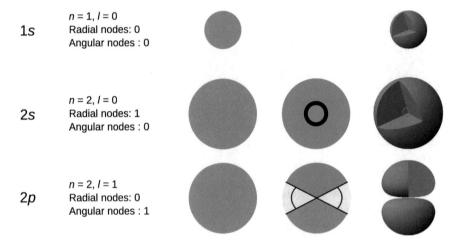

1s $n = 1, l = 0$
Radial nodes: 0
Angular nodes : 0

2s $n = 2, l = 0$
Radial nodes: 1
Angular nodes : 0

2p $n = 2, l = 1$
Radial nodes: 0
Angular nodes : 1

Fig. 10.5 Conceptual construction of atomic orbitals (final eigenfunctions Ψ) from the radial eigenfunction R (sphere), consideration of radial nodes, and application of angular nodes provided by the spherical harmonics (angular eigenfunctions Y). The three-dimensional graphs in the last panel are shown with cut-away wedges to reveal the interior of the functions. The colours indicate positive (blue) and negative (green) values of the wave function Ψ

electron cannot exist. The angular eigenfunction provides the overall shape information. As illustrated in Fig. 10.5, one might think of atomic orbitals as a sphere, from which particular regions—defined by the nodes—are cut off.

In the three-dimensional graphs of the final eigenfunctions, the areas where the wave functions assume positive and negative values (phase of the wave function) are typically mapped in different colours (see Fig. 10.5).

The real eigenfunctions, i.e. atomic orbitals, are summarised in Table 10.4 for the principal quantum numbers $n = 1, 2, 3$. The three-dimensional graphs of these orbitals are shown in Fig. 10.6. The s orbitals ($l = 0$) possess an overall spherical shape and are nested shells for $n > 1$. The p orbitals ($l = 1$) for $n = 2$ consist of two ellipsoids arranged either along the x-, y- or z-axis. This make-up is very similar for

Subshells

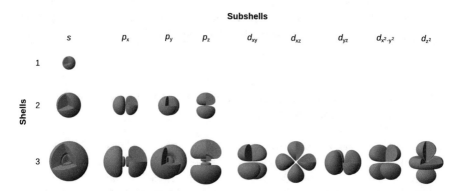

Fig. 10.6 Three-dimensional graphs of the atomic orbitals for the principal quantum numbers $n = 1, 2, 3$ as summarised in Table 10.4. The atomic orbitals are shown with cut-away wedges. The colours indicate positive (blue) and negative (green) values of the wave function Ψ

the p orbitals in the shell $n = 3$, however, an additional radial node leads to four final ellipsoids arranged along the major axis of extension. The shell $n = 3$ is the first shell with d orbitals, four of which appear very similar and take the form of four drop-like lobes arranged orthogonally around the central nucleus. The centres of all four lobes lie in one plane. These planes are the xy-, xz-, and yz-planes, where the lobes are located between the pairs of the cartesian coordinate system; one of these orbitals has the centres along the x- and y-axes. The fifth d orbital consists of a torus with two pear-shaped lobes placed symmetrically on the z-axis.

It is important to remember that the graphically depictions in Fig. 10.6 are the wave functions Ψ of the individual atomic orbitals. The probability to find the electron within the orbital is given by $|\Psi|^2$, which is a probability density function. Therefore, the shapes depicting the probability is different from those that illustrate Ψ; with exception of the s orbitals, all other orbitals become more prolate when considering $|\Psi|^2$. Also, due to the square operation, there is no longer a difference between positive and negative regions in the orbitals, thus making the probability density functions uniformly positive.

10.2   The Quantum Numbers

The solution of the Schrödinger equation for the hydrogen atom provided three quantum numbers, n, l and m. Importantly, the energy of the electron of atomic hydrogen only depended on the primary quantum number n (Eq. 10.9) and a degeneracy of n^2 (Eq. 10.8). These quantum mechanical findings are in agreement with the energy levels in Bohr's atomic model as well as the atomic spectrum of hydrogen.

However, when observing the atomic spectrum of hydrogen in the presence of an outside magnetic field, the degeneracy of energy levels disappears resulting in splitting of individual spectral lines. This effect was discovered in 1896 by the Dutch physicist Pieter Zeeman and is thus called the Zeeman effect.

10.2.1 The Quantum Numbers of the Orbital Angular Momentum and Magnetic Momentum

The discussion in the previous sections showed, that the type of rotation of an electron around the atomic nucleus is described by the spherical harmonics $Y_{l,m}(\theta,\phi)$ (Eq. 9.28) as described for the rigid rotor with space-free axis. Therefore, the quantum number l characterises the orbital angular momentum $\bar{l}$ of the orbiting electron. The orbital angular momentum is a vector and thus has a length and a direction. Notably, electrons in s orbitals ($l = 0$) possess no orbital angular momentum, but electrons in other orbitals ($p, d, f, \ldots$) do.

The quantum number m becomes important, when the electron rotates around a fixed axis, for example, when the atom is brought into a magnetic field. The quantum number m (or m_l) is thus called the magnetic quantum number and determines the orientation of the orbital angular momentum in space.

Since an electron orbiting the atomic nucleus is a moving moving charge, it represents an electrical current, which, in turn, induces a magnetic field. In the presence of an outside magnetic field, the magnetic field of the orbiting electron is aligned with the external magnetic field. More specifically, the orbital that the electron is in is aligned by the outside field. For illustration, we consider the d orbital where $l = 2$ and $|m_l| \leq l$ (see Eq. 9.27), so that there are five $(2 \cdot l + 1)$ degenerate energy levels. Since all five levels possess the same energy, the orbital angular momentum $\bar{l}$ has no preferred orientation. This changes when an external magnetic field is applied. The five energy levels of the d orbital are no longer degenerate and split into discrete levels that also enforce particular directions of the orbital angular momentum $\bar{l}$—this phenomenon is known as space quantisation (Fig. 10.7). The value of the orbital angular momentum (the length of the vector $\bar{l}$) is different for the different subshells and given by:

$$|\bar{l}| = \frac{h}{2\pi} \cdot \sqrt{l \cdot (l+1)}, \tag{10.14}$$

so for $l = 2$, this yields a value of $|\bar{l}| = \frac{h}{2\pi} \cdot \sqrt{6}$. In the presence of a magnetic field $\bar{B}$, the component parallel to $\bar{B}$ then depends on the magnetic quantum number m_l:

$$|\bar{l}_{\bar{B}}| = \frac{h}{2\pi} \cdot |m_l|. \tag{10.15}$$

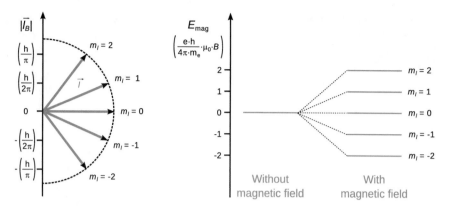

Fig. 10.7 Space quantisation illustrated for an electron in the d orbital. Left: Possible directions of the orbital angular momentum $\vec{l}$ with respect to an external magnetic field $\vec{B}$. Right: Space quantisation in the presence of an external magnetic field leads to a potential magnetic energy E_{mag}

10.2.2 The Spin Quantum Number

Based on the considerations in the previous section, one would expect that the spectral lines of atomic hydrogen split into three ($l = 1$) or five ($l = 2$) lines in the presence of an external magnetic field. However, experimentally it is observed that the spectral lines split into even numbered sets when a magnetic field is present. The pivotal discovery by George Uhlenbeck and Samuel Goudsmit in 1925 was the fact that electrons not only orbit the atomic nucleus but also rotate around their own axis, leading to the term electron spin. In addition to the orbital angular momentum $\vec{l}$, one thus also needs to consider a spin momentum $\vec{s}$, which is characterised by a spin quantum number s. In contrast to all other quantum numbers, the spin quantum number only assumes one value:

$$s = \frac{1}{2}.$$

(10.16)

In analogy to the magnetic quantum number m_l (Eq. 9.27), it follows for the magnetic spin quantum number m_s:

$$m_s = -s, s,$$

(10.17)

and therefore assumes the values of $-1/2$ or $1/2$.

The general considerations we introduced for the orbital angular momentum (Eqs. 10.14 and 10.15) are also valid for the spin momentum and thus lead to the equations:

Fig. 10.8 Possible directions
of the electron spin with
respect to an external
magnetic field

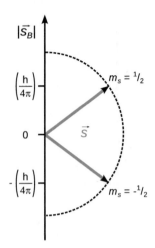

$$|\bar{s}| = \frac{h}{2\pi} \cdot \sqrt{s \cdot (s+1)}, \text{and} \qquad (10.18)$$

$$|\bar{s}_{\bar{B}}| = \frac{h}{2\pi} \cdot |m_s|. \qquad (10.19)$$

With s only assuming one value ($s = 1/2$), this means that the spin momentum for all electrons is the same, regardless of the orbital they occupy:

$$|\bar{s}| = \frac{h}{2\pi} \cdot \sqrt{\frac{3}{4}}. \qquad (10.20)$$

Also, there are only two possible states with respect to an external magnetic field $\bar{B}$ (Fig. 10.8).

10.2.3 Spin–Orbit Coupling

We mentioned earlier (Sect. 10.2.1) that orbiting of an electron around the atomic nucleus gives rise to a magnetic field. If an electron spins around its own axis, as we have just introduced in the preceding section, then this also constitutes the motion of a charge which also induces a magnetic field. Both of those magnetic fields couple and result in a total angular momentum $\bar{j}$

$$\bar{j} = \bar{l} + \bar{s}, \qquad (10.21)$$

which is characterised by the quantum number j of the total angular momentum. The values assumed by j combine the angular quantum number l and the spin quantum number s:

$$j = (l+s), (l+s-1), \ldots |l-s|. \tag{10.22}$$

For example, for an electron in a p orbital where $l = 1$ and $s = 1/2$, this yields:

$$j = (1 + \tfrac{1}{2}), (1 - \tfrac{1}{2}) = \tfrac{3}{2}, \tfrac{1}{2}.$$

The magnetic quantum number of the total angular momentum, m_j, can generally assume the values:

$$m_j = -j, (-j+1), \ldots, (j-1), j \tag{10.23}$$

and specifically for the above example of the electron in the p orbital:

$$j = \tfrac{1}{2} \qquad m_j = -\tfrac{1}{2}, \tfrac{1}{2}$$
$$j = \tfrac{3}{2} \qquad m_j = -\tfrac{3}{2}, -\tfrac{1}{2}, \tfrac{1}{2}, \tfrac{3}{2}.$$

This shows that through the spin–orbit coupling, there are already two energetically different states in the p orbital: one with $j = 1/2$, and one with $j = 3/2$. If an external magnetic field is applied, these states split up further, characterised by the magnetic quantum number of the total angular momentum (Fig. 10.9). The external magnetic field causes the disappearance of the degeneracy in the two j-states, resulting in two energy levels for $j = 1/2$ and four energy levels for $j = 3/2$, in agreement with the observed splitting of lines in the atomic spectra.

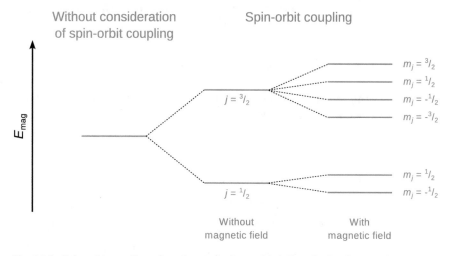

Fig. 10.9 Spin–orbit coupling of an electron in the p orbital. Even in the absence of an external magnetic field, there are two different states of the total angular momentum. In the presence of an external magnetic field, these two states split into further states (Zeeman effect)

10.3   Multi-Electron Atoms

10.3.1 The Schrödinger Equation for Multi-Electron Atoms

We have seen in the previous sections that the Schrödinger equation can be solved for the hydrogen atom, which consists of a positively charged nucleus and one electron. The same is true for any one-electron species, such as H_2^+, He^+, Li^{2+}, Be^{3+}, etc. However, atoms that possess more than one electron present additional challenges such that the Schrödinger equation can no longer be solved in an exact fashion.

If more than one electron is present, each electron experiences an attractive force by the nucleus, but a repelling force by all other electrons. The potential energy in the Schrödinger equation therefore not only depends on the distance of an electron from the nucleus, but also from the distance to all other electrons. As this requires consideration of all individual interactions between the electrons, it becomes impossible to determine an exact solution of the Schrödinger equation.

An approximation suggested by Douglas Hartree in 1927 replaces the individual inter-electronic interactions by the interaction of an electron with a mean field which may be assumed to be of spherical symmetry. The non-linear equations evolving from this approach are solved in an iterative fashion, and the methodology has become known as the self-consistent field method or Hartree-Fock method. Even though more accurate methods have been developed since, the Hartree-Fock method remains the starting point for almost all methods that describe multi-electron systems. Its inherent shortcomings stem from that fact that the mean field of other electrons are assumed to be of spherical symmetry. However, it has become apparent that especially for heavy atoms those errors are indeed very small.

10.3.2 Electronic Configuration of Atoms

We have seen in the previous discussion of the hydrogen atom, that the allowed energy levels can be calculated and only depend on the principal quantum number n (Eq. 10.9), such that the atomic orbitals can be arranged in the order of increasing energy:

$$1s < 2s = 2p < 3s = 3p = 3d < 4s = 4p = 4d = 4f < \ldots$$

For multi-electron systems, calculations show that the order differs from that of the one-electron system above; specifically:

$$1s < 2s < 2p < 3s < 3p < 4s < 3d < 4p < 5s < 4d \ldots$$

The reason for this difference is in the fact that electrons for example in the $2s$ orbital are closer to the nucleus than those in the $2p$ orbital, given the spherical distribution of the former (see Sect. 10.1.3). The $2s$ electrons thus experience a stronger attractive force by the nucleus than the $2p$ electrons which are shielded

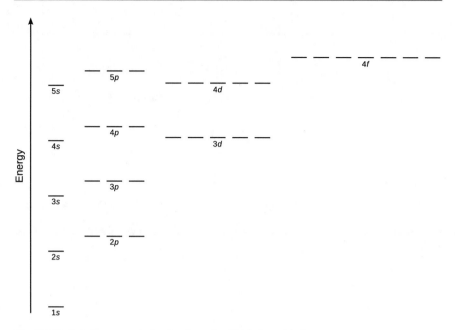

Fig. 10.10 Relative energetic levels of atomic orbitals in multi-electron systems

(by the 2s electrons). The shielding makes it a less stable arrangement and hence leads to a higher potential energy. When arranging the individual atomic orbitals relative to each other on an energy scale, one arrives at scheme as illustrated in Fig. 10.10.

The order number Z of the periodic system indicates the number of protons in the nucleus and hence the number of positive charges of an atom that need to be balanced by the surrounding electrons (when considering an element). Clearly, when populating the scheme in Fig. 10.10 with electrons for atomic hydrogen, the one electron needs to go into the lowest lying orbital (1s). When considering the helium atom, a first complication arises due to the fact that the 1s state is a two-fold degenerate state (see Sect. 10.2.2), with two possibilities for the magnetic spin quantum number $m_s = -1/2, +1/2$. The question is whether both electrons will assume the same (parallel) spin direction ($-1/2$ and $-1/2$, or $+1/2$ and $+1/2$), or whether they assume different (anti-parallel) directions ($-1/2$ and $+1/2$). Analysis of the atomic spectrum of helium shows that only the second scenario agrees with the observed spectrum, therefore, the two electrons occupying the 1s orbital assume anti-parallel spins (Fig. 10.11).

For the alkali metal lithium, the atomic spectrum indicates a one-electron system, despite the total number of three electrons that need to be considered. This can only be achieved if two electrons occupy the 1s orbital, as in the case of the preceding element (helium), and the third electron occupies the next higher orbital in the scheme (2s). Since the energy of the 2s orbital is higher than that of the 1s orbital, the electrons in the 2s orbital are in a less stable state, and therefore easier to remove from the atom. The ionisation energy of lithium, which can be determined from the

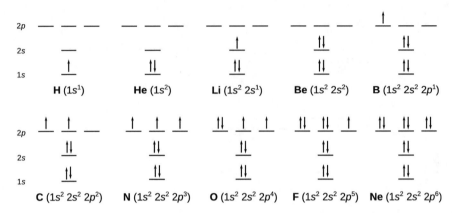

Fig. 10.11 Population of atomic orbitals with electrons for the first ten elements in the periodic system

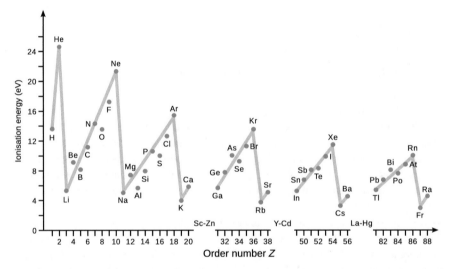

Fig. 10.12 The first ionisation energies of elements as a function of their order number (Z) in the periodic system

atomic spectrum (see Sect. 8.2.1), would thus be expected to be less than the ionisation energy of hydrogen or helium, and this is indeed the case (Fig. 10.12).

Another challenge is then observed with carbon. The $2p$ orbital has already been populated by boron with one electron. The question arises whether the next electron goes into the same p orbital or into another one. And what spin does it assume? Figure 10.11 shows that a new p orbital is occupied, and the spin of the added electron is the same as the one in the first p orbital.

These rules for populating atomic orbitals are known as the Aufbau principle, Pauli exclusion principle and Hund's rules:

▶ The Pauli exclusion principle states that in atomic and molecule systems, two electrons cannot possess the exact same four quantum numbers.

▶ The Aufbau principle is a consequence of the Pauli principle which requires electrons to occupy higher energy levels once lower levels are filled.

▶ According to Hund's rules, if there are multiple orbitals at similar energy levels, electrons populate individual orbitals before pairing up. Unpaired electrons that populate orbitals of the same energy level assume the same spin direction.

The Pauli exclusion principle, formulated by the Austrian physicist Wolfgang Pauli in 1925, is a general quantum mechanical principle that applies to all particles with half-integer spin (so-called fermions). Like the laws of thermodynamics, it is a fundamental principle of observation that cannot be proven.

Hund's rule is also known as the rule of maximum multiplicity. Since electrons are charged particles and pairing them up is energetically costly due to the repelling force between particles of like charges, this is avoided as long as possible. In general, the achievement of half-filled orbitals is preferred, even if an orbital of slightly lower energy is only partially filled as a consequence (see e.g. transition metals).

Above rules enable the prediction of electron configurations of individual atoms (and molecules). When applying these rules to the elements, arranged in increasing order of their nuclear charge (order number Z), the periodicity becomes apparent. These quantum mechanical rules therefore explain the periodic system of the elements which was originally assembled based on chemical and physical properties of the individual elements (and credited to Dmitri Mendeleev in 1869).

Table 10.6 The electron configuration and relationships of quantum numbers explain the periodicity in the periodic table of the chemical elements

Principal quantum number n	Shell	Electron symbol based on angular momentum quantum number l	Possible quantum numbers m_l	Possible spin states	Max. number of electrons for given	
					l	n
1	K	$1s$	1	2	2	2
2	L	$2s$	1	2	2	8
		$2p$	3	2	6	
3	M	$3s$	1	2	2	18
		$3p$	3	2	6	
		$3d$	5	2	10	
4	N	$4s$	1	2	2	32
		$4p$	3	2	6	
		$4d$	5	2	10	
		$4f$	7	2	14	

Table 10.6 summarises the relationships between the quantum numbers, electron configurations and shells.

An obvious illustration of the periodicity caused by the quantum mechanical electron configuration is given by the first ionisation energies, which describe the energy required to remove one (the outermost) electron from the atom (see Fig. 10.12). Their comparison shows that moving from a noble gas to the following alkali metal is accompanied by a drastic lowering of the binding energy of the outermost electron; the binding energy then rises again up to a maximum at the next noble gas. This implies phenomenologically that a new shell of electrons is being filled with each alkali metal.

10.4   Exercises

1. Calculate at which distances from the nucleus the $3s$ orbital possesses radial nodes.

2. Derive the electron configuration for (a) vanadium, and (b) chromium in their ground states.

3. Name the four quantum numbers of the electron and state their relationships/ possible values.

4. Describe the overall shape of the orbitals characterised by the angular momentum quantum number $l = 0, 1, 2$.

5. Determine the ground-state electronic configuration of the following species: S^-, Zn^{2+}, Cl^-, Cu^+.

6. (a) Using the knowledge about the allowed energy states (eigenvalues; Eq. 10.9), calculate the ionisation potential of the hydrogen atom.
   (b) The equation yielding the allowed energy states of a multi-electron atom is often modified to $E_n = -\dfrac{m_e \cdot e^4 \cdot (Z - \sigma)^2}{8 \cdot \varepsilon_0^2 \cdot h^2 \cdot n^2}$ to accommodate the shielding of the nuclear charge by electrons in the various orbitals; the parameter σ is thus called the screening constant. Calculate the value of σ for helium, assuming its first ionisation potential is 24.5 eV.

The Chemical Bond

11

After considering the make-up of isolated atoms, we now want to have a closer look at how atoms interact with each other to form compounds and molecules. Practical experiences with different substances tell us that the type of bonds between atoms can be quite different.

When combining two atoms of substantially different ionisation energies, such as for example observed with alkali halides (see Fig. 10.12), one can expect a tendency of an electron being transferred from one atom to another, resulting in the generation of ions. These substances are thus likely to form bonds based on electrostatic interactions, i.e. ionic bonds. A different type of bond is to be expected between atoms that have similar or the same ionisation energies, such as e.g. gaseous hydrogen or diamond. These types of substances are formed by covalent bonds. Yet a different type of substance are metals. While also being formed by interactions of atoms of the same element, the main difference of metals when compared to covalently bonded substances is the fact that they can conduct electric current. The metallic bond therefore must have particular features. Last, we know that noble gases generally resist chemical reactions and, for large parts, do not engage in compound formation. However, even noble gases form liquid and solid phases. In those states, the noble gas atoms are held together by van der Waals interactions.

In the following sections, we will discuss these four types of chemical bonds, but appreciate that these are particularly distinct instances; there will also be many cases where the type of chemical bond is a mixture of two of these distinct bonding types.

11.1 The Ionic Bond

As an example for an ionic substance we will consider the alkali halide NaCl which adopts a cubic close-packed crystal structure where every individual ion is surrounded by six counter-ions sitting on the vertices of an octahedron (Fig. 11.1).

© Springer International Publishing AG, part of Springer Nature 2018
A. Hofmann, *Physical Chemistry Essentials*,
https://doi.org/10.1007/978-3-319-74167-3_11

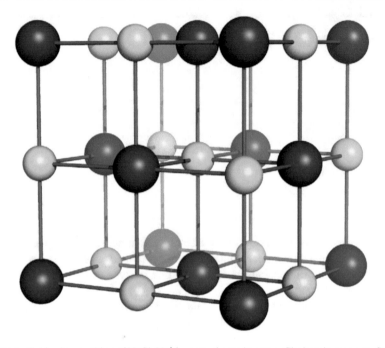

Fig. 11.1 Cubic close packing of NaCl. Na$^+$ ions are shown in green, Cl$^-$ ions in magenta. Sphere sizes have been decreased to allow drawing of the crystal lattice

The force describing the electrostatic interaction between two ions of charge Q_1 and Q_2 is given by Coulomb's law:

$$F_{\text{Coulomb}} = \frac{Q_1 \cdot Q_2}{\varepsilon \cdot r^2}, \qquad (11.1)$$

where ε is the dielectric constant of the medium. Since energy is the product of force and distance:

$$E = F \cdot r, \qquad (11.2)$$

the potential energy between two ions is due to the Coulomb force of electrostatic attraction and given by

$$V_C = \frac{Q_1 \cdot Q_2}{\varepsilon \cdot r}. \qquad (11.3)$$

For two ions with the charge numbers z_+ and z_- in a crystal ε lattice, this yields the following expression:

$$V_C = \frac{z_+ \cdot z_- \cdot e^2}{4\pi \cdot \varepsilon_0 \cdot r}, \qquad (11.4)$$

where we replaced the dielectric constant ε with the vacuum permittivity ε_0 (more precisely the product $4\pi\cdot\varepsilon_0$) introduced in Sect. 8.2.2. Note that the numerical value of the Coulomb potential energy V_C in Eq. 11.4 is negative, since z_+ has a positive and z_- a negative sign. V_C thus assumes an indefinite negative value at $r = 0$, i.e. when both ions collapse. This is certainly not what we observe. The Coulomb potential energy thus describes the real phenomenon only when the two ions are at sufficient distance from each other.

As two particles get spatially very close, the short-range repulsion forces gain weight. Therefore, we also need to consider a potential energy V_R describing this repulsion which may be modelled in this case by

$$V_R = A \cdot e^{-\frac{r}{\rho}}, \tag{11.5}$$

where A and ρ are constants.

The total potential energy between two ions in the lattice is thus a composite of the attractive (Coulomb) and the repulsive terms:

$$E_{\text{pot}} = V_C + V_R = \frac{z_+ \cdot z_- \cdot e^2}{4\pi \cdot \varepsilon_0 \cdot r} + A \cdot e^{-\frac{r}{\rho}}. \tag{11.6}$$

As illustrated in Fig. 11.2, the total potential energy E_{pot} possesses a minimum at a particular distance r_{eq} which is the equilibrium distance between the two ions.

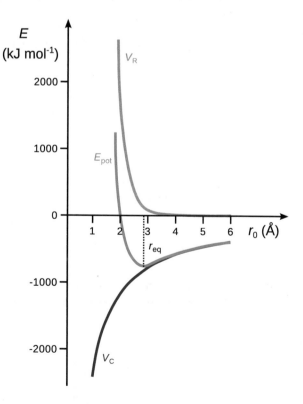

Fig. 11.2 Potential energy of the ionic bond, illustrated for NaCl

Since there are more than two ions in an ionic crystal, we need to consider all possible inter-ionic interactions. Therefore, we first focus on one cation which has a charge z_+ and the distance of this ion to each other ion with running index i shall be $r_{+,i}$. We thus obtain for the Coulomb potential energy for this cation:

$$V_{C+} = \frac{z_+ \cdot e^2}{4\pi \cdot \varepsilon_0} \cdot \sum_i \frac{z_i}{r_{+,i}}.$$

Similarly, this approach yields for the Coulomb potential energy of a particular anion:

$$V_{C-} = \frac{z_- \cdot e^2}{4\pi \cdot \varepsilon_0} \cdot \sum_i \frac{z_i}{r_{-,i}}.$$

Since the distances do not vary continuously, but systematically due to the regular arrangement of ions in the lattice, $r_{+,i}$ and $r_{-,i}$ can be expressed as multiples of a smallest distance r_0:

$$r_{+,i} = a_{+,i} \cdot r_0 \quad \text{and} \quad r_{-,i} = a_{-,i} \cdot r_0.$$

Further, we need to make sure that the interactions between cat- and anions are not counted twice when combining V_{C+} and V_{C-}, and thus introduce a factor of $^1/_2$. Considering 1 mol of substance, we obtain for the molar Coulomb potential energy in the ionic crystal:

$$V_C = \frac{1}{2} \cdot (V_{C+} + V_{C-}) \cdot N_A = \frac{z_+ \cdot z_- \cdot e^2 \cdot N_A}{4\pi \cdot \varepsilon_0 \cdot r_0} \cdot \frac{1}{2} \cdot \sum_i \frac{z_i}{z_- \cdot a_{+,i}} + \frac{z_i}{z_+ \cdot a_{-,i}}.$$

$$(11.7)$$

The factor

$$\frac{1}{2} \cdot \sum_i \frac{z_i}{z_- \cdot a_{+,i}} + \frac{z_i}{z_+ \cdot a_{-,i}} = M \qquad (11.8)$$

is called the Madelung constant M, named after the German physicist Erwin Madelung. This yields for the attractive (Coulomb) potential energy (Eq. 11.7) in the ionic crystal

$$V_C = \frac{1}{2} \cdot (V_{C+} + V_{C-}) \cdot N_A = \frac{z_+ \cdot z_- \cdot e^2 \cdot N_A}{4\pi \cdot \varepsilon_0} \cdot \frac{M}{r_0}$$

where z_+, z_- and r_0 are characteristic constants of a substance, and the Madelung constant M is a characteristic parameter capturing the geometric arrangements in a particular lattice type. Using the above expression, the total potential energy (Eq. 11.6) becomes

$$E_{pot} = \frac{z_+ \cdot z_- \cdot e^2 \cdot N_A}{4\pi \cdot \varepsilon_0} \cdot \frac{M}{r_0} + A \cdot e^{-\frac{r_0}{\rho}} \qquad (11.9)$$

The lattice energy $\Delta U_{lattice}$ of a crystal describes the energy required to break the bonds of 1 mol of an ionic solid under standard conditions; it equals the potential energy at the equilibrium distance r_{eq}, i.e. the minimum of the function $E_{pot}(r_0)$ illustrated in Fig. 11.2. At the minimum of a function $f(x)$, the derivative $\frac{df(x)}{dx}$ equals zero.

For the potential energy function in Eq. 11.9 this results in

$$\frac{dE_{pot}}{dr_0} = \frac{-z_+ \cdot z_- \cdot e^2 \cdot N_A}{4\pi \cdot \varepsilon_0} \cdot \frac{M}{r_0^2} - \frac{A}{\rho} \cdot e^{-\frac{r_0}{\rho}} = 0$$

and allows calculation of the parameter A:

$$A = \frac{-z_+ \cdot z_- \cdot e^2 \cdot N_A \cdot M \cdot \rho}{4\pi \cdot \varepsilon_0 \cdot r_0^2} \cdot e^{\frac{r_0}{\rho}}.$$

If we substitute this expression for A in Eq. 11.9 we obtain the following expression for the lattice energy of an ionic crystal:

$$\Delta U_{lattice} = E_{pot}(r_{eq}) = \frac{z_+ \cdot z_- \cdot e^2 \cdot N_A \cdot M}{4\pi \cdot \varepsilon_0 \cdot r_0} \cdot \left(1 - \frac{\rho}{r_0}\right). \qquad (11.10)$$

As mentioned above, the charge numbers z_+ and z_- are characteristic parameters of a substance, r_0 is available from the lattice parameters for this substance and M is a geometric constant for a particular lattice type. The extent of the repulsion potential is expressed by ρ. We appreciate that this parameter is directly linked to the compressibility of the ionic crystal: the longer the range of the repulsion potential, the larger is ρ and the less the lattice can be compressed. The experimental determination of compressibility of solids thus allows determination of the parameter ρ.

11.2 Electronegativity

The tendency of an atom to attract electrons when being part of a molecule is called electronegativity. The distribution of electrons in a heteronuclear molecules therefore can be seen as a direct result of differences in electronegativity. Accordingly, in an ionic bond, as discussed in the previous section, the difference in electronegativity between the two atoms is thus at its most extreme.

The most commonly used method to calculate electronegativity is the one proposed by Linus Pauling (Pauling 1932). In its original form, the Pauling scale of relative electronegativity values (arbitrarily) assigns a value of 4.0 to fluorine. The values of a subsequently refined scale are summarised in Table 11.1. Generally,

Table 11.1 Electronegativities of the elements in their most common and stable oxidation states

1	2	3	4	5	6	7	8	9	10	11	12	13	14	15	16	17	18
H 2.20																	He
Li 0.98	Be 1.57											B 2.04	C 2.55	N 3.04	O 3.44	F 3.98	Ne
Na 0.93	Mg 1.31											Al 1.61	Si 1.90	P 2.19	S 2.58	Cl 3.16	Ar
K 0.82	Ca 1.00	Sc 1.36	Ti 1.54	V 1.63	Cr 1.66	Mn 1.55	Fe 1.83	Co 1.88	Ni 1.91	Cu 1.90	Zn 1.65	Ga 1.81	Ge 2.01	As 2.18	Se 2.55	Br 2.96	Kr 3.00
Rb 0.82	Sr 0.95	Y 1.22	Zr 1.33	Nb 1.6	Mo 2.16	Tc 1.9	Ru 2.2	Rh 2.28	Pd 2.20	Ag 1.93	Cd 1.69	In 1.78	Sn 1.96	Sb 2.05	Te 2.1	I 2.66	Xe 2.60
Cs 0.79	Ba 0.89	(La)	Hf 1.3	Ta 1.5	W 2.36	Re 1.9	Os 2.2	Ir 2.20	Pt 2.28	Au 2.54	Hg 2.00	Tl 1.62	Pb 1.87	Bi 2.02	Po 2.0	At 2.2	Rn 2.2
Fr 0.7	Ra 0.9	(Ac)	Rf	Db	Sg	Bh	Hs	Mt	Ds	Rg	Cn	Uut	Fl	Uup	Lv	Uus	Uuo

(La)	La 1.1	Ce 1.12	Pr 1.13	Nd 1.14	Pm 1.13	Sm 1.17	Eu 1.2	Gd 1.2	Tb 1.1	Dy 1.22	Ho 1.23	Er 1.24	Tm 1.25	Yb 1.1	Lu 1.27
(Ac)	Ac 1.1	Th 1.3	Pa 1.5	U 1.38	Np 1.36	Pu 1.28	Am 1.13	Cm 1.28	Bk 1.3	Cf 1.3	Es 1.3	Fm 1.3	Md 1.3	No 1.3	Lr 1.3

caesium is the least electronegative element in the periodic table (0.79), while fluorine is the most electronegative (3.98).

11.3 The Covalent Bond

The occurrence of ionic bonds appears as a fairly straightforward phenomenon, since we can easily appreciate its coming-about due to the attraction between two particles of opposite charge. A more complex situation arises when bonds occur between rather similar atoms, i.e. such that exhibit rather small differences in their electronegativity values. Obviously, the formation of particular molecules must result in stable arrangements, meaning that these molecules correspond to a minimum in the overall energy function.

As we have previously seen, even in the case of multi-electron atoms, it is not possible to find an exact solution of the Schrödinger equation. The same problem arises when one considers molecules which comprise of multiple atoms and thus many electrons. Different ways of approximation of such more complex systems have been developed, including the valence bond (VB) method and the molecular orbital (MO) method.

The molecular orbital method, introduced by Friedrich Hund and Robert Mulliken in 1928, considers the nuclear scaffold of a molecule (e.g. the two nuclei of a di-atomic molecule) with varying distances. First, the molecular orbitals are determined; these orbitals localise to multiple atoms (i.e. they are poly-centric), as opposed to the atomic orbitals which localise to just one atom. Then, the molecular orbitals are filled with electrons according to the Aufbau principle as well as the Pauli principle and Hund rules.

11.3.1 The Born-Oppenheimer Approximation

Since atoms comprise of the atomic nucleus as well as electrons, the kinetic energies of both nuclei and electrons need to be considered in a rigorous treatment of energies in a molecule. The Born-Oppenheimer approximation poses that the heavier nuclei remain static at time scales during which the electrons are in movement (see also Franck-Condon principle, Sect. 13.4.1). In other words, this approximation allows separation of the motion of electrons from that of atomic nuclei. In mathematical terms, the wave function of a molecule can thus be thought of as being composed of an electronic and a nuclear function:

$$\Psi = \Psi_{\text{electrons}} \cdot \Psi_{\text{nuclei}}. \tag{11.11}$$

At the time scale of electron motion, the atomic nuclei only affect electrons by an electrostatic potential.

11.3.2   Linear Combination of Atomic Orbitals

If we consider a di-atomic molecule consisting of atoms 1 and 2, electrons travel on paths that wrap around both atoms. However, if they are spatially closer to atom 1, they will experience mainly the forces by the nucleus of atom 1 as well as the electrons in its vicinity. The forces originating from the nucleus and electrons of atom 2 are comparably small, due to atom 2 being located further away. The situation electrons find around atom 1 will therefore be similar (but not the same) as if they were part of an isolated atom 1 that is not part of a molecule. The same argument can be made for the situation around atom 2.

It is thus conceivable that the molecular orbital Ψ will have strong characteristics of the individual atomic orbitals Ψ_1 and Ψ_2 in vicinity of either atom 1 or atom 2. Mathematically, this can be modelled by a linear combination of the individual atomic orbitals Ψ_1 and Ψ_2 which constitute the molecular orbital Ψ. The contributions of the individual atomic orbitals—which are now called base functions—don't have to be the same and are thus expressed in the general form by weighting factors (c_1, c_2):

$$\Psi = c_1 \cdot \Psi_1 + c_2 \cdot \Psi_2 \tag{11.12}$$

This approach is called the linear combination of atomic orbitals (LCAO) and can alternatively be formulated as:

$$\Psi = \Psi_1 + \lambda \cdot \Psi_2, \tag{11.13}$$

where λ is a measure of the polarity of the molecular orbital.

Conceptually the LCAO describes an interference of the two base functions Ψ_1 and Ψ_2. Keeping in mind that the base functions are indeed wave functions, we remember from Sect. 8.1.2 that waves can interfere in a constructive and a destructive manner. In a similar fashion we need to consider a two types of interactions between the base functions; they can either add or subtract, giving rise to two different molecular orbitals, a bonding one Ψ and an anti-bonding one Ψ^*:

$$\Psi = c_1 \cdot \Psi_1 + c_2 \cdot \Psi_2$$

$$\Psi^* = c_1 \cdot \Psi_1 - c_2 \cdot \Psi_2.$$

As an example, we consider the formation of the hydrogen molecule H_2. The $1s$ orbitals of each of the individual H atoms combine according to the linear combination of atomic orbitals to yield two molecular orbitals, the σ and the σ^* orbital (Fig. 11.3). Since the σ orbital shows build-up of electron density in the space between two the atomic nuclei, it is called a bonding orbital, since its population with electrons keeps the two atomic nuclei together. In contrast, the σ^* orbital shows low electron density between the two nuclei, hence its population with electrons would lead be destabilising for the assembly of the two atoms; it is called anti-bonding orbital.

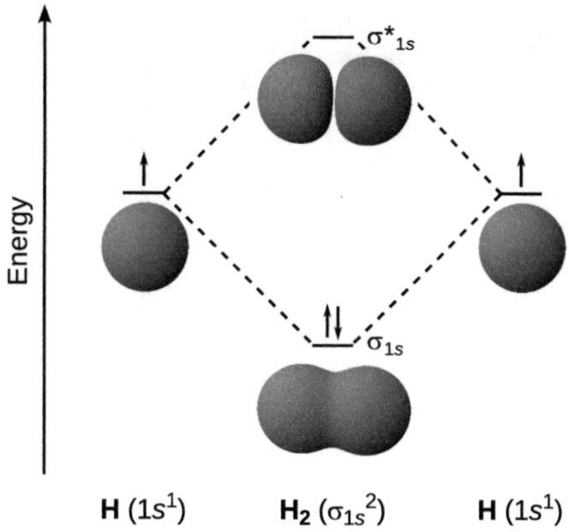

Fig. 11.3 Linear combination of the two $1s$ orbitals in hydrogen leads to bonding and anti-bonding molecular orbitals of H_2

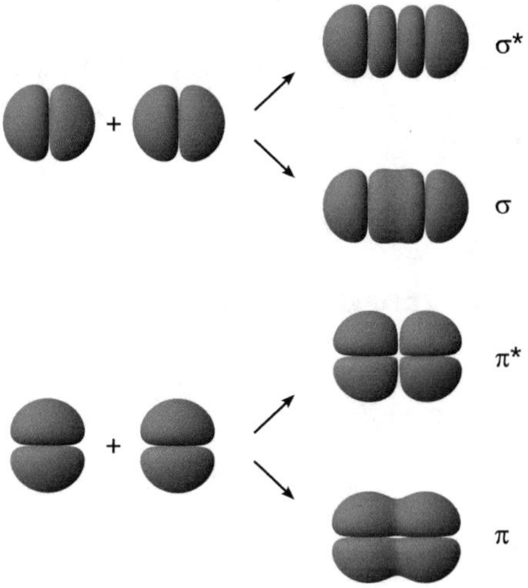

σ^*_{1s}

σ_{1s}

H ($1s^1$) H_2 (σ_{1s}^2)   H ($1s^1$)

Fig. 11.4 The interaction of two p orbitals can result in formation of either σ/σ^* or π/π^* molecular orbitals

σ^*

σ

π^*

π

When considering linear combinations of the p orbitals of two atoms engaging in an interaction, it becomes obvious that there are two types of overlap (Fig. 11.4): In one case, the p orbitals extending along the inter-nuclear axis can overlap and this gives rise to electron density between the two atoms along the inter-nuclear axis. If we think of the inter-nuclear distance as a rotation axis, we appreciate that the resulting orbital has rotational symmetry, just like the σ_s orbital. This type of interaction therefore results in a σ_p and a σ^*_p orbital.

The other two p orbitals have their lobes extending either above/below or in front/behind of the internuclear axis. Overlap of those p orbitals therefore leads to electron density between the two atoms above/below or in front/behind of the inter-nuclear axis. Such orbitals are called π orbitals. As before, constructive and destructive interference is possible, giving rise to two molecular orbitals upon interaction, π and

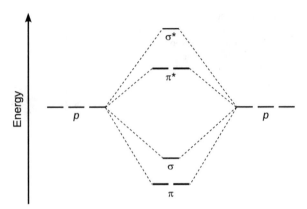

Fig. 11.5 Energy diagram
for linear combination of the
p orbitals in the second period

π^*. Notably, the π/π^* orbitals do not possess rotational symmetry, as the sign of those orbitals changes when rotating $180°$ around the inter-nuclear axis. The relative order of σ- and π orbitals arising from linear combination of $2p$ orbitals is illustrated in Fig. 11.5.

Generally, when choosing atomic orbitals for linear combination, three pre-requisites need to be fulfilled:

- The energies of the two atomic orbitals need to be at a comparable level.
- Both atomic orbitals need to be able to produce sufficient overlap.
- The two atomic orbitals need to have the same symmetry with respect to the inter-atomic axis.

11.3.3 Bond Order

We have mentioned earlier that the molecular orbitals are populated with electrons according to the Aufbau principle. Of course, just as in the case of atomic orbitals, the Aufbau principle requires knowledge of the relative energetic levels of the individual molecular orbitals, so that we fill the different orbitals from the bottom to the top. As illustrated for O_2 in Fig. 11.6, this requires that anti-bonding orbitals are populated and we remember that electrons in anti-bonding orbitals result in destabilisation of the bond between two atoms.

In order to obtain an overall measure of the strength of a bond, the bond order is defined as:

$$\text{bond order} = \frac{1}{2} \cdot \left(N_{\text{bonding}} - N_{\text{anti-bonding}} \right) \qquad (11.14)$$

Fig. 11.6 MO scheme for oxygen (O_2)

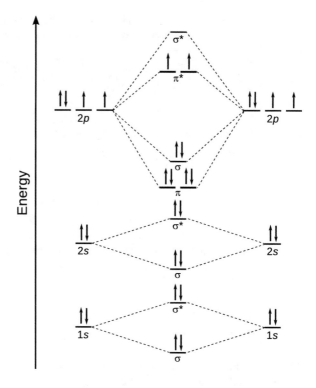

where $N_{bonding}$ is the number of electrons in bonding and $N_{anti\text{-}bonding}$ the number of electrons in anti-bonding orbitals. For the O_2 molecule (see Fig. 11.6), the border is calculated as per:

$$\text{bond order} = \frac{1}{2} \cdot (10 - 6) = 2.$$

11.3.4 Magnetic Properties of Diatomic Molecules

The molecular orbital scheme for oxygen highlights that the O_2 molecule has two unpaired electrons. Because unpaired electrons can orient in either direction, they exhibit magnetic moments that can align with an external magnetic field; such molecules are called paramagnetic. In contrast, molecules that do not possess unpaired electrons are called diamagnetic; they possess no net magnetic moment and are thus not attracted into a magnetic field. Indeed, diamagnetic materials exhibit a weak repulsion to external magnetic fields.

The prediction of the magnetic properties of molecules is a major advantage of molecular orbital theory as compared to the concept of Lewis structures. The concept of Lewis structures, introduced by Gilbert N. Lewis in 1916 (Lewis 1916) for covalently bonded molecules, predicts the bonding between and lone pair electrons of atoms in molecules based on the number of electrons in their outermost shell (the

Table 11.2 Properties of select diatomic gases. The magnetic properties and bond order are predicted by molecular orbital theory. The experimental values of bond lengths and bond energies reflect the predicted bond order

Molecule	Magnetic properties	Bond order	Bond length (Å)	Bond energy (kJ mol^{-1})
H_2	diamagnetic	1	0.74	436.0
He_2	–	0	–	–
N_2	diamagnetic	3	1.10	941.4
O_2	paramagnetic	2	1.21	498.8
F_2	diamagnetic	1	1.42	150.6
Ne_2	–	0	–	–

valence shell). For the O_2 molecule, the Lewis structure does not inform about the presence of unpaired electrons (and hence paramagnetism).

Table 11.2 summarises properties of select diatomic gases. Importantly, the bond order and electron configuration determined by the molecular orbital theory correctly predicts the magnetic properties and even the fact that some diatomic molecules (such as for example He_2, Ne_2) do not exist, since their bond order is zero.

11.3.5 Dipole Moment

The considerations in the previous chapters assumed that the linear combination of atomic orbitals occurs between two atoms of the same element, which results in formation of homonuclear di-atomic molecules. In the illustrations (Figs. 11.3, 11.5, and 11.6), this manifests itself in the fact that the two interacting atomic orbitals possess the same energy. Moreover, at the level of the quantum mechanical wave function, this results in equality of the two coefficients c_1 and c_2 (Eq. 11.12) or $\lambda = \pm 1$ (Eq. 11.13).

Whereas the same methodology can be applied to hetero-nuclear di-atomic molecules, we appreciate that the two atoms no longer belong to the same element, hence the energies of the interacting atomic orbitals are no longer equal. Similarly, the weighting coefficients c_1 and c_2 are not the same, and $\lambda \neq \pm 1$.

As mentioned in Sect. 11.3.2, the coefficient λ is a measure of the polarity of the molecular orbital, and if its value deviates from 1, the bond possesses a permanent electric dipole momentum (i.e. a bond moment). This is a result of one atom in the molecule attracting electrons more strongly than the other, leading to a formal partial negative charge on one atom and a formal positive charge on the other. In general, a dipole moment μ arises from a charge separation in space, and is thus defined as

$$\mu = Q \cdot r, \tag{11.15}$$

where Q is the separated charge (e.g. $1 \cdot e = 1.602 \cdot 10^{-19}$ C) and r the distance between the positive and negative centres. Since such separations happen at the scale of a chemical bond, the dipole moment is measured in multiples of $3.338 \cdot 10^{-30}$ C m which is called the debye:

Bond	Bond moment (D)	Electronegativity difference
HF	1.9	1.8
HCl	1.1	1.0
HBr	0.80	0.8
HI	0.42	0.5
H–O	1.5	1.2
H–N	1.3	0.8
H–P	0.40	0
C–H	0.40	0.4
C–F	1.4	1.5

Table 11.3 Examples of bond moments and electronegativity difference of the constituting atoms

$$[\mu] = 3.338 \cdot 10^{-30} \text{ C m} = 1 \text{ D}$$

and leads to the dipole moment for the charge separation due to one electron displaced by $1 \text{ Å} = 0.1 \text{ nm} = 10^{-10}$ m of

$$\mu = 1.602 \cdot 10^{-19} \text{ C} \cdot 10^{-10} \text{ m} = 1.602 \cdot 10^{-29} \text{ C m} = 4.8 \text{ D}.$$

When assessing the overall dipole moment of a molecule, one considers the individual bond moments as vectors (i.e. they have a value/length and a direction) and estimates the overall dipole moment by vector addition.

Table 11.3 compares bond moments of some select heteronuclear bonds with the electronegativity difference between the two atoms. If the dipole moment in a bond becomes larger, this results in the bond adopting more and more ionic character.

Dipole Moment of Ionic Compounds

The electronegativity difference between Na and Cl of 2.1 suggests that NaCl forms an ionic bond. If the bonding in crystalline NaCl was 100% ionic, the charge on the sodium atom was $+1 \cdot e$, and the charge on the chlorine $-1 \cdot e$. With an inter-nuclear distance of 2.36 Å, this results in a dipole moment of $\mu = 11.34$ D. The experimental value of the dipole moment can be obtained by microwave spectroscopy and yields $\mu = 9.001$ D. The ratio of the experimental and theoretical dipole moments

$$\frac{\mu_{exp}}{\mu_{theor}} = \frac{9.001 \text{ D}}{11.34 \text{ D}} = 0.794$$

indicates that the bonding between sodium and chlorine in the ionic solid is ~80% ionic.

11.4   The Metallic Bond

The characteristic property of metals is their ability to conduct electric current. In contrast to the covalent bond where electrons are localised, the metallic bond is characterised by delocalised electrons. Notably, this delocalisation does not result in weaker bonding; the bond energies of metals are indeed of the same order as those of ionic crystals or covalent di-atomic molecules.

The ability to conduct electrons requires a reasonable spatial extension of the metal (as opposed to, for example, fairly isolated di-atomic gas molecules). In other words, in order for a metallic bond to exist, one requires a bulk assembly of atoms in a condensed phase. In order to understand the metallic bond, it will thus be necessary to consider the arrangement of large numbers of atoms in space. For illustration (Fig. 11.7), we start with a di-atomic molecule, say Li_2 (which can be observed in the gas phase). Focussing on the outer shell only, the molecular orbital scheme for Li_2 can be established by linear combination of the two $2s$ orbitals, leading to a σ and a σ^* molecular orbital. Upon addition of a third and a fourth Li atom, the $2s$ orbitals of those atoms combine with the $2s$ orbitals of the initial Li atoms which leads to three and four molecular orbitals of different energies, respectively. For a large number of atoms (N), N molecular orbitals with slightly differing energies arise. If $N = 6.022 \cdot 10^{23}$, then the metal has 1 mol atoms and $6.022 \cdot 10^{23}$ molecular orbitals. Whereas the energies of these individual orbitals are different, they are so densely situated that an energy band (here: s-band) is established in which the energy levels are pseudo-continuous.

In the case of a Li atom, there is one electron in the $2s$ orbital. This means that in metallic lithium, there are as many electrons as molecular orbitals arising from the $2s$ combination. These electrons populate the s-band; but because every orbital can harbour a maximum of two electrons, the band is only half-populated. Due to the energy levels in the band being pseudo-continuous, electrons can be take up indefinitely small amounts of energy and thus become highly mobile. The energy can be provided by an electric field, established by a potential difference, which leads to electrons travelling and thus conduction of an electric current.

Fig. 11.7 The assembly of discrete atoms into bulk condensed matter in metals leads to generation of an electron band

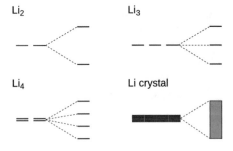

11.5   The van der Waals Bond

In the previous sections, we have seen that condensed matter can arise from different types of bonds, namely ionic (e.g. NaCl), metallic (e.g. Fe) and covalent (e.g. diamond). But what about the condensed phases such as for example the liquid or solid states of the noble gases, where the electronic configuration does not allow for covalent or ionic bonds?

The electrostatic potential of the nucleus is fully balanced by the electrons surrounding it. In the case of the noble gases, the electrons are distributed around the nucleus with spherical symmetry, due to full occupancy of all shells. However, this scenario describes the situation only in an averaged time window. Since the electrons are orbiting the nucleus, at any particular time, there may be a distribution that is not of spherical symmetry, and therefore leads to a temporary dipole momentum (see also Sect. 12.2). This dipole moment establishes an electric field whose value is given by

$$E = \frac{\mu_A}{4\pi \cdot \varepsilon_0 \cdot r^3} \qquad (11.17)$$

where μ_1 is the dipole moment in atom A, r is the distance from the dipole and ε_0 is the permittivity *in vacuo*. The electric field E then leads to an induced dipole moment in a neighbouring atom (atom B):

$$\mu_B = \alpha \cdot E = \frac{\alpha \cdot \mu_A}{4\pi \cdot \varepsilon_0 \cdot r^3}; \qquad (11.18)$$

α is called the polarisability and discussed in more detail in Sect. 12.2.1.

The potential energy of a dipole in an electric field is given by the product between the two quantities and given a negative sign since it is an attractive interaction:

$$V_{attr} = -\mu_B \cdot E = -\alpha \cdot E^2 = -\frac{\mu_A^2 \cdot \alpha}{4\pi^2 \cdot \varepsilon_0^2 \cdot r^6}. \qquad (11.19)$$

The important relationship in above equation is that the attractive interaction between the two atoms (= temporary dipoles) varies with r^{-6}:

$$V_{attr} \sim -\frac{1}{r^6}. \qquad (11.20)$$

This attractive potential is called the van der Waals potential (see Fig. 11.8).

If the distance between two atoms is made less and less, the two atoms approach each other and their electron shells start to overlap. This universal repulsive force becomes stronger, the closer the two atoms get. Empirically, it was found that this potential varies with r^{-12} and is known as the Pauli repulsive potential:

$$V_{rep} \sim \frac{1}{r^{12}}. \qquad (11.21)$$

Fig. 11.8 The Lennard-Jones potential (E_{pot}) as a combination of the attractive van der Waals potential (V_{attr}) and the repulsive potential (V_{rep}) between two atoms. The minimum in the potential curve occurs at the equilibrium distance r_{eq}.

$\sigma = \sqrt[6]{\frac{a}{b}}$

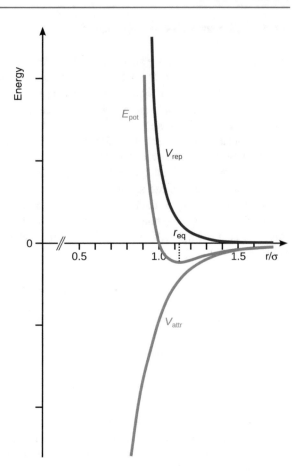

The sum of the above attractive and repulsive terms yields the potential between two particles based on temporary dipoles. The combination of the van der Waals and Pauli terms is called the Lennard-Jones potential:

$$E_{pot} = \frac{b}{r^{12}} - \frac{a}{r^6}. \tag{11.22}$$

The universal repulsive force arises from two quantum mechanical principles and prevents two atoms from occupying the same space. It is indeed of universal importance since it gives rise compressibility and hardness of solids, was famously linked by Victor Weisskopf to the heights of mountains, lengths of ocean waves, and even sizes of stars (Weisskopf 1975). On the one hand, due to the uncertainty relationship by Heisenberg (see Sect. 8.1.6), the position of the electrons surrounding an atomic nucleus cannot be exactly located. They possess kinetic energy which results in a pressure that would drive them off the nucleus if they were not held back by the attractive force of the positively charged nucleus. If the volume occupied by the electrons was decreased by an approaching atom intruding in that space, the electron pressure would increase dramatically and thus resist this change in volume. Ultimately, this is also the reason why condensed matter can hardly be compressed.

On the other hand, according to the Pauli exclusion principle (see Sect. 10.3.2), the electrons in an atom must all have different quantum numbers. In the event of an approaching atom intruding into the space of another atom, some electrons are forced into higher quantum states to fulfil the exclusion principle. However, higher quantum states occupy larger volumes, thereby counteracting the attempted volume decrease by intruding atoms.

11.6 Crystal Field and Ligand Field Theory

As we have seen earlier, a further level of complexity arises in transition metals since they populate the fivefold degenerate d orbitals. We found that four of the d orbitals possess a very similar shape and differ mainly in the orientation of the different lobes (see Fig. 11.9). The fifth d orbital was unique as it extends its two lobes along the z-axis and has a torus around its equator.

In an isolated transition metal atom, the five d orbitals are fully degenerate and possess the same energy. However, when the transition metal engages in a complex

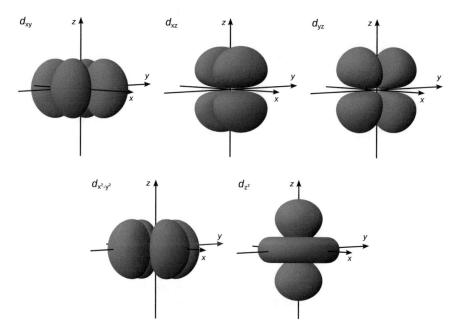

Fig. 11.9 Three-dimensional graphs of the five d orbitals of the third shell (principal quantum number $n = 3$) with respect to a Cartesian coordinate system

where it interacts with ligands that are distribution around it in a particular spatial distribution, ligands perturb the d orbitals that extend into their direction, thereby affecting the energies of the individual d orbitals. Those d orbitals that do not extend into the direction of a ligand are stabilised (their energy decreases) and the other d orbitals whose lobes extend into the direction of the ligands are destabilised (their energies increase). This concept is known as the crystal field theory and the energy difference between the stabilised and destabilised d orbitals is called the crystal field splitting Δ.

11.6.1 Crystal Field Splitting

Obviously, the way in which the degenerate d orbitals split upon complex formation depends on the geometry of the complex. The geometries of transition metal complexes mainly consist of tetrahedral, octahedral and square planar arrangements; the crystal field splitting for these cases is illustrated in Fig. 11.10.

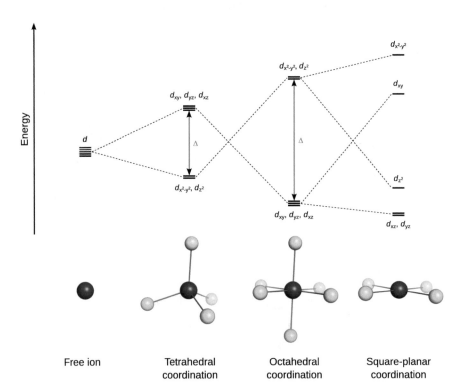

Fig. 11.10 Energetic splitting of d orbitals in response to different coordination geometries. Those orbitals that extend along a direction projects towards a ligand are destabilised and move to higher energy; the others are stabilised and move to lower energy

Importantly, the magnitude of Δ, which can be measured spectroscopically (d-d-transitions), depends on the identity and charge of the metal ion, as well as the type of ligand. By characterising a range of different ligands, a spectrochemical series can be established which lists the ligands in the order of increasing Δ:

$$I^- < Br^- < S^{2-} < SCN^- < Cl^- < NO_3^- < F^- < OH^-$$
$$< oxalate^{2-} < H_2O < NCS^- < H_3C - CN <$$
$$\leftarrow \textbf{weak field} - \textbf{strong field} \rightarrow$$
$$NH_3 < ethylenediamine < 2,2^{'} - bipyridyl < o - phenantrolin < NO_2^-$$
$$< CN^- < CO$$

Given the same geometry of coordination and identical ligands, the crystal field splitting Δ for different metals increases in the following order:

$$Mn^{2+} < Ni^{2+} < Co^{2+} < Fe^{2+} < Fe^{3+} < Cr^{3+} < Co^{3+} < Rh^{3+} < Ir^{3+} < Pt^{4+}$$
$$\leftarrow \textbf{weak field} - \textbf{strong field} \rightarrow$$

Because a complex with tetrahedral geometry has fewer ligands than a complex with octahedral geometry, the magnitude of the crystal field splitting observed with tetrahedral coordination (Δ_t) is smaller than with octahedral coordination (Δ_o):

$$\Delta_t = \frac{4}{9} \cdot \Delta_o. \tag{11.23}$$

11.6.2 Low-Spin and High-Spin Complexes

The magnitude of the crystal field splitting Δ can affect the electronic configuration, and thus the magnetic properties of a complex; in particular, if there are several electrons populating the d orbitals, such as for example in Fe^{3+}.

As illustrated in Fig. 11.11, electrons are filled into the orbitals according to the principles we have established before. One electron is added to each of the degenerate orbitals until each orbital has one electron with the same spin. In an octahedral complex, this works straightforward up to the third electron. The fourth electron could either be placed as a paired electron in the lower set of d orbitals (low-spin complex) or as an unpaired electron in the energetically higher set of d orbitals (high-spin complex). Whether the first or the second scenario happens depends on the magnitude of crystal field splitting Δ. Therefore, high-spin complexes are typically found with metals/ligands at the weak-field end of the spectrochemical series (such as e.g. $[FeCl_6]^{3-}$); vice versa, low-spin complexes are expected in complexes that comprise of metals and ligands at the high-field end of the spectrochemical series (such as e.g. $[Fe(CN)_6]^{3-}$). Since the formation of either a high-spin or a low-spin complex affects the pairing of electrons, the crystal field splitting has repercussions in the magnetic properties of a complex.

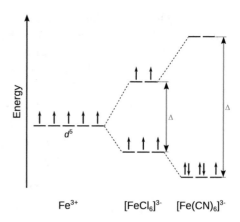

Fig. 11.11 Comparison of crystal field splitting for two Fe^{3+} complexes

Importantly, with octahedral geometry, the options of forming either low-spin or high-spin complexes only exist for systems with 4–7 electrons in the d orbitals (d^4, d^5, d^6, d^7 complexes). For d^1, d^2 d^3 and d^8, d^9, d^{10} complexes, there is only one possible electron configuration.

11.6.3 Ligand Field Theory

Properties such as the absorption of visual light due to d-d-transitions and magnetic susceptibility of metal complexes can be macroscopically observed, and the crystal field theory delivers predictions that are in good agreement with macroscopic observations for some complexes. However, the concept has a fundamental shortcoming in that it only considers the electrostatic interactions between metal and ligands, and ignores any covalent character of metal-ligand bonds; you may have noticed that the ligand orbitals are not actually featured in Fig. 11.11.

The ligand field theory overcomes this defect by also taking into account covalent contributions of metal-ligand interactions. To apply this concept, we construct molecular orbitals as introduced earlier (Sect. 11.3.2). The d orbitals that do not extend into the direction of the coordinated ligands are now deemed to not take part in bond formation and thus called non-bonding orbitals. Accordingly, their energy does not change as compared to the set of degenerate d orbitals in the free metal ion.

This concept is illustrated for $[FeCl_6]^{3-}$ in Fig. 11.12. The Fe^{3+} ion possesses five electrons in the $3d$ orbitals and has empty $4s$ and $4p$ orbitals. Energetically, the $3d$, $4s$ and $4p$ orbitals are at a level that allows molecular orbital formation with the atomic orbitals provided by the six chloride ligands. In this scheme, a total of 17 electrons need to be filled, twelve of which occupy the bonding orbitals. The three non-bonding d orbitals (coloured red in Fig. 11.12) and the two anti-bonding orbitals (upper pair of yellow coloured orbitals in Fig. 11.12) arising from the linear combination of two $3d$ orbitals with the two ligand orbitals results in an energetic scheme we have already seen in the discussion of the crystal field theory. Therefore, the crystal field splitting Δ is also observed in the ligand field theory.

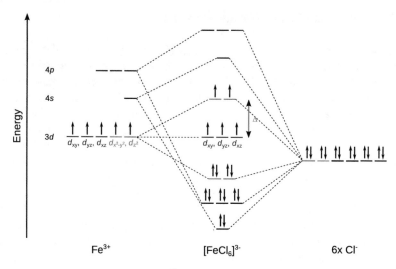

Fig. 11.12 Molecular orbitals for $[FeCl_6]^{3-}$

11.7   Valence Bond Theory and Hybridisation of Atomic Orbitals

The valence bond (VB) approach considers the overlap of the half-filled valence atomic orbitals of each atom containing one unpaired electron, thereby assuming that all bonds are localised bonds. This approach builds on the Lewis structures of molecules which have been introduced in introductory chemistry courses. Historically, the VB theory is a further development of the Lewis structures that also accounts for the geometric shape of molecules. Its main advantage is indeed the description of molecular shapes, but shortcomings remain in the correct prediction of electronic structures in some cases (e.g. molecular oxygen, O_2). However, despite those deficits, the VB method is still frequently used in qualitative descriptions of bond formation in molecules.

Conceptually, the VB method assumes that due to the spatial closeness of the electrons and nuclei of two atoms engaging in a bond, their orbitals become distorted as a consequence of the electrostatic interactions. Therefore, these orbitals no longer adopt their pure forms but rather a mix of the features of the pure orbitals; the resulting orbitals are thus called hybrid orbitals. Importantly, the mixing must not be confused with the linear combination of atomic orbitals in Sect. 11.3.2 which resulted in molecular orbitals. The hybrid orbitals are still atomic orbitals, since they arise from a single atom.

11.7.1 Hybridisation of Atomic Orbitals

To illustrate the concept, we consider methane (CH_4). All experimental studies show that the four C–H bonds in methane are identical in length and energy; the four hydrogen atoms are indistinguishable and the molecule possesses tetrahedral symmetry, resulting in the angle of 109.5° between any pair of C–H bonds. Recalling the three-dimensional arrangements of the $2s$ and the three $2p$ orbitals, it is obvious that the tetrahedral symmetry cannot be explained by overlap of those 'pure' atomic orbitals with the four $1s$ orbitals of the hydrogen atoms. Since four identical bonds require four identical orbitals, one can suggest four identical atomic orbitals which may be derived from one s and three p orbitals by mixing them into four hybrid orbitals (sp^3) that extend into the four directions of a tetrahedron. The mathematical representation of the mixing is given in Table 11.4. Each sp^3 orbital has the same shape (see Fig. 11.13) and its direction is determined by the signs. The C–H bonds are now accomplished by overlap of the four $1s$ orbitals of the hydrogen atoms with the four sp^3 hybrid orbitals of carbon (see Fig. 11.14).

Schematic drawings such as the one in Fig. 11.14 qualitatively depict the shape of atomic orbitals and frequently used in pen-and-paper discussions of the binding situation in molecules.

In contrast to methane, ethylene (C_2H_4) is a planar molecule. The angle between two C–H bonds is 120°, which can be explained by the formation of three sp^2 hybrid orbitals from the $2s$ and two $2p$ orbitals (for example, p_x and p_y; see Fig. 11.13 and Table 11.4) of carbon, each occupied with one electron, ready to pair up with the one electron in a bonding orbital provided by another atom ($2\times 1s$ orbital of hydrogen and $1\times sp^2$ hybrid orbital from the other carbon; see Fig. 11.14). The fourth remaining electron occupies a pure 'left-over' p orbital; since p_x and p_y have been mixed, the remaining pure p orbital is p_z. Overlap of the two p_z orbitals of the two

Table 11.4 Mixing rules of to obtain hybrid from pure atomic orbitals. There are as many hybrid orbitals as pure orbitals used for mixing

Hybrid orbital	Mixing of atomic orbitals	Topology	Angle between hybrid orbitals
sp	$t_1 = \sqrt{\frac{1}{2}} \cdot s + \sqrt{\frac{1}{2}} \cdot p_x$ $t_2 = \sqrt{\frac{1}{2}} \cdot s - \sqrt{\frac{1}{2}} \cdot p_x$	linear	180°
sp^2	$t_1 = \sqrt{\frac{1}{3}} \cdot s + \sqrt{\frac{2}{3}} \cdot p_x$ $t_2 = \sqrt{\frac{1}{3}} \cdot s - \sqrt{\frac{1}{6}} \cdot p_x + \sqrt{\frac{1}{2}} \cdot p_y$ $t_3 = \sqrt{\frac{1}{3}} \cdot s - \sqrt{\frac{1}{6}} \cdot p_x - \sqrt{\frac{1}{2}} \cdot p_y$	planar	120°
sp^3	$t_1 = \frac{1}{2} \cdot (s + p_x + p_y + p_z)$ $t_2 = \frac{1}{2} \cdot (s + p_x - p_y - p_z)$ $t_3 = \frac{1}{2} \cdot (s - p_x + p_y - p_z)$ $t_4 = \frac{1}{2} \cdot (s - p_x - p_y + p_z)$	tetrahedral	109.5°

Fig. 11.13 Shapes and directions of sp, sp^2 and sp^3 hybrid orbitals obtained by mixing pure s and p orbitals as specified in Table 11.4

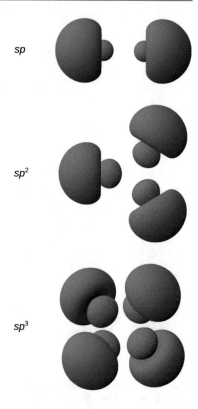

neighbouring carbon atoms provides a further bond, in addition to the bond arising from overlap of two sp^2 hybrid orbitals. Earlier (Sect. 11.3.2), we have introduced the distinction between electron density arising between atoms on the inter-nuclear axis (σ bond) and above/below or in front/behind the inter-nuclear axis (π bond). As illustrated in Fig. 11.14, the overlap of the pure p orbitals in ethylene occurs above and below the C–C axis and therefore constitutes a π bond.

Bonding in the linear acetylene (C_2H_2) molecule can be explained by mixing of the $2s$ and one $2p$ orbital of carbon, resulting in the formation of two sp hybrid orbitals (see Fig. 11.13 and Table 11.4). As illustrated in Fig. 11.14, the sp orbitals of the two neighbouring carbon atoms overlap on the inter-nuclear axis and thus form a σ bond. The second sp hybrid orbital on each of the carbon atoms overlaps with the $1s$ orbital of the hydrogen atoms, forming another σ bond. The remaining pure p orbitals p_y and p_z on each of the neighbouring carbon atoms also possess one electron each and overlap above/below (p_z) and in front/behind (p_y) of the inter-nuclear axis and therefore give rise to two π bonds.

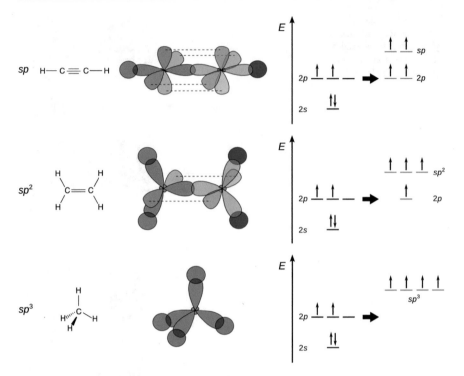

Fig. 11.14 Schematic illustration of bond formation in acetylene, ethylene and methane. The hybrid atomic orbitals of carbon are coloured turquoise, the $2p$ orbitals grey and the $1s$ orbitals of hydrogen are coloured magenta

11.7.2 Valence Shell Electron Pair Repulsion

The hybridisation of atomic orbitals can be successfully applied to many molecules comprising of covalent bonds. In ammonia (NH_3), for example, one might think that due to the electronic configuration of $1s^2\,2s^2\,2p^3$ of nitrogen the three hydrogen atoms might be bound by overlap of the three $2p$ orbitals of nitrogen (with one electron each) with the 1s orbitals of the hydrogen atoms (also one electron each). This would result in an angle of 90° between any two N–H bonds. However, experimentally, the angle between two N–H bonds is observed with 107.3°.

Alternatively, we can consider formation of four sp^3 hybrid orbitals on the nitrogen; three of those hybrid orbitals are occupied by one electron and overlap with the hydrogen 1s orbitals. The fourth sp^3 hybrid orbital is populated with two electrons of anti-parallel spin and forms a so-called lone pair. Repulsion between the lone pair electrons and the electrons in the N–H σ bonds 'pushes' the three N–H bonds closer together and thus causes a decrease of the expected bond angle from

109.5° to 107.3°. This concept forms the central idea of the so-called valence shell electron pair repulsion (VSEPR).

The repulsive effect is more pronounced in the presence of more electrons. For example, in the water (H_2O) molecule, the bond angle between the two O–H bonds is 104.5° and therefore even further decreased from tetragonal angle of 109.5°. With oxygen having an electron configuration of $1s^2\, 2s^2\, 2p^4$, the orbitals of the second shell can again hybridise to form four sp^3 hybrid orbitals. However, two of those hybrid orbitals are occupied by two electrons each, thus forming two lone pairs which exert a stronger repulsive effect towards the electrons in the σ bonds than the one lone pair in the case of ammonia.

The VSEPR approach remains a popular method for qualitative description of covalent bonding in for lighter elements, but does not predict correct geometry for some compound groups involving heavier elements (such as e.g. calcium, strontium and barium halides).

11.7.3 Resonance Structures and Electron Delocalisation

Both the molecular orbital (MO) theory and the valence bond (VB) theory are useful approaches to describe chemical bonding. One aspect of the VB method that is very appealing to many chemists is the fact that it allows depiction of molecules in connectivity diagrams which remains the by far most frequently used method to denote molecular structures.

However, it is often impossible to denote the properties of particular molecules with a single structure. The most prominent example is certainly benzene (C_6H_6) for which a cyclic structure was first proposed by Kekulé in 1865. A ring structure can be accomplished with $6 \cdot 4 = 24$ carbon valence electrons, if one assumes three alternating double bonds (see Fig. 11.15). Such a structure would require that there are two types of C–C bond lengths in the ring: 1.54 Å for a single, and 1.33 Å for a double bond. In contrast, the experimentally determined C–C bond length in benzene is 1.40 Å. We realise that the assignment of the C–C double bond in the ring is entire arbitrary and can formally suggest a second structure where the double bonds are localised differently. More realistically, though, the molecule resonates between the two depicted structures which depict two possible extremes. The possible extreme structures are those we can depict with localised bonds and are called resonance structures. In the depiction of such structures, a double-headed arrow is used to indicate that the structure(s) shown are indeed resonance structures.

Fig. 11.15 Depiction of resonance and aromatic structures of benzene

Resonance structures

Aromatic structure

Aromaticity

Cyclic planar molecules that possess increased chemical stability when compared to linear molecules with the same number of atoms are called aromatic compounds. The increased stability is a direct consequence of the electron delocalisation within the molecule, i.e. the existence of resonance structures.

The properties of aromatic systems include:

- Delocalised π electrons, typically a result of alternating single and double bonds
- Coplanar structure
- Atoms are part of one or more ring systems
- Hückel's rule: the number of π electrons is $4 \cdot N + 2$, with $N = 0, 1, 2, \ldots$

When depicting chemical structures, the existence of aromaticity is often shown as a circular bond (see Fig. 11.15).

Resonance structures arise when the valence bond method is used to describe molecules; this is typically the case when constructing connectivity diagrams such as in Fig. 11.15. An alternative explanation for phenomena such as aromaticity is provided by the molecular orbital (MO) theory. Focussing on the π electron system, we need to consider the six p_z orbitals which need to be combined in accordance with the linear combination of atomic orbitals. Therefore, six different linear combinations of the wave functions of the six p_z orbitals have to be generated. This results in three bonding and three anti-bonding molecular orbitals. The phase distribution in the different molecular orbitals are indicated in Fig. 11.16 by colours and the boundaries define the nodal planes. The (bonding) MO with the lowest energy shows that these electrons are delocalised over the entire ring. With increasing energy, the number of nodal planes increase and thus the compartmentalisation of the orbitals. Electrons populating the higher energy MOs are therefore increasingly localised.

Fig. 11.16 The molecular orbitals and schematic representation of wave functions of the π system of benzene

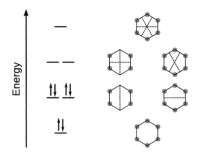

11.8 Exercises

1. NaCl crystallises in the lattice type sodium chloride ($M = 1.747565$) with $r_0 = 2.82$ Å and $\rho = 0.321$ Å. Calculate the lattice energy of NaCl.

2. Discuss the bonding of carbon monoxide (CO) using the valence bond theory without and with hybridised atomic orbitals as well as the molecular orbital theory. Compare the bond order derived by the three approaches.

3. Repeat exercise 10.5 and determine for each of the species S^-, Zn^{2+}, Cl^- and Cu^+ whether they are diamagnetic or paramagnetic.

4. Using an appropriate molecular orbital scheme, explain why $[Co(NH_3)_6]^{3+}$ is a diamagnetic low-spin complex, whereas $[CoF_6]^{3-}$ is a paramagnetic high-spin complex.

Intermolecular Interactions

12

In the preceding chapter, we discussed the various types of chemical bonds that are holding individual atoms together such as to build up new entities—molecules (or metals). These new entities possess different characteristics and functions than the individual atoms they comprise of. Molecules can engage in further interactions, which is the subject of this chapter.

The strength of interaction decreases in the following three phenomena in the order of their discussion: the interactions of permanent dipoles is stronger than interactions involving induced dipoles. The weakest interaction thus arises from interactions exclusively between induced dipoles, the so-called London dispersion force. The combined interactions involving dipoles constitute the van der Waals interactions (or van der Waals bond) which has been discussed in Sect. 11.5. For consistency, we continue to denote the potential energy as V in the following sections in order to avoid confusion with the electric field E. The dipole moment has been introduced in Sect. 11.3.5.

The hydrogen bond gives rise to a much stronger, but still non-covalent, interaction between two molecules. This type of interaction occurs when a hydrogen atom is bound to a highly electronegative (see Sect. 11.2) atom such as nitrogen, oxygen or fluorine. Coulomb interactions are electrostatic interactions and have been discussed when introducing the ionic bond (Sect. 11.1).

12.1 Interactions of Permanent Dipoles

12.1.1 Dipole–Dipole Interactions

Polar molecules are characterised by a localisation of charges in different parts of the molecule. Frequently, the value of charges separated is smaller than the fundamental charge e, i.e. the localised charges are partial charges (denoted by $\delta+$ and $\delta-$). This charge separation constitutes a permanent dipole and such molecules may interact with each either through attractive electrostatic interactions.

© Springer International Publishing AG, part of Springer Nature 2018
A. Hofmann, *Physical Chemistry Essentials*,
https://doi.org/10.1007/978-3-319-74167-3_12

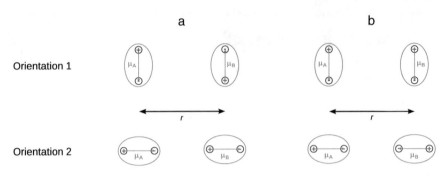

Fig. 12.1 Extreme orientations of two permanent dipoles for attractive (a) and repulsive (b) interactions

For two molecules with dipole moments μ_A and μ_B, respectively, many different relative orientations (with attractive as well as repulsive interactions) of the two dipoles are possible. Four extreme cases of orientations with attractive/repulsive interactions are shown in Fig. 12.1.

In orientation 1A, the interaction energy (potential energy) is

$$V = -\frac{\mu_A \cdot \mu_B}{4\pi \cdot \varepsilon_0 \cdot r^3} \qquad (12.1)$$

and in orientation 2A, the energy is:

$$V = -\frac{2 \cdot \mu_A \cdot \mu_B}{4\pi \cdot \varepsilon_0 \cdot r^3}, \qquad (12.2)$$

whereby ε_0 is the permittivity *in vacuo* and r the distance between the two dipoles. Note that the negative sign in Eqs. 12.1 and 12.2 relates to the directions of the dipole moments as shown in Fig. 12.1. The interaction energy assumes negative values and hence describes an attraction. If the direction of one of the dipole moments is reversed (orientations 1B and 2B in Fig. 12.1), the negative sign in above equations needs to be reversed to, which makes the interaction energy positive and thus describes a repulsion.

If we consider a bulk system of above molecules, it would appear at a first glance that all possible orientations are populated and the mean interaction energy was zero, since the attractive and repulsive interactions cancel each other in the sum. However, due to the Maxwell–Boltzmann statistics (see Sect. 5.1.2), orientations with favourable interactions outweigh those with unfavourable interactions with a factor that is proportional to

$$e^{-\frac{E}{kT}} = e^{-\frac{\mu_A \cdot \mu_B}{4\pi \cdot \varepsilon_0 \cdot r^3 \cdot k \cdot T}}.$$

Averaged over all rotational orientations of the two dipole moments μ_A and μ_B, the interaction energy of two permanent dipoles, the interaction energy between the two dipoles has been derived by Willem Hendrik Keesom (1915). The interactions between two permanent dipoles are thus also known as Keesom interactions and their interaction (potential) energy is inversely proportional to the sixth power of r, i.e. it falls of rapidly with increasing distance:

$$V = -\frac{2}{3} \cdot \frac{\mu_A^2 \cdot \mu_B^2}{(4\pi \cdot \varepsilon_0)^2 \cdot r^6} \cdot \frac{1}{k \cdot T}. \tag{12.3}$$

The above relationship also shows that the interaction energy decreases with increasing temperature. This is to be expected, since the molecules (=permanent dipoles) possess higher kinetic energy at higher temperatures which prevents the spatial alignment of dipole moments required for favourable interactions.

12.1.2 Ion–Dipole Interactions

The interactions between ions and permanent dipoles are an important characteristic of aqueous solutions. Water is a highly polar molecule with a dipole moment of 1.8 D. In the presence of cations (e.g. Na^+), the water dipoles are arranged around the cations such that their negative (oxygen) ends point towards the cation. Similarly, around an anion (e.g. Cl^-), the dipoles are oriented with their positive (hydrogen) ends pointing towards the anion. Due to these attractive interactions, ions in aqueous solutions are always hydrated.

12.2 Interactions of Temporary Dipoles

Neutral, non-polar atoms or molecules possess no localised electrical charge or permanent dipole moment (i.e. no separated partial charges). However, the approach of an ion or a dipole induces a temporary dipole in the non-polar species. The interaction between the ion or dipole with the induced dipole results in an attractive interaction.

12.2.1 Ion–Induced Dipole Interactions

In Fig. 12.2, we consider an atom that possesses an electron density of spherical symmetry around the nucleus. The approach of a charged particle (e.g. a cation) distorts the electron density of the atom and induces a dipole. The magnitude of the induced dipole moment μ_{ind} depends on the properties of the atom (or chemical moiety) as well as the electric field elicited by the approaching charged particle. The ease with which an outside electric field can distort the electron density of the neutral species is called polarisability α. The interaction energy of ion–induced dipole interactions is then given by:

Fig. 12.2 Attractive
interactions of ion-induced
and dipole-induced temporary
dipoles in a neutral atom. The
dot indicates the position of
the nucleus

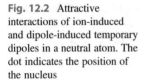

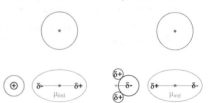

Table 12.1 Polarisability and related quantities for H_2O

Polarisability	α	$1.65 \cdot 10^{-40}$ C m^2 V^{-1}	$=$	$1.65 \cdot 10^{-40}$ C^2 m^2 J^{-1}
Polarisability volume	α'	$1.48 \cdot 10^{-30}$ m^3	$=$	1.48 Å^3
Molar polarisability volume	α'_m	0.89 cm^3 mol^{-1}		

$$V = -\frac{1}{2} \cdot \frac{\alpha \cdot e^2}{4\pi \cdot \varepsilon_0 \cdot r^4} = -\frac{1}{2} \cdot \frac{\alpha' \cdot e^2}{r^4}, \qquad (12.4)$$

where ε_0 is the vacuum permittivity and r the distance between the charged ion and
the neutral species. We recall that the polarisability is the factor of proportionality
between the field strength E and the induced dipole momentum μ_{ind} (see Sect. 11.5):

$$\mu_{ind} = \alpha \cdot E. \qquad (11.18)$$

Often, the polarisability is expressed as a related quantity that is more suitable to
the context in which it is determined. Frequently, it is the polarisability volume α'
which is given by

$$\alpha' = \frac{\alpha}{4\pi \cdot \varepsilon_0}, \qquad (12.5)$$

and has the units of a volume. Some polarisability-related quantities for H_2O are
summarised in Table 12.1.

In larger molecules, polarisability is typically assessed for individual groups, such
as bonds or chemical functionalities. In general, larger groups with diffuse electron
densities possess a higher polarisability than smaller groups with strongly located
electrons. Highly polarisable groups therefore include

- anions,
- groups with π electron systems (e.g. phenyl group, nucleic acids, etc.),
- unsaturated bonds (e.g. C=C, C=N, nitro groups).

Generally, the polarisability of atoms depends on two factors. First, the
polarisability of atoms with a large number of electrons is higher than those of
atoms with a smaller number of electrons (the nuclear charge has less control on

charge distribution the more electrons there are). Second, the more distant electrons are from the atomic nucleus, the less their localisation can be controlled by the nuclear charge; therefore, the polarisability increases the more distant electrons are located from the nucleus. The combination of both factors results in the observation that heavier atoms possess a higher polarisability, since heavier atoms possess more electrons and occupy more distant orbitals.

In molecules with an extended shape (this excludes tetrahedral, octahedral and icosahedral molecules), the orientation with respect to an electric field can also affect the polarisability. Higher polarisability in such molecules (e.g. 2,4-hexadiene) is achieved when the electric field is applied parallel rather than perpendicular to the molecule.

12.2.2   Dipole–Induced Dipole Interactions

A permanent dipole can also distort the electron distribution of a neutral atom or non-polar molecule, and thereby induce a temporary dipole. This leads to an attractive interaction as illustrated in Fig. 12.2 and the interaction energy as derived by Peter Debye:

$$V = -\frac{\alpha \cdot \mu^2}{(4\pi \cdot \varepsilon_0)^2 \cdot r^6}. \tag{11.19}$$

In above equation, ε is again the dielectric constant of the medium, r the distance between the two dipoles, α the polarisability of the non-polar molecules and μ the dipole moment of the permanent dipole.

The interaction between permanent and induced dipoles is also known as the Debye force. Importantly, in contrast to the interactions of permanent dipoles (Keesom interactions), the interactions involving induced dipoles is not dependent on the temperature. Since the temporary dipoles can be induced instantaneously, the thermal motion of the molecules does not affect this interaction.

12.3   London Dispersion Force

Whereas the Keesom interactions and Debye forces require either a charged or polar species to be present, we now consider matter that entirely consists of neutral/non-polar atoms or molecules, such as e.g. helium or nitrogen gas. Since such gases can be condensed into liquids, there must be attractive interactions between those atoms/molecules in the absence of charged or polar species.

As illustrated in Fig. 12.2, an atom with spherical electron density is non-polar because it possesses no permanent dipole moment. However, this is only the view on average. At any individual moment in time, the electron density may not be spherically uniform but exhibit localised peaks. Such deformation of the spherical symmetry is mainly due to collisions between individual atoms. The non-uniform distribution of electrons constitutes a temporary dipole which can induce another temporary dipole in a neighbouring atom at appropriate distance and thus give rise to an attractive interaction, named the London dispersion force after the physicist Fritz London (1930).

For a pure substance, London showed that the interaction (potential) energy is given by

$$V = -\frac{3}{4} \cdot \frac{I \cdot \alpha^2}{r^6}, \tag{12.6}$$

where α is the polarisability, and I the first ionisation potential. Similarly, for a mixture of substances A and B, the interaction is given by:

$$V = -\frac{3}{2} \cdot \frac{I_A \cdot I_B}{I_A + I_B} \cdot \frac{\alpha_A \cdot \alpha_B}{r^6}. \tag{12.7}$$

In contrast to the other dipole-based interaction types mentioned before (collectively termed van der Waals interactions), the London dispersion forces are always attractive. Independent of the relative orientation of two non-polar molecules, the induced dipoles will always possess compatible directions (since they are induced). The London dispersion forces are the weakest interactions among the van der Waals interactions. This is in agreement with macroscopic observations: the above-mentioned liquid helium and nitrogen boil at 4.2 and 77 K, respectively. These low temperatures suggest that only weak forces are holding the atoms/molecules together in the liquid state.

12.4 van der Waals Interactions

Many molecules do not just engage one of the interactions discussed above, but their condensed states are held together by a combination of interactions involving dipoles. These combined interactions are termed van der Waals interactions, referring to all weaker forces between molecules and thus contrasting the stronger intermolecular interactions (Coulomb attraction, hydrogen bond). The attractive van der Waals interactions are often modelled by a potential that varies with sixth power of the distance—a relationship we have observed several times in above discussions (e.g. Debye force, London force).

Balanced by the universal repulsive force, which varies with the twelfth power of the distance, the van der Waals interactions result in the bonding interaction between two non-polar atoms/molecules. This interaction is often described by a Lennard–Jones potential and has been introduced in Sect. 11.5:

Table 12.2 Contributions of the various types of van der Waals forces in select molecules

	Boiling point (°C)	Dipole moment (D)	Polarisability (10^{-40} C m^2 V^{-1})	Interaction		
				Dipole–induced dipole (%)	Dipole–dipole (%)	Dispersion (%)
Ar	−186	0	1.85	0	0	100
CO	−190	0.117	2.20	0	0	100
HCl	−84	1.08	2.93	4.2	14.4	81.4
NH$_3$	−33	1.47	2.47	5.4	44.6	50.0
H$_2$O	100	1.85	1.65	4.0	77.0	19.0

$$V = \frac{b}{r^{12}} - \frac{a}{r^6}. \tag{11.22}$$

This superposition of the energies resulting from the attractive and repulsive forces results in a potential energy function that depends on the distance between the two molecules (or atoms) and that possesses a minimum at the equilibrium distance (see Fig. 11.8). This distance is the average distance the two particles maintain if there are no other forces acting on them.

By way of example, the potential energy of two argon atoms approaching other decreases as they are brought closer together. However, the minimum energy attained at the equilibrium distance is approx. −1.3 kJ mol^{-1} the value of which is less than the thermal energy at ambient temperature ($E_{\text{therm}} = R \cdot T \approx 2.5$ kJ mol^{-1}), and thus not enough to hold the two atoms together. These non-bonding attractions enable argon to exist as a liquid and solid at low temperatures (when the potential energy is larger than the thermal energy). However, at ambient temperature, the potential energy due to van der Waals interactions is not enough to withstand disruptions caused by thermal energy, so argon exists as a gas under ambient conditions.

Table 12.2 shows estimates of the contributions of the various types of van der Waals forces that act between different types of molecules. This comparison highlights the importance of the ubiquitous dispersion forces on the one hand, even in cases of polar molecules (high dipole moment).

12.5 The Hydrogen Bond

With hydrogen only possessing one electron, the nucleus becomes partially unshielded when the atom engages in a covalent bond. The 'partial proton' can then interact directly with a nearby atom that possesses a lone pair of electrons, such as oxygen, nitrogen, fluorine or chlorine. The existence of hydrogen bonds has drastic effects on properties of substances and is of tremendous importance for

Table 12.3 Comparison of boiling points of a homologue series. The polarisability increases with the heavier group VI atoms. This leads to stronger dispersion forces and thus higher boiling points. However, the strong hydrogen bonds in water give rise to a substantial increase in interaction energy between H_2O molecules, leading to an unusually high boiling point

Liquid	Boiling point (K)	Polarisability α	Dipole moment μ (D)	H-bonds possible
H_2O	373		1.8	+
H_2S	213		1.1	−
H_2Se	231		0.4	−
H_2Te	271		0.2	−

the folding and properties of biological macromolecules. Depending on geometry and environmental conditions, the potential energy of a hydrogen bond is between 5 and 30 kJ mol^{-1}, which makes it much stronger than van der Waals interactions, but weaker than covalent or ionic bonds. The hydrogen bond extends from the hydrogen bond acceptor (the atom that has a lone pair of electrons) to the hydrogen bond donor (the electronegative atom to which the hydrogen atom is covalently bound).

For example, the boiling points of liquids in a series of homologues increases with polarisability (which results in stronger London dispersion forces; see Sect. 12.3). However, in the series shown in Table 12.3, H_2O possesses a substantially higher boiling point than the homologous compounds, while indeed it would be expected to have the lowest boiling point based on the London dispersion forces. Despite the low polarisability of H_2O, the existence of intermolecular hydrogen bonds leads to much stronger interactions between water molecules and thus results in the much higher boiling point.

Each H_2O molecule contains two hydrogen atoms and two lone pairs. In water, the number of hydrogen bonds is therefore maximised by a tetrahedral arrangement of hydrogen atoms around the oxygen atoms. In the hexagonal structure of ice, each oxygen atom is surrounded by a distorted tetrahedron of hydrogen atoms that form bridges to the oxygen atoms of adjacent water molecules. Importantly, the hydrogen atoms are not located at the same distance from the two oxygen atoms they connect (see Fig. 12.3); the shorter distance is indicative of a covalent O–H bond, and the longer distance constitutes the hydrogen bond. The hexagonal structure of ice is the form of all natural snow and ice on Earth (note the sixfold symmetry in ice crystals grown from water vapour). The packing of H_2O molecules in this cage-like structure is expanded as compared to the packing of molecules in liquid water. Therefore, ice is less dense than liquid water, which explains why it floats on the liquid.

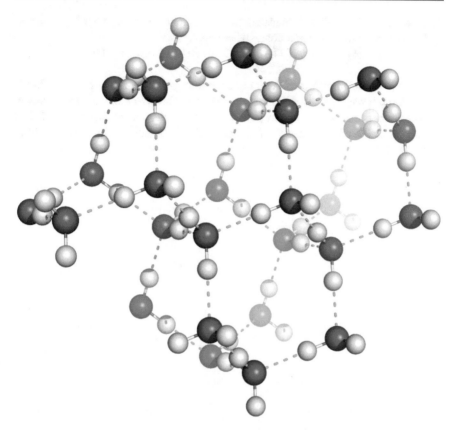

Fig. 12.3 The hexagonal structure of ice. Oxygen atoms are shown in red and hydrogen atoms in grey. Hydrogen bonds are indicated by green dashed lines. The distance of the covalent O–H bond measures 1.01 Å and that of the hydrogen bond 1.74 Å, respectively

12.6 Exercises

1. The dipole moment of HF is 1.92 D.

    (a) Calculate the potential energy of the attractive dipole–dipole interaction between two HF molecules oriented along the x-axis in a plane, separated by 5 Å.

    (b) What is the potential energy of the dipole–dipole interaction for 1 mol HF?

    (c) Calculate the average thermal energy of bulk matter at room temperature. Can the dipole–dipole interaction of HF be sustained at room temperature?

2. Explain why the boiling point of the two isomers of butane, n-butane and i-butane, are different. Which isomer has the higher boiling point?

3. In which of the following substances are molecules held together by hydrogen bonds? Draw the hydrogen bonds where applicable.

    (a) XeF_4, (b) CH_4, (c) H_2O, (d) NaH, (e) BH_3, (f) NH_3, (g) HI.

4. What is the relationship of the interaction energy between two particles due to Coulomb forces and dispersion forces with the distance? Compare the falloff of these energies with distance by calculating their ratio for the distances 1, 2 3, 4 and 5 Å. Based on these results, assess the implications for ideal/non-ideal behaviour of solutions.

Interactions of Matter with Radiation **13**

In Sect. 8.2.1, we already had a look at the interaction of atoms and electromagnetic radiation. Specifically, the atomic spectra were used to obtain insights into the inner fabric of atoms and their energetic states. Importantly, the spectral data provided the experimental verification of theoretical models such as the atomic model by Bohr (Sect. 8.2.2) and the quantum mechanics of the hydrogen atom by the Schrödinger equation (Sect. 10.1).

Since detailed information about structure, bonding and intra-/inter-molecular processes can be obtained from analysis of the interaction between matter and electromagnetic radiation, the area of spectroscopy is of fundamental importance for a wide variety of chemical disciplines.

13.1 General Spectroscopic Principles

In the study of matter and its interaction with electromagnetic radiation, two general processes can be distinguished:

- absorption describes the uptake of energy by an atom or molecule from incident electromagnetic radiation
- emission describes the release of electromagnetic radiation from an atom or molecule.

Either process is due to transitions between different energetic states; we have used this concept already in Sect. 8.2.2. The fundamental equation is therefore:

$$\Delta E = E_{\text{end}} - E_{\text{start}} = h \cdot \nu, \tag{13.1}$$

which means that in order to enable a transition from the lower (start) to the higher (end) energy level, electromagnetic radiation is required that possesses just the right quantum of energy ΔE. This is called the resonance condition.

© Springer International Publishing AG, part of Springer Nature 2018
A. Hofmann, *Physical Chemistry Essentials*,
https://doi.org/10.1007/978-3-319-74167-3_13

When irradiating matter with light of appropriate energies, such transitions between discrete energy states occur as a consequence of either induced absorption or emission. Additionally, if a molecule has populated an excited state and it transitions back into the ground state, such transition might be accompanied by spontaneous emission of photons.

13.1.1 Intensity

A fundamental question of the interaction between radiation and matter addresses the loss of intensity of an incident beam as it travels through a sample of interest (Fig. 13.1). Practically, in most cases the sample will be contained in cuvette and potential interactions of the incident radiation with the cuvette material thus needs to be corrected for when conducting the experiment.

If the intensity of the beam at any point x on its way through the sample is I, the loss of intensity dI is proportional to the intensity at that point. Furthermore, dI is also proportional to the sample thickness dx penetrated:

$$-dI \sim I, \; -dI \sim dx \Rightarrow -dI \sim I \cdot dx \Rightarrow -dI = \mu \cdot I \cdot dx$$

Conversion of the proportionality to an equality relation requires the introduction of the proportionality constant μ (also known as the linear attenuation coefficient; see Sect. 13.6.2). Considering that the overall thickness of the sample shall be l and the intensity of the incident beam be I_0, the above equation can be integrated between the boundaries of $x = 0$ and $x = l$ after separating the integration variables I and x:

$$\int_{I_0}^{I} \frac{-dI}{I} = \int_{0}^{l} \mu \cdot dx$$

$$-(\ln I - \ln I_0) = \mu \cdot (l - 0)$$

$$\ln \frac{I_0}{I} = \mu \cdot l$$

(13.2)

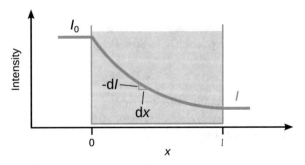

Fig. 13.1 Loss of intensity of an incident beam due to absorption

The natural logarithm of above equation is converted into a decadic logarithm, which yields:

$$\lg\frac{I_0}{I} = a \cdot l \tag{13.3}$$

known as the Bouger-Lambert law where a is the linear absorption coefficient. If the sample consists of a solution where the solvent shows no absorption of the incident radiation, the absorption coefficient depends on the molar concentration c of the radiation-absorbing solute:

$$a = \varepsilon \cdot c$$

and introduces the molar absorption coefficient ε, from which the law of Lambert-Beer is obtained:

$$A = \lg\frac{I_0}{I} = \varepsilon \cdot c \cdot l \tag{13.4}$$

where A is called absorbance and the ratio of transmitted versus incident intensity is called transmittance: $T = \frac{I}{I_0}$. The absorbance has no units and can in principle take infinite values; however, in experimental designs, only values of $A < 2$ deliver useful results, owing to the fact that the law of Lambert-Beer is only valid for comparably low concentrations (<0.5 M).

13.1.2 Intensity and Absorption Strength at the Molecular Level

A spectrum observed for a particular sample represents the superposition of a large number of transitions in individual atoms or molecules. The intensity of the observed absorption peaks therefore depends on the difference in the number of atoms/ molecules occupying the two different energetic states. The ratio of these populations is given by the Boltzmann expression (see also Sect. 5.1.2)

$$\frac{N(E_2)}{N(E_1)} = e^{-\frac{E_2 - E_1}{k_B \cdot T}} \tag{13.5}$$

where $N(E_2)$ and $N(E_1)$ are the number of atoms or molecules in the excited (energy E_2) and ground states (energy E_1), respectively. The larger the difference in the population of the two states, the more intense the spectral peak will be.

From physics and radio communications it is known that in order to transmit or receive radio waves one requires an oscillating dipole and a dipole antenna, respectively. In addition, since radio waves emitted by an oscillating dipole are linearly polarised, the receiving dipole antenna needs to be oriented such that is in alignment with the plane of polarisation of the emitted signal. The emission and absorption of light (or generally electromagnetic radiation) by molecules follows the very same

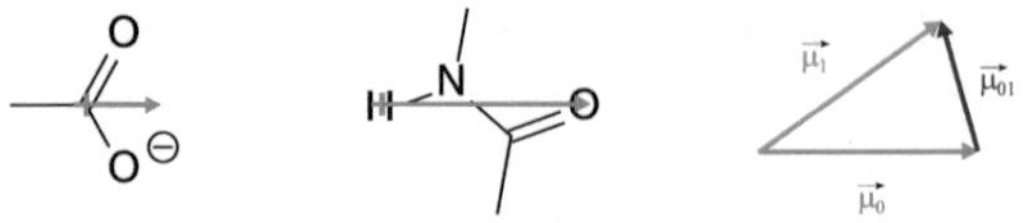

Fig. 13.2 Permanent dipole moment of the carboxyl group (left) and the amide bond (centre). Right: The permanent dipole momentum changes when the molecule transitions into an excited state. The difference vector between the permanent dipole momentum in the ground state ($\vec{\mu}_0$) and the excited state ($\vec{\mu}_1$) is called transition dipole momentum ($\vec{\mu}_{01}$)

physical concepts. One can thus only expect absorption or emission of light by molecules that possess a dipole momentum. Typically, these will be permanent dipoles, but in some applications dipoles may also be induced in molecules that have no permanent dipole momentum. In cases involving linearly polarised light, one also needs to consider orientation effects.

Chromophores are the light absorbing moieties within a molecule. Due to differences in electronegativity between individual atoms, they possess a spatial distribution of electric charge. This results in a dipole momentum $\vec{\mu}_0$ (ground state), such as for example the permanent dipole momentum of a carboxylic acid or an amide group (Fig. 13.2).

When light is absorbed by the chromophore, the distribution of electric charge is altered and the dipole momentum changes accordingly ($\vec{\mu}_1$; excited state). The transition dipole momentum $\vec{\mu}_{01}$ is the vector difference between the dipole momentum of the chromophore in the ground and the excited state. This transition dipole momentum is a measure for transition probability, and its dipole strength, D_{01}, is defined is the squared length of the transition dipole momentum vector:

$$D_{01} = |\vec{\mu}_{01}|^2 \tag{13.6}$$

The strength of this transition dipole momentum is directly related to the probability with which a transition occurs, i.e. the strength of an absorption band. Spectroscopic data can thus be analysed to obtain numerical values for a transition dipole momentum which connects the absorption spectrum to the quantum mechanical wave function of a molecule.

13.1.3 Selection Rules

Owing to the underlying quantum mechanics, transitions between different energy states have to follow particular selection rules. Whereas allowed transitions possess a high probability of occurring, forbidden transitions are unlikely to occur. Forbidden transitions can be classified into spin-forbidden and symmetry-forbidden transitions.

Spin-forbidden Transitions
As we have seen earlier, the electronic states of atoms and molecules can be described by orbitals which contain up to two electrons paired with anti-parallel

Table 13.1 Spin multiplicity of atoms and molecules

Number of unpaired electrons	Total spin S	Spin multiplicity M	
0	0	1	Singlet state
1	½	2	Doublet state
2	1	3	Triplet state
3	³⁄₂	4	Quartet state

spin orientation. The total spin S is calculated as the sum of the individual electron spins (s_i):

$$S = \sum_i s_i. \tag{13.7}$$

The spin multiplicity M is given by

$$M = 2 \cdot S + 1 \tag{13.8}$$

and informs about the number of different possible arrangements of the unpaired spins in an external magnetic field (Table 13.1).

For a transition to be allowed, the spin multiplicity must not change, i.e.

$$M = \text{const. and therefore } \Delta S = 0. \tag{13.9}$$

Thus, a transition from a singlet to a triplet state and vice versa is normally forbidden.

Symmetry-forbidden Transitions

In Sect. 13.1.2 above, we have introduced the transition dipole momentum μ_{01} the strength of which is a measure for transition probability. Quantum-mechanically, the transition dipole momentum is described as

$$\mu_{01} = \int \Psi_0^* \cdot \left(\sum_i Q_i \cdot r_i \right) \cdot \Psi_1 d\tau \tag{13.10}$$

which tracks the positions (r_i) of the electron charges (Q_i) in the molecule. Ψ_1 is the wave function describing the molecular orbital of the excited state and Ψ_0^* is the complex conjugate wavefunction of the ground state; $d\tau = dx\, dy\, dz$ is the volume element. The wavefunctions inherently describe the symmetry of the molecular orbital. Unless the product $\Psi_0^* \cdot \left(\sum_i Q_i \cdot r_i \right) \cdot \Psi_1$ is of a certain symmetry, the integral in Eq. 13.10 will be zero and therefore the strength of the transition dipole momentum $D_{01} = |\vec{\mu}_{01}|^2$ will be zero, i.e. the transition will not occur. Transitions with $D_{01} \rightarrow 0$ are called forbidden transitions, the probability of their occurrence is low. If $D_{01} \rightarrow 1$, the transition is called allowed and occurs with high probability.

Symmetry rules and wavefunctions

When comparing the three wave functions in Fig. 13.3, it becomes obvious that Ψ_0 and Ψ_2 are symmetric with respect to their centre plane (mirror plane), but Ψ_1 possess no mirror symmetry (instead it possesses an inversion centre). The symmetry properties of wavefunctions are also known as parity.

Ψ_0 and Ψ_2 are thus called 'symmetric' or 'even', and Ψ_1 is called 'anti-symmetric' or 'odd'. In algebraic terms, this means:

- if $f(x) = f(-x)$, then f is an even function;
- if $f(x) = -f(-x)$, then f is an odd function.

When multiplying functions, the following rules apply:

- An even function times an even function yields an even function;
- an odd function times an odd function yields an even function;
- an even function times an odd function yields an odd function.

When dealing with wavefunctions, there are two general rules to be considered:

- The integral over all space for an even function is non-zero.
- The integral of an odd function over all space yields zero.

Using ethylene as an example, we can now use Eq. 13.10 to appraise a potential transition from the π to the π^* orbital of ethylene (Fig. 13.4):

Ψ^*_0: The π orbital of ethylene is even.

Ψ_1: The π^* orbital of ethylene is odd.

$(Q_i \cdot r_i)$: The distribution of charge develops with distance and is therefore an odd function.

The product this yields: 'even'·'odd'·'odd' = 'even'. The integral over the product is thus non-zero and the transition from the π to the π^* orbital in ethylene is allowed.

In contrast, if the molecular orbitals of the excited state had the same symmetry as those of the ground state, such as e.g. in butadiene, then transitions between those orbitals are forbidden based on the symmetry rules.

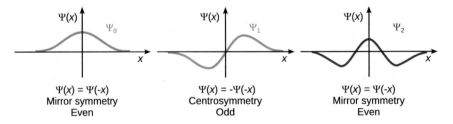

Fig. 13.3 The symmetry (parity) of wavefunctions

Fig. 13.4 The π and π* orbitals of ethylene

Ethylene

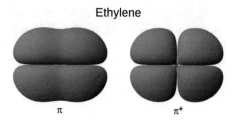

π   π*

13.1.4 Line Width and Resolution

Theoretically, the two energy levels involved in a transition have discrete energy values which should result in a sharp spectral line. However, since the life time of the excited state of an atom or molecule is limited and much shorter than the life time in the ground state, the peak width broadens due to the Heisenberg uncertainty relationship (see also Sect. 8.1.6). Spectral lines thus have a non-zero width. Applied to transitions between energy levels, the uncertainty relationship is

$$\Delta E \cdot \Delta t \geq \frac{h}{4\pi} \tag{13.11}$$

where ΔE is and Δt are the uncertainties in determining the energy difference of a transition and life time of a particular state, respectively. The uncertainty relationship as given in Eq. 13.11 causes a natural line width $\Delta \lambda_0$ for a particular spectral transition. The uncertainty relationship enables us to calculate a minimum natural line width as per

$$\Delta \lambda_0 \cdot \Delta t \geq \frac{h}{4\pi} \Rightarrow \Delta \lambda_0 \geq \frac{h}{4\pi \cdot \Delta t} \tag{13.12}$$

where Δt is the observation time.

In the recorded spectrum, the line shape follows a Lorentzian function. If the maximum intensity (=transition probability) for a transition is recorded at the central wavelength λ_c, then there will be a distribution of intensities given by the relationship

$$I = \frac{\text{const.}}{(\lambda - \lambda_c)^2 + \left(\frac{\Delta \lambda_0}{2}\right)^2} . \tag{13.13}$$

Additionally, the width of spectral lines may also be broadened by collisional processes or chemical reactions. For example, the absorption spectrum of a substance in its liquid state typically shows broader peaks than a spectrum obtained in the gas phase. Since the likelihood of collisions is larger in the liquid than in the gas phase, excited molecules are more readily deactivated and thus have a shorter life time.

As illustrated in Fig. 13.5, the peak width is typically defined as the width at half maximum (FWHM, full width at half maximum). From a practical perspective, it is important that two neighbouring peaks in a spectrum can be distinguished. This separation of two neighbouring peaks in a spectrum is called resolution. It is quantitatively measured by the resolving power, which in case of a wavelength-based spectrum is calculated as per:

$$RP = \frac{\lambda}{\Delta\lambda}. \tag{13.14}$$

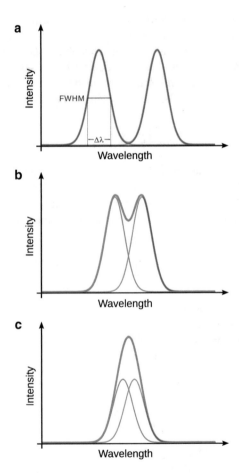

Fig. 13.5 Schematic explanation of full width at half maximum as well as the effect of resolution. (**a**) Two fully resolved peaks; (**b**) two overlapping peaks; (**c**) two non-resolved peaks. The red line indicates the observed spectrum which is the sum of the individual peaks

Here, $\Delta\lambda$ is the difference between the wavelengths of the two neighbouring peaks $\lambda_1 = \lambda$ and $\lambda_2 = \lambda + \Delta\lambda$.

13.1.5 The Electromagnetic Spectrum

Figure 13.6 shows the spectrum of electromagnetic radiation, organised in increasing energy from left to right. We have previously introduced the energy of a photon (Sect. 8.1.4) and can thus introduce the relationship between wavelength λ and wavenumber $\tilde{\nu}$ and the photon energy as per:

$$E = h \cdot \nu = \frac{h \cdot c}{\lambda} = h \cdot c \cdot \tilde{\nu}. \tag{13.15}$$

From this equation, we immediately appreciate that the frequency ν and the wavenumber $\tilde{\nu}$ increase with increasing energy, whereas the wavelength gets shorter the higher the energy of the radiation.

Different types of interactions happen between electromagnetic radiation of certain energy regions with matter. The various interactions with matter and the appropriate types of spectroscopy are mapped to the electromagnetic spectrum in Fig. 13.6.

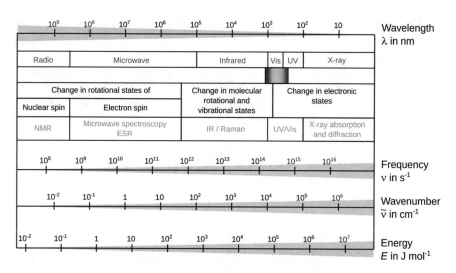

Fig. 13.6 The electromagnetic spectrum and the use of particular regions for spectroscopic applications

13.2   Magnetic Resonance Spectroscopy

13.2.1 General Principles

As we have seen in Sect. 10.2.2, an electron rotating around its own axis (spin) constitutes a charged particle in motion, leading to a magnetic spin momentum $\bar{s}$ which is characterised by the magnetic spin quantum number m_s. As a consequence, the electron spin is oriented with respect to an external magnetic field (see Fig. 10.8).

If we consider atomic nuclei, it is obvious that these are charged particles, too, and, upon rotation around their own axis there is a resulting nuclear spin momentum $\bar{I}$ the value of which is given as

$$|\vec{I}| = \frac{h}{2\pi}\sqrt{I\cdot(I+1)},\qquad(13.16)$$

introducing the nuclear spin quantum number I. Whereas the electron spin quantum number s assumes only value ($\frac{1}{2}$), the value of the nuclear spin quantum number I is not as tightly restricted and, in general, cannot be predicted; values between 0 and 8 have been observed so far. Whereas protons and neutrons (the components constituting the atomic nucleus; also called nucleons) each have net spins of $\frac{1}{2}$, there are also spin interactions among the elementary particles (quarks) that, in turn, compose the nucleons. As a result of this complexity, there is no simple formula to predict I based on the number of protons and neutrons within an atom. However, there are three main rules:

- If the number of protons (Z) as well as the number of neutrons (N) is even, the nuclear spin quantum number I is zero. For example, ^{12}C and ^{16}O possess no nuclear spin.
- If Z is even and N odd, or vice versa, then I is half-integral. For example, the nuclear spin quantum number of ^{1}H, ^{13}C, ^{15}N, ^{19}F and ^{31}P is $I = \frac{1}{2}$.
- If both Z and N are odd, then the nuclear spin quantum number takes an integral value. For example, ^{2}H and ^{14}N have $I = 1$.

If either an electron or a nucleus with their respective spins are exposed to an external magnetic field $\vec{B}$, then the spins need to orient themselves such that the space component of the electron spin ($\vec{S}_{\vec{B}}$) or nuclear spin ($\vec{I}_{\vec{B}}$) can only take the following values (Table 13.2):

Table 13.2 Comparison of electron and nuclear spin directions in an external magnetic field $\vec{B}$

	Electron spin	Nuclear spin
Space component parallel to magnetic field	$\|\vec{s}_{\vec{B}}\| = \dfrac{h}{2\pi}\|m_s\|$	$\|\vec{I}_{\vec{B}}\| = \dfrac{h}{2\pi}\|m_I\|$
	with	
Spin quantum number	$m_s = -s, +s = -\frac{1}{2}, +\frac{1}{2}$	$m_I = -I, -I+1, \ldots, I-1, I$

An external magnetic field will exert a torque on a magnetic dipole and, therefore, the potential energy of both electrons and nuclei includes a magnetic component which is given as the product between the magnetic momentum $\vec{m}_{\vec{B}}$ of either electron or nucleus and the magnetic field $\vec{B}$ itself:

$$E_{\text{pot}} = -\vec{m}_{\vec{B}} \cdot \vec{B}. \tag{13.17}$$

The magnetic momentum $\vec{m}_{\vec{B}}$ is derived from the spin quantum numbers m_s (electron) and m_I (nucleus):

We appreciate that in this derivation, the use of a gyromagnetic ratio (also called g-factor) is introduced. The g-factor for an orbital is $g_l = 1$, but for the electron spin a factor of $g_e \approx 2$ has been found. For nuclei, the g-factor depends on the individual element; measurements for the protein and neutron yielded that

$$g_{\text{proton}} = 5.5856947, \text{ and}$$
$$g_{\text{neutron}} = -3.8260837.$$

Surprisingly, the g-factor for the neutron is far from zero, despite the neutron does not carry a charge! This indicates that inside the neutron there is an internal structure involving the movement of charged particles (the elementary particles called quarks).

Table 13.3 also introduced the magnetons which are units of the magnetic momentum. For the electron, this yields the Bohr magneton μ_B, and for nuclei units of nuclear magnetons μ_N are commonly used. The quantisation of electron and nuclear magnetic momentum thus yields equally spaced energy levels.

As an illustration, we will calculate the the potential magnetic energy difference for the two magnetic spin states of the electron ($m_s = \pm^1/_2$) as well as the the two nuclear spin states of the proton ($m_I = \pm^1/_2$) in the presence of an external magnetic field $\vec{B}$. For the electron, this yields:

$$E\left(m_s = +\frac{1}{2}\right) = +\frac{1}{2} \cdot 2.002 \cdot 9.285 \cdot 10^{-24} \text{JT}^{-1} \cdot \left|\vec{B}\right| = 9.294 \cdot 10^{-24} \text{JT}^{-1} \cdot \left|\vec{B}\right|$$

$$E\left(m_s = -\frac{1}{2}\right) = -\frac{1}{2} \cdot 2.002 \cdot 9.285 \cdot 10^{-24} \text{JT}^{-1} \cdot \left|\vec{B}\right| = -9.294 \cdot 10^{-24} \text{JT}^{-1} \cdot \left|\vec{B}\right| \quad (13.18)$$

$$\Delta E = \left|E\left(m_s = +\frac{1}{2}\right) - E\left(m_s = -\frac{1}{2}\right)\right| = g_e \cdot \mu_B \cdot \left|\vec{B}\right| = 18.59 \cdot 10^{-24} \text{JT}^{-1} \cdot \left|\vec{B}\right|$$

The same evaluation yields for the proton:

Table 13.3 Comparison of the potential magnetic energy of electrons and nuclei

	Electron spin	Nuclear spin
Potential magnetic energy	$E_{\text{pot}} = g_e \cdot \mu_B \cdot m_s \cdot \left|\vec{B}\right|$	$E_{\text{pot}} = -g_N \cdot \mu_N \cdot m_I \cdot \left|\vec{B}\right|$
	where	
Gyromagnetic ratio	$g_e = 2.0023193134$	g_N depends on the nucleon
Magneton	$\mu_B = 9.284832 \cdot 10^{-24}$ J T^{-1}	$\mu_N = 5.050824 \cdot 10^{-27}$ J T^{-1}

$$E\left(m_I = +\frac{1}{2}\right) = -\frac{1}{2} \cdot 5.586 \cdot 5.051 \cdot 10^{-27} \mathrm{JT}^{-1} \cdot \left|\vec{B}\right| = -14.105 \cdot 10^{-27} \mathrm{JT}^{-1} \cdot \left|\vec{B}\right|$$

$$E\left(m_I = -\frac{1}{2}\right) = +\frac{1}{2} \cdot 5.586 \cdot 5.051 \cdot 10^{-27} \mathrm{JT}^{-1} \cdot \left|\vec{B}\right| = 14.105 \cdot 10^{-27} \mathrm{JT}^{-1} \cdot \left|\vec{B}\right| \quad (13.19)$$

$$\Delta E = \left| E\left(m_I = +\frac{1}{2}\right) - E\left(m_I = -\frac{1}{2}\right) \right| = g_N \cdot \mu_N \cdot \left|\vec{B}\right| = 28.21 \cdot 10^{-27} \mathrm{JT}^{-1} \cdot \left|\vec{B}\right|$$

We see, firstly, that whereas for the electron the state $m_s = -\frac{1}{2}$ is energetically favoured, for the proton it is the state with $m_I = +\frac{1}{2}$. Second, assuming the same strength of the externally applied magnetic field, the difference between both energy levels is much larger for electrons than for protons.

In the introduction to this chapter, we discussed that in order to elicit a transition from a just the difference in energy between the two states the spin momentum component, i.e. satisfies the resonance condition. Assuming an external magnetic field of 2.0 T, the energy required for proton spin resonance would thus be:

$$h \cdot \nu = 28.21 \cdot 10^{-27} \text{ J T}^{-1} \cdot 2.0 \text{ T} \Rightarrow \nu = 85.1 \text{ MHz} \Rightarrow \lambda = 3.5 \text{ m}.$$

which is in the region of radio waves.

We can thus conclude that in order to satisfy the nuclear magnetic resonance condition, electromagnetic radiation in the range of radio waves will be required. Since the energy difference of the magnetic electron spin levels is three orders of magnitude higher (at the same magnetic field strength), radiation in the microwave range will satisfy the resonance condition for electron spin resonance. In principle, the resonance condition can in both cases be achieved by either applying a constant magnetic field and variation of the frequency ν of the incident electromagnetic radiation, or a constant frequency and variation of the magnetic field $\left|\vec{B}\right|$.

Using the Boltzmann expression introduced in Sect. 13.1.2, it is possible to calculate the ratio of populations of the lower and the higher energy states of electron and nuclear spins in the presence of an external magnetic field (Table 13.4). Owing to the small energy differences between higher and lower states, the differences in population of both states are extremely small. Provision of the resonance energy ΔE by incident electromagnetic radiation leads to a change of the population ratio as the energetically higher spin states are increasingly populated after absorption of ΔE.

For the comparison in Table 13.4, we have used the field strength of the external magnetic field supplied by the instrumentation used for electron spin resonance or

Table 13.4 Comparison of the ratio of populations of lower and higher states for electron and nuclear spin using the energy differences from Eqs. 13.18 and 13.19 at a temperature of $T = 300$ K

	Electron spin	Nuclear spin
External magnetic field strength $\left|\vec{B}\right|$	0.5 T	2.0 T
Energy difference ΔE between lower and higher state	$9.30 \cdot 10^{-24}$ J	$56.4 \cdot 10^{-27}$ J
Ratio of population of lower and higher spin states $\frac{N_{high}}{N_{low}} = e^{-\frac{\Delta E}{k_B T}}$	0.9977572	0.9999864

Table 13.5 Comparison of T_1 and T_2 relaxation

T_1 relaxation	T_2 relaxation
Longitundinal relaxation Spin-lattice relaxation	Transversal relaxation, Spin-spin relaxation
Requires energy transfer from spins to environment ("lattice").	May occur with or without energy transfer.
Source of fluctuating field is molecular motion of a nearby electron or nucleus.	Anything causing T_1 relaxation also causes T_2 relaxation.
	T_2 relaxation can also occur without T_1 relaxation. Major causes include de-phasing by static local field disturbances and flip-flop exchanges between spins.

nuclear magnetic resonance spectrometry. A closer look at the observed phenomena shows that the field strength experienced by a particular electron or nucleus in a molecule does not exactly equal the the field strength of the external magnetic field but is slightly different. Since the electron density varies with the type of chemical bond, the observed resonance is shifted with respect to the position expected based on the external magnetic field strength. This shift is also called chemical shift and constitutes an important parameter when deducing the chemical structure from magnetic resonance spectra.

After the absorption of the resonance energy caused an increase in the population of the higher spin states, the reverse process occurs whereby spins return to the lower energy state—this is known as relaxation. The population ratio N_{high}:N_{low} returns to values determined by the presence of the external magnetic field (see Table 13.4). The energy released during those transitions is dissipated into the environment ('lattice') as heat. This process happens with a kinetics of first order and has a rate known as spin-lattice relaxation time or longitudinal relaxation time T_1. The rate constant is, accordingly, T_1^{-1}.

A related phenomenon is observed for the phases of the spins. After turning on the magnetic field, the movement of all spins are in phase, but the exchange of energy between spins, inhomogeneous/disturbed local magnetic fields as well as the spin-lattice relaxation all lead to a decrease in the phase correlation of spin movements. This process also happens with a kinetics of first order and is described by the transversal (sometimes also called spin-spin) relaxation time T_2.

Both T_1 and T_2 are characteristic parameters for the life time of the excited state and thus affect the peak width observed in magnetic resonance spectra. A brief comparison is summarised in Table 13.5 (Fig. 13.7).

13.2.2 Nuclear Magnetic Resonance (NMR) Spectroscopy

NMR spectroscopy requires a strong homogeneous magnetic field which is provided in contemporary spectrometers by superconducting magnets (currently up to 23.5 T equivalent to a resonance frequency of 1 GHz). Historically, the early spectrometers

a T_1 Relaxation

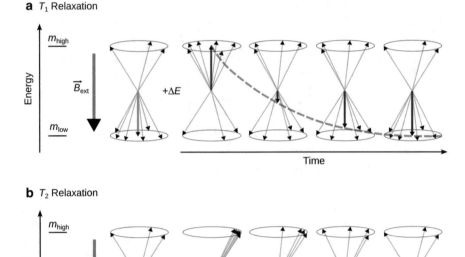

b T_2 Relaxation

Fig. 13.7 Illustration of T_1 and T_2 relaxation. '$+\Delta E$' indicates provision of the resonance energy. (**a**) T_1 or longitudinal relaxation leads to a decrease of the magnetisation anti-parallel to the external magnetic field. (**b**) T_2 or transversal relaxation results in decrease of magnetisation orthogonal to the external magnetic field

operated in the so-called continuous wave method and used electromagnets. A sweep generator was used for sweeping either the magnetic or radio frequency field through the resonance frequencies of the sample. Contemporary spectrometers are operated as Fourier Transform NMR spectrometers where all required radio frequencies are transmitted at once in a radiation pulse and cause nuclei in the magnetic field to flip into the higher-energy alignment (so-called pulse-acquire method). In the following time T, the nuclei return to their original states and emit a radio frequency signal called the free induction decay (FID). The FID contains all the resonance signals of the sample, but they are coded in the time domain. With the computational process of Fourier transformation, these data can be converted into a conventional spectrum where the resonance intensities are ordered per frequency. For illustration, we will focus on proton (^{1}H) NMR to discuss some general principles of this spectroscopic method.

When considering the energy difference between the higher and lower proton spin states as derived in Eq. 13.19, one needs to be aware that the value of $|\vec{B}|$ in that equation is not the strength of the external magnetic field $|\vec{B}_{ext}|$, but the local

Fig. 13.8 The field of small magnets experiencing an external magnetic field opposes the direction of the external field

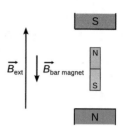

magnetic field experienced at the position of the nucleus. The local field strength $|\vec{B}|$ is less than $|\vec{B}_{\text{ext}}|$, since the orbiting electrons generate a magnetic field counteracting the external magnetic field. The orbital motion of electrons around a nucleus creates constitutes current loops, which produce magnetic fields of their own. In the presence of an external magnetic field, these current loops will align in such a way as to oppose the applied field. Macroscopically, this effect may be illustrated by a tiny magnet that is brought into an external magnetic field; the small magnet will align itself such that its field is opposed to the external magnetic field (Fig. 13.8). The lesser magnetic field exhibited by a particular nucleus therefore leads to a lesser resonance energy which is called the chemical shift. Since the electron density in a molecule varies with location, depending on the chemical bonds, different chemical shifts are observed for different bonding situations.

Since the registered chemical shifts are extremely small compared to the external magnetic field, NMR spectra are not plotted against the effective magnetic field strength but rather compared to an internal standard; for ^{1}H-NMR this is typically tetramethylsilane (TMS) which produces a single peak in a proton NMR experiment. The chemical shift δ is defined as

$$\delta = \frac{B_{\text{sample}} - B_{\text{standard}}}{B_{\text{standard}}} \cdot 10^6 = \frac{\nu_{\text{sample}} - \nu_{\text{standard}}}{\nu_{\text{standard}}} \cdot 10^6 \qquad (13.20)$$

and therefore yields $\delta = 0$ ppm for the standard. Figure 13.9 shows the ^{1}H-NMR spectrum of 1-chloropropane which features three clusters of peaks. The step-type curves show the integration of the individual peak clusters (i.e. peak areas). The integration values form the ratio 2:2:3 which indicates that the peak clusters originate from the $ClCH_2-$, the $-CH_2-$ and the CH_3-group with 2, 2 and 3 protons, respectively. The fine structure in each peak cluster demonstrates the substantial impact of the local environment and is thus an extremely important part of structure determination.

As evident from Fig. 13.9, the ^{1}H-NMR spectrum of 1-chloropropane shows a triplet peak for the methyl group ($\delta = 1.02$ ppm). This is due to the coupling of the spin of methyl protons with the spins of neighbouring methylene protons (spin-spin coupling). With two possible spins for each methylene proton, the spin of a methyl

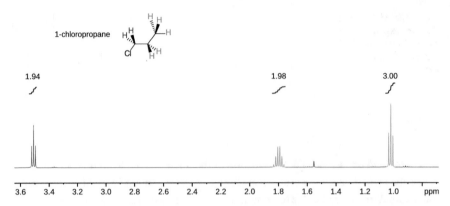

Fig. 13.9 ^{1}H-NMR spectrum of 1-chloropropane acquired in CDCl$_3$ at room temperature on a Bruker 500 MHz Avance III NMR spectrometer at room temperature. Each peak cluster has been integrated and the numerical values are shown above the integration curves. (Spectrum courtesy of Dr Siji Rajan.)

proton can couple with four different combinations of methylene proton spin orientations; however, two of those combinations are indistinguishable and thus collapse into one peak (Fig. 13.9).

The methylene proton spins experience the the spin configurations of the methyl group of which there are eight different. However, there are three indistinguishable configurations each that possess a total spin of $+^1/_2$ and $-^1/_2$, respectively. The splitting pattern of the methylene protons thus yields a quartet. Additionally, the methylene proton spins can also couple with the spins of the chloromethyl protons (three distinct possible orientations) which leads to a triplet splitting of each of the quartet peaks. This results in a complex multiplet that is difficult to resolve and appears as a sextet ($\delta = 1.80$ ppm). The two protons of the chloromethyl group split into a triplet pattern ($\delta = 3.51$ ppm) as their spins experience the spin combinations of the two methylene protons.

The separation between the lines of a multiplet yields the spin-spin coupling constant J. The magnitude of the coupling constant is determined by the extent of the magnetic interaction between two nucleic spins. There are three important observations regarding proton spin-spin coupling:

- Nuclei that have the same chemical shift (i.e. they constitute equivalent nuclei) do not interact with each other in this manner.
- Coupling occurs primarily between protons that are separated by three bonds; the number of bonds between two coupling nuclei is indicated by a subscript on the coupling constant (e.g. ^2J, ^3J, ...).
- Coupling is most noticeable with protons bonded to carbon (Fig. 13.10).

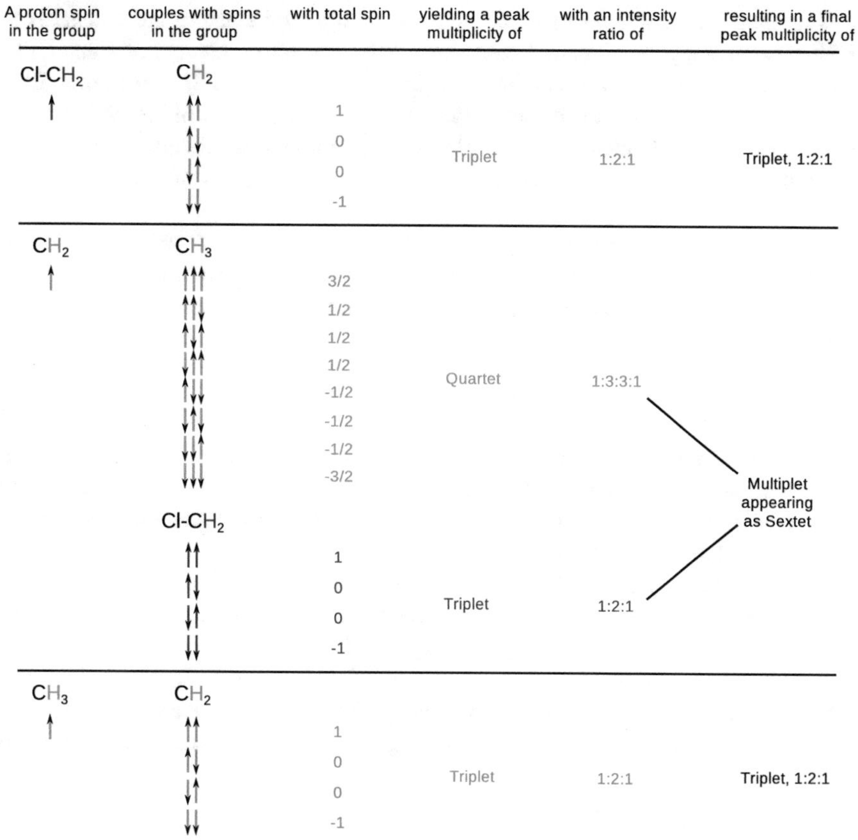

A proton spin in the group	couples with spins in the group	with total spin	yielding a peak multiplicity of	with an intensity ratio of	resulting in a final peak multiplicity of
Cl-CH$_2$ ↑	CH$_2$ ↑↑	1			
	↑↓	0			
	↓↑	0	Triplet	1:2:1	Triplet, 1:2:1
	↓↓	-1			
CH$_2$ ↑	CH$_3$ ↑↑↑	3/2			
	↑↑↓	1/2			
	↑↓↑	1/2			
	↓↑↑	1/2			
	↑↓↓	-1/2	Quartet	1:3:3:1	
	↓↑↓	-1/2			
	↓↓↑	-1/2			
	↓↓↓	-3/2			
					Multiplet appearing as Sextet
Cl-CH$_2$ ↑	↑↑	1			
	↑↓	0			
	↓↑	0	Triplet	1:2:1	
	↓↓	-1			
CH$_3$ ↑	CH$_2$ ↑↑	1			
	↑↓	0			
	↓↑	0	Triplet	1:2:1	Triplet, 1:2:1
	↓↓	-1			

Fig. 13.10 Proton spin-spin couplings in 1-chloropropane

13.2.3 Electron Spin Resonance (ESR) Spectroscopy

ESR (also called electron paramagnetic resonance or EPR) spectroscopy shares the general concepts with NMR spectroscopy and both methods therefore have many features in common. In ESR spectroscopy, a sample is exposed to a homogeneous external magnetic field which leads to a slight over-population of the lower spin state (see Table 13.4). The energy difference ΔE between the lower and higher spin state is overcome by provision of electromagnetic radiation that satisfiers the resonance condition. In Sect. 13.2.1 we have seen, that the energy difference between the two spin states of the electron is about three orders of magnitude larger than in case of the nuclear spin states. Using Eq. 13.18 and assuming an external magnetic field of 0.5 T, the energy required for electron spin resonance is:

$$\text{h} \cdot \nu = 18.59 \cdot 10^{-24} \text{ J T}^{-1} \cdot 0.5 \text{ T} \Rightarrow \nu = 14.0 \text{ GHz} \Rightarrow \lambda = 2 \text{ cm}$$

which is in the region of microwaves.

Importantly, only molecules that possess an unpaired electron are amenable to ESR spectroscopy, since paired electrons cannot undergo a change of spin as this would violate the Pauli exclusion principle (see Sect. 10.3.2). Therefore, ESR spectroscopy is limited to the investigation of radical species of which there are naturally only few. However, the use of so-called spin-labels—chemical groups with a stabilised radical that can be covalently attached to molecules of interest—allows the investigation of many processes by means of the attached reporter group carrying an unpaired electron. ESR spectroscopy with spin labels is a frequently used method in the biosciences to study processes on and around proteins and membranes. Only two transitions are allowed for the unpaired electron, though, as is reflected by the selection rules:

$$\Delta m_s = \pm 1, \Delta m_I = 0. \tag{13.21}$$

The rule demanding conservation of the nuclear spin ($\Delta m_I = 0$) can be explained if one assumes that the motion of a nucleus is much slower than that of an electron. In other words, in the time it takes for the electron to change its spin ($\Delta m_s = \pm 1$), the nuclear spin has no time to reorient.

For practical reasons, ESR spectra don't register the absorption peak but rather its first derivative (Fig. 13.11). For free electrons, the ESR spectrum contains just one peak. If we consider the hydrogen atom, which constitutes the simplest radical consisting of one proton and one electron, we find that the ESR spectrum shows two peaks with the same intensity; the distance between both peaks is measured at 50.7 milli-tesla (mT). For comparison, the signal of the free electron comes to lie in

Fig. 13.11 (a) ESR absorption signal of an unpaired electron, and (b) its first derivative. (c) The hyperfine splitting of the ESR signal of the unpaired electron in the hydrogen atom shows a doublet with a coupling constant of $a^H = 50.7$ mT

a

b

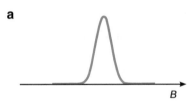

c

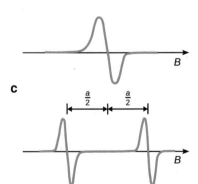

the centre between the two peaks of the hydrogen atom. Similar to the observations made in nuclear magnetic resonance, the spin signal of the hydrogen electron is split into a doublet.

The splitting is the result of an interactions between the spin of the unpaired electron and the nuclear spin, known as the Fermi contact. This interaction only arises when the unpaired electron is inside the nucleus—a phenomenon that can only be explained using the probability density (introduced in Sect. 8.3.2):

$$\rho(x) = |\Psi(x)|^2 = \Psi^*(x) \cdot \Psi(x). \tag{8.32}$$

The effect of the nuclear magnet on the spin of the unpaired electron can be summarised in form of an additional magnetic field $\vec{B}_{contact}$ (contact field), which depends on the magnetic moment of the nucleus (given by the quantum number m_I), the g-factor of the nucleus (g_N), the nuclear magneton (μ_N), as well as the spin density of the unpaired electron in the nucleus ($|\Psi|^2$):

$$\left|\vec{B}_{contact}\right| = \frac{8\pi}{3} \cdot g_N \cdot \mu_N \cdot |\Psi|^2 \cdot m_I = a^N \cdot m_I \tag{13.22}$$

thereby defining the hyperfine coupling constant a^N.

For the proton, the probability density of the electron in the $1s$ orbital can be calculated, thereby allowing the computation of a theoretical hyperfine coupling constant for the hydrogen atom. The calculated value of $a^H = 50.8$ mT is in excellent agreement with the experimentally observed value of 50.7 mT.

If the unpaired electron of a radical interacts with two different (non-equivalent) nuclei, for example protons H1 and H2, a more complex splitting pattern arises due to independent nuclear spin orientations of both protons. Interaction of the electron with H1 leads to a doublet pattern with a coupling constant of a^{H1} as discussed above. Due to the presence of H2, one needs to consider that each of those situations can also show coupling with the nuclear spin orientations of H2, which possesses a different coupling constant a^{H2}. The hyperfine splitting therefore results in 'doublet of doublet' pattern where all peaks possess the same intensity (Fig. 13.12 left). Notably, if the electron was to interact with two equivalent protons (e.g. in the formyl radical H_2CO^-), then the coupling constants a^{H1} and a^{H2} have the same value and the two centre peaks collapse into one peak. The resulting pattern of the hyperfine splitting thus appears as a triplet with intensity ratio 1:2:1 (Fig. 13.12 right). Generally, the number of hyperfine lines can be predicted as per

$$\text{number of lines} = 2 \cdot N \cdot I + 1 \tag{13.23}$$

where N is the number of chemically equivalent nuclei and I is the nuclear spin.

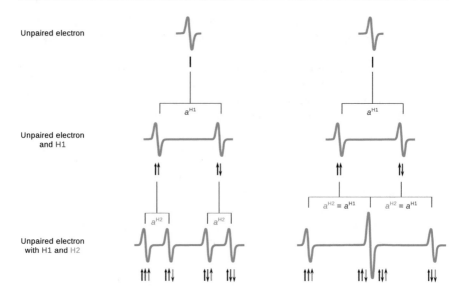

Fig. 13.12 Hyperfine splitting by coupling of an unpaired electron with two non-equivalent protons (left) and two equivalent protons (right)

13.3 Rota-vibrational Spectroscopy

13.3.1 The Rotational Spectrum

In Sect. 9.2.2 we discussed the rigid rotor with space-free axis which we now consider as a model to describe the rotation of a two-atomic molecule. This model assumes that the two atoms are bonded to each other at a fixed distance ('rigid') and that the axis of the two-atomic molecule is allowed to take any orientation in an outside coordinate system ('space-free axis') while the centre of gravity rotates around the origin. Equation 9.29 delivered the allowed energy levels of such a rotor as

$$E(J) = h \cdot c \cdot B \cdot J \cdot (J+1), \tag{9.29}$$

whereby the rotational constant

$$B = \frac{h}{8\pi^2 \cdot c \cdot I} \tag{13.24}$$

is dependent on the momentum of inertia I; h and c are the Planck constant and the speed of light, respectively. In order for a transition between different rotational levels to occur, the resonance condition needs to be met; i.e. electromagnetic radiation of the matching energy needs to be provided to (absorption spectrum) or is released from (emission spectrum) the molecule. Second, we need to consider relevant selection rules as there are many different rotational levels between which transitions could potentially occur.

The selection rules define that only transitions with $\Delta J = \pm 1$ are possible. If we assume a starting level of J_{start} and an end level of $J_{end} = J_{start} + 1$, we can calculate the resonance energy for a transition:

$$
\begin{aligned}
\Delta E &= E(J_{end}) - E(J_{start}) \\
\Delta E &= h \cdot c \cdot B \cdot J_{end} \cdot (J_{end} + 1) - h \cdot c \cdot B \cdot J_{start} \cdot (J_{start} + 1) \\
\Delta E &= h \cdot c \cdot B \cdot [(J_{start} + 1) \cdot (J_{start} + 2) - J_{start} \cdot (J_{start} + 1)] \\
\Delta E &= h \cdot c \cdot B \cdot [J_{start}^2 + 3 \cdot J_{start} + 2 - J_{start}^2 - J_{start}] \\
\Delta E &= h \cdot c \cdot B \cdot (2 \cdot J_{start} + 2) \\
\Delta E &= 2 \cdot h \cdot c \cdot B \cdot (J_{start} + 1).
\end{aligned} \tag{13.25}
$$

Using the relationship in Eq. 13.15, we can thus calculate the wavenumber of a transition between the two rotational levels J and $(J+1)$ as per:

$$
\widetilde{\nu} = 2 \cdot B \cdot (J + 1). \tag{13.26}
$$

This equation allows a conclusion as to the appearance of a rotational spectrum. For the transition of rotational level 0 to level ($0 \rightarrow 1$) we are expecting a peak in a plot of absorbance versus wavenumber that is spaced at a distance of $2 \cdot B$ from $\widetilde{\nu} = 0$. For transition $1 \rightarrow 2$ the peak is spaced $4 \cdot B$ from $\widetilde{\nu} = 0$, etc. In other words, Eq. 13.26 predicts a spectrum with peaks with a distance of $2 \cdot B$ between them (see Fig. 13.14). In reality, the distance between the absorption peaks decreases as the wavenumber increases, owing to the fact that the molecule is indeed not a rigid rotor as assumed in the above model. At higher rotational states, the distance between the two atoms (and thus the momentum of inertia $I = m \cdot r^2$, Eq. 9.7) increases due to the centrifugal force. As per Eq. 13.24 above, an increase in I prompts a decrease in B (see analysis of rota-vibrational spectra in Sect. 13.3.4).

To appreciate the type of electromagnetic radiation that fulfils the resonance condition for a rotational spectrum, we consider the molecule HCl which possesses a bond length of 1.29 Å. This yields a momentum of inertia of

$$
I = \mu \cdot r^2 = \frac{1 \cdot 35}{1 + 35} u \cdot 1.67 \cdot 10^{-27} \frac{kg}{u} \cdot (1.29 \cdot 10^{-10} m)^2 = 2.70 \cdot 10^{-47} kg\, m^2.
$$

The rotational constant of HCl is then

$$
\begin{aligned}
B &= \frac{h}{8\pi^2 \cdot c \cdot I} = \frac{6.626 \cdot 10^{-34} \, J\, s}{8\pi^2 \cdot 2.99 \cdot 10^8 \, m\, s^{-1} \cdot 2.70 \cdot 10^{-47} \, kg\, m^2} \\
B &= \frac{6.626}{8\pi^2 \cdot 2.99 \cdot 2.70} \cdot 10^{-34-8+47} \cdot \frac{kg\, m^2\, s^{-2}\, s}{kg\, m^3\, s^{-1}} \\
B &= 0.0104 \cdot 10^5 \, m^{-1} = 10.4 \, cm^{-1}.
\end{aligned}
$$

For the transition from the rotational ground to the first excited state ($0 \rightarrow 1$), the following wavenumber can be computed by using Eq. 13.26:

$$\tilde{\nu} = 2 \cdot B \cdot (J+1) = 2 \cdot 10.4 \text{ cm}^{-1} \cdot (0+1) = 20.8 \text{ cm}^{-1}$$

and converted into a wavelength or frequency

$$\lambda = \frac{1}{\tilde{\nu}} = 0.048 \text{ cm} = 480 \text{ } \mu\text{m}$$

$$\nu = \frac{c}{\lambda} = \frac{2.99 \cdot 10^8 \text{ m s}^{-1}}{480 \cdot 10^{-6} \text{ m}} = 0.62 \cdot 10^{12} \text{ Hz} = 0.62 \text{ THz},$$

corresponding to a resonance energy of $\Delta E = 4.1 \cdot 10^{-22}$ J. These values indicate that rotational transitions in molecules require a resonance energy in the range of microwaves and far infrared. Purely rotational spectroscopy is thus often called microwave spectroscopy (Fig. 13.13).

If the transition moment is the same for all different starting levels J (which is indeed the case), then the intensity of individual peaks in the rotational spectrum should be dependent on the population of the starting level of the transition that elicits an individual peak. However, one needs to remember that the quantum mechanical discussion of the rigid rotor resulted in several degenerate states, whereby $(2J + 1)$ states of the same energy were possible for each state J (see Sect. 9.2.2). This degeneracy has to be accounted for in addition to the Boltzmann expression that describes the ratio of population of a rotational state J with respect to all states:

$$\frac{N_J}{N} = \frac{(2 \cdot J + 1) \cdot e^{-\frac{h \cdot c \cdot B \cdot J \cdot (J+1)}{k_B \cdot T}}}{\sum_i (2 \cdot J_i + 1) \cdot e^{-\frac{h \cdot c \cdot B \cdot J_i \cdot (J_i+1)}{k_B \cdot T}}}$$

Using the above equation, the population ratio of N_J/N and $N_{J=0}/N$ can be evaluated, and the quotient of both ratios yields the ratio of populations of state J and state $J = 0$:

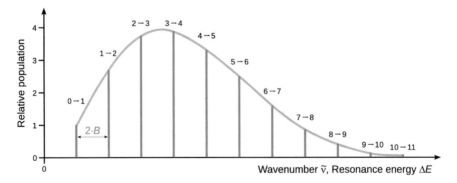

Fig. 13.13 Wavenumbers for transitions in a rotational spectrum (see Eq. 13.26). The absorption of individual transitions is based on the relative absorption of the starting states as given in Eq. 13.27

$$\frac{\frac{N_J}{N}}{\frac{N_{J=0}}{N}} = \frac{N_J}{N_{J=0}} = (2 \cdot J + 1) \cdot e^{-\frac{h \cdot c \cdot B \cdot J \cdot (J+1)}{k_B \cdot T}} \tag{13.27}$$

For HCl at $T = 293$ K with $B = 10.4$ cm^{-1} the relative population of state $J = 1$ with respect to $J = 0$ is then:

$$\frac{N_{J=1}}{N_{J=0}} = 3 \cdot e^{-\frac{0.412 \cdot 10^{-21} J}{4.05 \cdot 10^{-21} J}} = 2.7,$$

which means that the population of the rotational level $J = 1$ is 2.7 times higher than the rotational ground state $J = 0$. The peak intensity in the rotational spectrum thus progresses through a maximum value with increasing J, as indicated in Fig. 13.14.

13.3.2 The Vibrational Spectrum

The quantum mechanical background for vibrational modes of molecules are given by the harmonic oscillator which we discussed in Sect. 9.3 where we derived the discrete energy levels in dependence of the vibrational quantum number v as

$$E = h \cdot \nu_0 \cdot \left(v + \frac{1}{2} \right) \tag{9.38}$$

whereby the oscillation frequency ν_0 is a function of the force constant k and the reduced mass μ:

Fig. 13.14 Comparison of the potential energy of the harmonic (blue) and anharmonic (red) oscillator

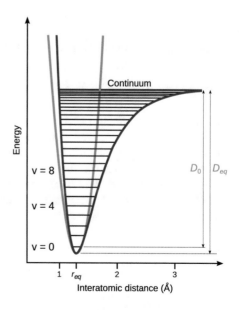

$$\nu_0 = \frac{1}{2\pi} \cdot \sqrt{\frac{k}{\mu}} \tag{9.34}$$

The selection rule for vibrational transitions is $\Delta v = \pm 1$, allowing only transitions between neighbouring vibrational levels. The resonance energy required for a transition from v_{start} to $v_{end} = v_{start} + 1$ is thus:

$$
\begin{aligned}
\Delta E &= E(v_{end}) - E(v_{start}) \\
\Delta E &= h \cdot \nu_0 \cdot \left(v_{end} + \frac{1}{2} \right) - h \cdot \nu_0 \cdot \left(v_{start} + \frac{1}{2} \right) \\
\Delta E &= h \cdot \nu_0 \cdot \left(v_{start} + 1 + \frac{1}{2} - v_{start} - \frac{1}{2} \right) \\
\Delta E &= h \cdot \nu_0
\end{aligned}
\tag{13.28}
$$

The frequency of the electromagnetic radiation fulfilling the resonance condition is thus the same as the frequency of the vibrational mode of the molecule. Using again HCl ($k = 480.6 \text{ N m}^{-1}$) as an example, the type of energy required for vibrational resonance can be calculated:

$$\Delta E = h \cdot \nu_0 = \frac{h}{2\pi} \cdot \sqrt{\frac{k}{\mu}} = \frac{h}{2\pi} \cdot \sqrt{\frac{k \cdot [m(H) + m(Cl)]}{m(H) \cdot m(Cl)}}$$

$$\Delta E = \frac{6.626 \cdot 10^{-34} \text{ J s}}{2\pi} \cdot \sqrt{\frac{480.6 \text{ N m}^{-1} \cdot 36 \text{ u}}{35 \text{ u}^2 \cdot 1.67 \cdot 10^{-27} \text{ kg u}^{-1}}}$$

$$\Delta E = 1.06 \cdot 10^{-34} \text{ J s} \cdot \sqrt{296 \cdot 10^{27} \cdot \frac{\text{kgms}^{-2} \text{ m}^{-1} \text{ u}}{\text{u}^2 \text{ kg u}^{-1}}}$$

$$\Delta E = 1.06 \cdot 10^{-34} \text{ J s} \cdot 5.44 \cdot 10^{14} \text{ s}^{-1} = 5.77 \cdot 10^{-20} \text{ J}$$

This corresponds to a wavelength and wavenumber of:

$$\lambda = \frac{h \cdot c}{\Delta E} = 3.43 \times 10^{-6} \text{ m} = 3.43 \text{ } \mu\text{m} \quad \tilde{\nu} = \frac{1}{\lambda} = 0.291 \cdot 10^6 \text{ m}^{-1} = 2910 \text{ cm}^{-1},$$

thus indicating that the resonance energies are in the region of the infrared.

The resonance energy for a vibrational transition is thus two orders of magnitudes higher than that required for transition between two rotational states. This has substantial implications for the relative population of the individual vibrational states. In analogy to the discussion of the electron/nuclear magnetic states (Table 13.4) as well as the rotational states (Eq. 13.27), this can be calculated using the Boltzmann statistics:

$$\frac{N_{\text{v}}}{N_{\text{v}=0}} = e^{-\frac{h \cdot \nu_0 \cdot (v+1/2)}{k_B \cdot T}} \tag{13.29}$$

For HCl at $T = 293$ K with $\nu_0 = 0.87 \cdot 10^{14}$ Hz, the relative population of state $v = 1$ with respect to $v = 0$ is then:

$$\frac{N_{\text{v}=1}}{N_{\text{v}=0}} = e^{-\frac{1.5 \cdot 5.77 \cdot 10^{-20} \text{J}}{4.05 \cdot 10^{-21} \text{J}}} = 5.1 \cdot 10^{-10},$$

indicating that at room temperature virtually all molecules occupy the vibrational ground state.

13.3.3 Anharmonicity

A closer look at Eq. 13.29 shows that at increasing temperatures one expects that the population of vibrationally excited molecules will be continuously increasing. It should thus be possible to store ever higher energy in molecules which would undergo ever more intense vibrations. This situation is in contradiction to experimental observations: we know that molecules dissociate (i.e. the distance between atoms becomes very large) if they are exposed to high enough temperature.

We remember that the model we used to describe vibrational modes of a di-atomic molecule is based on the harmonic oscillator (see Eq. 9.33 and Fig. 9.9). Graphically, this is illustrated by a symmetric potential curve where the potential energy develops with the square of the distance between two bonded atoms:

$$E_{\text{pot}}(r) = \frac{1}{2} \cdot k \cdot \left(r - r_{\text{eq}}\right)^2 \tag{13.30}$$

where the equilibrium bonding distance between the two atoms is r_{eq}, and k is the force constant.

Two adjustments of the harmonic potential are required to describe the real behaviour of molecular vibration (see Fig. 13.14):

- Even at high energy levels, the distance between the two atoms cannot get arbitrarily small. The Coulomb repulsion between the two nuclei prevents very small inter-atomic distances. The left branch of the parabolic function thus needs to develop steeper than suggested by the harmonic potential.
- The right branch of the potential function needs to progress into a horizontal line at an energy level that represents the dissociation energy. When the molecule reaches this vibrational state, the distance between the two atoms can become arbitrarily large.

The resulting function describes a so-called anharmonic oscillator. At low energy levels, an anharmonic oscillator behaves very similar to a harmonic oscillator. At higher energies, substantial deviations from harmonicity are observed. The energy difference between the minimum of the potential energy curve and the horizontal branch equals the dissociation energy D_{eq}. Since any bond possesses a zero point energy

$$E(v = 0) = h \cdot \nu_0 \cdot \left(0 + \frac{1}{2}\right) = h \cdot \nu_0,$$

the first vibrational level is elevated by $h \cdot \nu_0$ from the minimum of the potential energy curve. The experimentally relevant dissociation energy is thus D_0 which is the difference between the vibrational ground state and the horizontal branch of the potential curve.

The potential curve of the anharmonic oscillator cannot be rigorously derived. The Lennard-Jones potential (see Sect. 11.5 and Fig. 11.8) is one model to describe the potential energy function of an anharmonic oscillator. Another very frequently used function is the Morse function

$$E_{pot}(r) = D_{eq}\left[1 - e^{-\beta \cdot (r - r_{eq})}\right]^2 \tag{13.31}$$

which features the dissociation energy D_{eq} and the constant β as a modified force constant (see also Eq. 9.37). Mathematically, the above function can be developed into a so-called Maclaurin series:

$$E_{pot}(r) = D_{eq} \cdot \beta^2 \cdot (r - r_{eq})^2 - D_{eq} \cdot \beta^3 \cdot (r - r_{eq})^3$$
$$+ \frac{7}{12} \cdot D_{eq} \cdot \beta^4 \cdot (r - r_{eq})^4 - \dots$$

which shows high similarity with the potential energy function for the harmonic oscillator above (Eq. 13.30); the anharmonicity correction results from additional terms with $(r - r_{eq})^3$, $(r - r_{eq})^4$, etc. Using this Morse potential to solve the Schrödinger equation, the following energy eigenvalues are obtained for the anharmonic oscillator:

$$E(v) = h \cdot \nu_0 \cdot \left(v + \frac{1}{2}\right) - h \cdot \nu_0 \cdot x_e \cdot \left(v + \frac{1}{2}\right)^2 + h \cdot \nu_0 \cdot y_e \cdot \left(v + \frac{1}{2}\right)^3$$
$$- \dots \tag{13.32}$$

with the anharmonicity constants x_e and y_e.

The anharmonicity has two major implications:

- The selection rules are different from those of the harmonic oscillator, owing to a different wavefunction. The selection rules for the anharmonic oscillator allow further transitions with $\Delta v = \pm 1, \pm 2, \pm 3, \dots$

- A transition with $\Delta v = \pm 1$ is called fundamental; transitions with $\Delta v = \pm 2, \pm 3,$ ... are called overtones.
- Due to the presence of additional terms with power 2 and higher in Eq. 13.32, the distance between consecutive vibrational levels is not constant but decreases with increasing quantum number v. Therefore, there is only a finite number of vibrational levels until the dissociation energy is reached.

13.3.4 The Rota-vibrational Spectrum: Infrared Spectroscopy

The discussion of the pure rotational and pure vibrational spectra in the previous sections showed that the resonance energy required for vibrational excitation is by about two orders of magnitudes larger than that required for rotational excitation. One can thus expect that upon vibrational excitation vibrational modes will also be excited. The observed vibrational spectra therefore consist of peaks rather than lines. If sufficient spectral resolution can be achieved, the vibrational peaks show a structure that can be analysed as to rotational transitions. The resulting spectra are thus called rota-vibrational spectra.

In infrared spectroscopy, the probing of rota-vibrational transitions is carried in an absorption experiment where samples are exposed to infrared light and the absorbance is monitored. Since vibrational as well as rotational transitions will be observed, one needs to consider the selection rules for the vibrational quantum number as well as those for the rotational quantum number; for the purpose of the following discussion, we are ignoring the possible overtones due to anharmonicity:

- $\Delta v = \pm 1$
- $\Delta J = \pm 1$.

Figure 13.15 shows two vibrational states $v = 0$ and $v = 1$ and overlayed on each the rotational states. It is obvious that when transitioning from $v_{start} = 0$ to $v_{end} = 1$, there are two ways to satisfy the rotational selection rule; one can start from a lower rotational state and transition to the next higher rotational state ($J_{start} = 0, J_{end} = 1$) or vice versa ($J_{start} = 1, J_{end} = 0$). This gives rise to two different blocks of transitions (called branches) which are grouped at lower (P-branch) and higher (R-branch) energies, respectively.

The wavenumbers for the two different branches can be calculated considering Eqs. 13.25 and 13.28:

In exceptional cases, e.g. when molecules exist in triplet state (where the total electron spin is not zero), the selection rules may also include $\Delta v = \pm 1$ and $\Delta J = 0$. In such cases, a third branch called Q-branch is observed.

A close look at the expressions in Table 13.6 shows that a wavenumber $\tilde{v}$ that corresponds to the pure vibrational transition ($\tilde{v}_0$) is not possible, i.e. the pure vibrational transition is not observed, since $\Delta J = 0$ is not allowed in the absence of a Q-branch. This gives rise to the so-called zero gap of the rota-vibrational spectrum; the distance between the two first peaks of the P- and R-branches is:

Fig. 13.15 Top: Rotational transitions for the excitation of the first vibrational state from the vibrational ground state. Bottom: The resulting schematic rota-vibrational spectrum

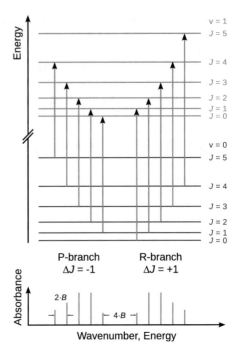

Table 13.6 Wavenumbers for rota-vibrational transitions

P-branch	R-branch
$\Delta v = +1,\ \Delta J = -1$	$\Delta v = +1,\ \Delta J = +1$
$\widetilde{\nu} = \widetilde{\nu} + 2 \cdot B \cdot J_{\text{start}}$	$\widetilde{\nu} = \widetilde{\nu}_o + 2 \cdot B \cdot (J_{\text{start}} + 1)$

$$\Delta J = -1 : J_{\text{start}} = 1, J_{\text{end}} = 0 \Rightarrow \widetilde{\nu} = \widetilde{\nu}_o - 2 \cdot B \cdot 1$$

$$\Delta J = +1 : J_{\text{start}} = 0, J_{\text{end}} = 1 \Rightarrow \widetilde{\nu} = \widetilde{\nu}_o + 2 \cdot B \cdot (0 + 1)$$

$$\Delta\widetilde{\nu} = \widetilde{\nu}_0 + 2 \cdot B - \left[\widetilde{\nu}_0 - 2 \cdot B\right] = 4 \cdot B.$$

Figure 13.16 depicts the rota-vibrational spectrum of HCl and shows two particular details. First, there are two peaks for each individual vibrational transition, owing to the fact that HCl is a mixture of two isotopes $H^{35}Cl$ and $H^{37}Cl$. $H^{37}Cl$ has a slightly larger reduced mass than $H^{35}Cl$ (but the same force constant). According to Eqs. 13.24 and 13.26, this results in a lower frequency and hence a smaller wavenumber.

Second, the distance between the vibrational peak clusters decreases with increasing wavenumber; this cannot be explained by an effect of higher vibrational levels as in that case, the effect should be symmetric in the P- and R-branch with respect to the zero gap. The explanation for this effect becomes obvious when comparing the harmonic and anharmonic oscillator potential curves (Fig. 13.14). The potential curve for the anharmonic oscillator is symmetric with respect to the equilibrium

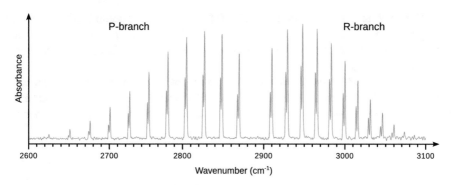

Fig. 13.16 Rota-vibrational infrared spectrum of HCl. The peaks of individual transitions are split into two with the less intense peaks (at lower energies) belonging to isotope $H^{37}Cl$ and the more intense peaks (at higher energies) arise from $H^{35}Cl$

distance r_{eq}; when moving to higher vibrational levels, the equilibrium distance remains constant. In case of the anharmonic oscillator this is different. Excitation of higher level vibrational modes leads to an increase in the bonding distance. With an increase in distance between the two bonded atoms, there is an increase in the momentum of inertia and thus a decrease of the rotational constant B (Eq. 13.24). The rotational constant is therefore dependent on the vibrational mode and its decrease with larger quantum numbers v is reflected in a decrease of the distance between peaks in the rota-vibrational spectrum.

We thus conclude that in order to explain all observations in the rota-vibrational spectrum, we not only need to replace the harmonic with the anharmonic oscillator model, but also the rigid rotor with the non-rigid rotor model.

13.3.5 The Rota-vibrational Spectrum: Raman Spectroscopy

In the discussion of general concepts of absorption spectroscopy in Sect. 13.1.2, we mentioned that an interaction of incident electromagnetic radiation with a molecule (= absorption) only happens if the molecule possesses a dipole momentum. This means that

- homonuclear two-atomic molecules (e.g. O_2, Cl_2, etc)
- symmetric multinuclear linear molecules (e.g. CO_2)
- fully symmetric multinuclear molecules (e.g. CH_4)

possess no rotational absorption spectrum due to the absence of a dipole momentum. Furthermore, homonuclear two-atomic molecules possess no vibrational absorption spectrum either.

However, in this discussion it was implied that the light-absorbing matter needed to possess a permanent dipole momentum which changes upon interaction with incident electromagnetic radiation, thus giving rise to a transition dipole momentum and hence absorption.

We now consider the possibility that the interaction of incident radiation can induce a dipole momentum in a molecule. This phenomenon results in a similar situation, namely that there has been a change in the dipole momentum. The induced dipole momentum μ_{ind} can be calculated as per

$$\mu_{ind} = \alpha \cdot E \tag{11.18}$$

where E is the electrical field strength and α is the polarisability, a quantity we have discussed earlier (see Sect. 12.2.1).

If the electrical field E at the molecule arises due to an electromagnetic wave (with energy $h \cdot \nu$), the field changes periodically with the frequency ν and so does the induced dipole momentum μ_{ind}. In turn, the dipole momentum μ_{ind} oscillating with frequency ν causes emission of light with the energy $h \cdot \nu$. Therefore, the molecule emits light of the same energy that it is being exposed to. This effect is known as Rayleigh scattering and macroscopically appears as an elastic scattering of light by an object.

However, if an energy transfer between incident radiation and molecule happens during this process, the resulting scattering is inelastic, i.e. the energy of the emitted light has different energy than the incident light; this is called the Raman effect. It had been predicted by Adolf Smekal (Smekal 1923) and experimentally proven by Sir Chandrasekhara Venkata Raman. Two situations can arise:

- The incident photon transfers energy onto the molecule which, in turn, transitions into a higher energy state. The emitted light thus has lesser energy than the incident light. This gives rise to the so-called Stokes lines (or S-branch).
- The incident photon gains energy from the molecule which transitions from a state of higher energy to a state of lower energy. The emitted light here has higher energy than the incident light, giving rise to the anti-Stokes lines (or O-branch).

The energy levels between which such transitions occur are the rotational levels as well as the vibrational levels. Therefore, the energy difference between incident and emitted light corresponds to the energy difference between the rotational (and vibrational) levels between which the transition occurs.

The selection rules for rota-vibrational transitions in the Raman effect are

- $\Delta v = \pm 1$
- $\Delta J = 0, \pm 2$

Since it is the energy difference between the incident light and the emitted light that contains the information about the rota-vibrational transitions in the molecule, Raman spectra are interpreted based on the ΔE (or $\Delta \nu$). This is called the Raman shift and obtained as difference between an individual peak and the energy or wavenumber of the incident light which features as the Rayleigh scattering peak in the Raman spectrum.

For a rotational transition $J_{start} \rightarrow J_{end}$ with $J_{end} = J_{start} + 2$, the wavenumber can be quantum mechanically calculated based on the rotational quantum number as per

$$\Delta \tilde{\nu} = \pm 4 \cdot B \cdot \left(J_{start} + \frac{3}{2} \right). \tag{13.33}$$

For $J_{start} = 0$, this yields for the first Stokes line:

$$\Delta \tilde{\nu} = -4 \cdot B \cdot \left(0 + \frac{3}{2} \right) = -6 \cdot B$$

and for the first anti-Stokes line:

$$\Delta \tilde{\nu} = +4 \cdot B \cdot \left(0 + \frac{3}{2} \right) = +6 \cdot B.$$

It is thus obvious, that the rotational Raman spectrum (Fig. 13.17) features the first Stokes and anti-Stokes lines in a distance of $6 \cdot B$ from the Rayleigh scattering peak. The peaks for the following transitions are then spaced at a distance of $4 \cdot B$ from each other.

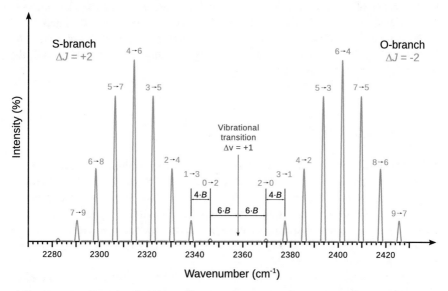

Fig. 13.17 Simulated rotational Raman spectrum of N_2. If light with a wavenumber of 2358 cm^{-1} is used as incident beam, the spectrum would feature a broad Rayleigh scattering peak at this wavenumber. The N-N triple bond vibration occurs at 2358 cm^{-1}

Fig. 13.18 The function $(1 + \cos^2\theta)$ oscillates between values of 1 and 2

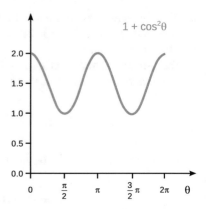

The angular dependency of the Raman scattering intensity is given by the relationship

$$I = I_0 \cdot \frac{8 \cdot \pi^2 \cdot \alpha^2}{\lambda^4 \cdot r^2} \cdot \left(1 + \cos^2\theta\right) \tag{13.34}$$

where I_0 is the intensity of the incident beam at wavelength λ, α is the polarisability, r the distance of the sample to the detector and θ the angle between the incident and the scattered ray. The function $\cos^2\theta$ oscillates between the values of 0 and 1 and assumes 0 at 90°, and the function $(1 + \cos^2\theta)$ thus has its lowest values at 90° and 270° (see Fig. 13.18). Therefore, the scattering intensity at right angles is half the scattering intensity in forward direction.

13.3.6 Conclusion

Infrared and microwave spectroscopy on the one hand and Raman spectroscopy on the other are useful tools that allow determination of force constants and atomic distances in bonds. The former methods require a change in the permanent dipole momentum of a substance during transition between different rotational or vibrational states. In contrast, Raman spectroscopy requires a change of the polarisability during the transition. Therefore, the two types of rota-vibrational spectroscopic methods are often complementary.

The vibrational spectra of molecules are typically rather complex and therefore assure that two different molecules have different vibrational spectra. Among other applications, this can be used very effectively in technique called fingerprinting. The vibrational spectrum of an unknown substance is acquired using either infrared or

Raman spectroscopy and then compared with a database of reference spectra. If a match is found, the unknown substance can be identified.

13.4   Electron Transition Spectroscopy

Electromagnetic radiation of sufficient energy may not only trigger excitation of rotational or vibrational modes of molecules but also cause transition of electrons into energetically higher orbitals. From the discussion of the hydrogen atom in Sect. 10.1, we know that the allowed electronic energy states vary with the principal quantum number n as per:

$$E_n = -\frac{m_e \cdot e^4}{8 \cdot \varepsilon_0^2 \cdot h^2 \cdot n^2}, \tag{10.9}$$

which yields for $n = 2$ and $n = 4$:

$$E_2 = -\frac{9.11 \cdot 10^{-31} \, kg \cdot \left(1.602 \cdot 10^{-19} \, C\right)^4}{8 \cdot \left(8.854 \cdot 10^{-12} \, A^2 \, s^4 m^{-3} kg^{-1}\right)^2 \cdot \left(6.626 \cdot 10^{-34} Js\right)^2 \cdot 2^2}$$

$$E_2 = -0.00054 \cdot 10^{-15} \frac{kg^3 \, m^6}{s^6 \, J^2} = -3.40 eV$$

$$E_4 = -\frac{9.11 \cdot 10^{-31} \, kg \cdot \left(1.602 \cdot 10^{-19} \, C\right)^4}{8 \cdot \left(8.854 \cdot 10^{-12} \, A^2 \, s^4 \, m^{-3} \, kg^{-1}\right)^2 \cdot \left(6.626 \cdot 10^{-34} \, Js\right)^2 \cdot 4^2}$$

$$E_4 = -0.00014 \cdot 10^{-15} \frac{kg^3 \, m^6}{s^6 \, J^2} = -0.85 \, eV$$

and thus for the difference between both electronic states:

$$\Delta E = E_4 - E_2 = 2.55 \, eV.$$

According to the resonance phenomenon, electromagnetic radiation of the same energy is required to enable a transition between both states. The wavelength of such radiation thus needs to be:

$$\lambda = \frac{h \cdot c}{\Delta E} = \frac{6.626 \cdot 10^{-34} \, Js \cdot 2.99 \cdot 10^8 \, ms^{-1}}{4.09 \cdot 10^{-19} \, J} = 4.84 \cdot 10^{-7} \, m = 484 \, nm,$$

which maps to the visual region of the electromagnetic spectrum. This estimation used the electronic orbitals of the hydrogen atom. In molecules, the energy difference between orbitals is typically found in the region of 1–10 eV. The light required to meet the resonance condition is this in the range of the visual and ultraviolet region of the spectrum.

From the discussion of rota-vibrational spectra in the previous sections, we recall that the vibrational excitation of a molecule causes a simultaneous excitation of rotational modes, i.e. the rotational spectrum is superimposed on any vibrational spectrum. Since the energy quanta meeting the resonance condition for electronic transitions (1–10 eV $\approx 10^{-19}$–10^{-18} J) are one to two orders of magnitude larger than those for vibrational transitions (10^{-20} J), rota-vibrational transitions will simultaneously occur with electronic transitions. Therefore, rota-vibrational spectra are superimposed on the electronic spectra. The fact that multi-nuclear molecules possess fairly large moments of inertia, the observation of rotational fine structure and vibrational transitions typically occurs as broad bands in the electronic spectra.

In the introduction to this chapter we considered the overall selection rules for transitions between different energy states: one was concerned with the overall electronic spin state and the other with the symmetry of wave functions.

- For the electronic spin states, it was stated that the multiplicity M has to remain constant, and therefore the total spin S of the molecule does not change during the transition:

$$\Delta S = 0. \tag{13.9}$$

- For the symmetry rules, one needs to consider the symmetry of the wavefunctions (also called parity) as well as the spatial properties of orbitals. If two orbitals involved in a transition do not possess large amplitudes in the same region of space, a transition will not be possible (see also Eq. 13.10).

These selection rules certainly need to be considered for electronic transitions, just as for the rota-vibrational transitions discussed in previous sections. However, further effects will need to be taken into account (see below).

13.4.1 UV/Vis Absorption Spectroscopy

If we consider a diatomic molecule, we can represent the individual electronic states (S_0, S_1, T_1, etc) by their potential energy curves, each of which features the set of vibrational and rotational energy levels. In general, the electronic excited states (S_1, T_1) possess a larger bond distance than the molecule in its ground state. The three potential curves are therefore not only positioned at different levels on the energy scale, but also at different positions with respect to the nuclear displacement scale (Fig. 13.19).

Since the time required for an electronic absorption (10^{-15} s) is much shorter than the time required to complete a vibration (10^{-14}–10^{-13} s), one assumes that the atoms remain at their current position while the electronic transition happens. This is the so-called Franck-Condon principle and requires that we draw an electronic transition as a strictly vertical arrow in the potential energy plot in Fig. 13.19.

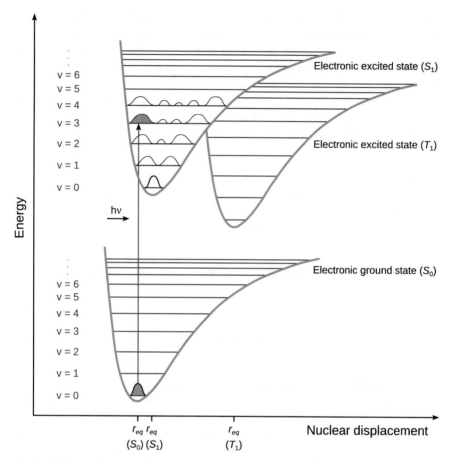

Fig. 13.19 Potential energy diagram sowing different electronic states of a di-atomic molecule and the process of absorption of a photon with energy $h \cdot \nu$

Conceptually, it is based on the Born-Oppenheimer approximation that assumes that the motion of atomic nuclei and electrons in a molecule can be separated.

From previous discussions, we know that the vibrational excited states are essentially not populated when the molecule is in its ground state. When absorbing UV/Vis light (which causes an electronic transition), one thus needs to start in the vibrational ground state of the electronic ground state ($S_0 v_0$). As per the resonance condition, an energy quantum will be absorbed only if it exactly matches the energy difference between $S_0 v_0$ and a vibrational excited state in the electronic excited state ($S_1 v_i$). Different vibrational levels will be populated to different extents, therefore producing a larger or smaller amplitude in the acquired spectrum. The extent to which a particular vibrational state in the electronic excited state is populated depends on the overlap integral between the probability density Ψ^2 (see Sect. 8.3.2) of the starting state ($S_0 v_0$) and that of the end state ($S_1 v_i$). The vibrational overlap integral is also known as the Franck-Condon factor. Graphically, this means that the most likely transition is the one where the probability densities of starting and end state share the largest area under their graphs when transitioning strictly vertically. In Fig. 13.19, the largest overlap is between the probability densities of S_0v_0 and S_1v_3.

We also note that because the start and end vibrational levels of an electronic transition belong to two different potential curves, the vibrational selection rule of $\Delta v = \pm 1$ is no longer valid. The Franck-Condon principle applies to electronic transitions (as opposed to pure rota-vibrational transitions) and thus requires a more complex mathematical treatment than the introductory discussion of the transition dipole momentum (Eq. 13.10), at the centre of which are the overall wavefunctions that are obtained as the product of the individual vibrational, electronic and spin wavefunctions.

13.4.2 Spontaneous Emission from Electronically Excited States

While the elevation of an electron from the ground into an excited state requires the absorption of UV/Vis light, systems may return from the electronic excited to the ground state, in which case there is a possibility that the energy difference is released as a photon. This is the process of emission. When investigating the fabric of atoms, we have seen that atomic absorption and emission spectra produced identical line spectra. In contrast to molecules, however, we did not need to consider rotational or vibrational quantum states. In the following, we will discuss the situation in molecules where the rota-vibrational quantum levels need to be considered.

As discussed in the previous section, upon absorption, electrons typically transition into the first excited singlet state S_1 and the molecule assumes an excited vibrational state (S_1v_i). The life time of such excited states is generally around 10^{-9}–10^{-5} s and therefore much longer than the time required to complete a vibration (10^{-14}–10^{-13} s). Most of the vibration al energy in the excited electronic state is thus dissipated as heat into the environment and the molecule subsequently relaxes into the vibrational ground state S_1v_0. This process is called internal conversion and is not accompanied by emission of radiation.

Whereas some molecules return to the electronic ground state S_0 by further non-radiative relaxation (internal conversion), others emit a photon in this process and transition into a vibrational excited state of the electronic ground state (S_0v_i). This spontaneous emission of light is called fluorescence. Note that the starting state for the fluorescence emission is always the vibrational ground state of the electronic excited state (here: S_1v_0). For this electronic transition, the Franck-Condon principle needs to be applied as introduced in the previous section, since the inter-atomic distances in the molecule remain constant during the electronic transition. The same considerations as for absorption apply, and it turns out that the population of the vibrational end states during the transition are mirrored to to those in the absorption process (see Fig. 13.20). In other words, if the state S_1v_3 was the preferred end state

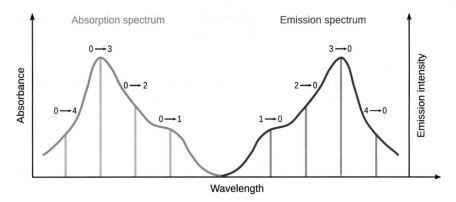

Fig. 13.20 Absorption and fluorescence emission spectra typically are mirror images of each other. The fluorescence spectrum always appears at longer wavelengths than the absorption spectrum (Stokes shift). The band structure of the absorption spectrum shows the vibrational structure of the excited state. In contrast, the fluorescence spectrum shows the vibrational structure of the ground state

Fig. 13.21 Jablonski diagram explaining transitions in and between electronic ground and excited states

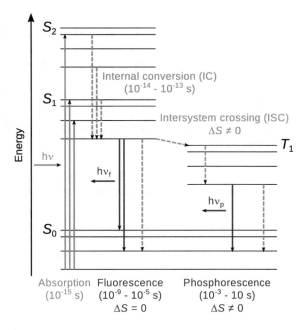

during absorption (Fig. 13.19), the most populated end state in the fluorescence emission process will be $S_0 v_3$.

When illustrating electronic transitions in molecules, the scheme showing potential energy curves (Fig. 13.19) are often replaced by so-called Jablonski diagrams (Fig. 13.21). Here, the individual states are depicted by horizontal lines that are grouped together based electronic states; along the y-axis, the states are arranged according to increasing energy. Non-radiative transitions are indicated by dotted arrows and radiative transitions by straight arrows.

Fluorescence
The spontaneous emission of fluorescence photons following light absorption follows a first-order kinetics and the intensity of light emitted is therefore given by

$$I = I_0 \cdot e^{-\frac{t}{\tau}} \tag{13.35}$$

where τ is the mean life time of the fluorescent state.

The efficiency of the fluorescence process is expressed by the quantum yield Φ:

$$\Phi = \frac{\text{Number of photons emitted}}{\text{Number of photons absorbed}}. \tag{13.36}$$

Fluorescent molecules (also called fluorophores) may not operate at 100% efficiency and emit less photons than they absorbed, since the excited state can be deactivated by other processes. These processes include:

- internal conversion (here: relaxation from electronic excited to electronic ground state without radiation);
- intersystem crossing (change from the singlet to a triplet state of same energy; see below);
- quenching (transfer of excess energy to a quenching molecule);
- fluorescence resonance energy transfer (transfer of excess energy to an acceptor molecule that subsequently emits a fluorescence photon);
- photoreaction.

The rate constant for all deactivation processes is then the sum of the radiative (fluorescence) and all non-radiative processes:

$$k = k_{\text{radiative}} + \sum k_{\text{non-radiative}}.$$

The reciprocal of this overall rate constant k is the mean life time τ of the fluorophore. The ratio of the rate constant describing the radiative process and the overall rate constant therefore equals the quantum yield:

$$\Phi = \frac{k_{\text{radiative}}}{k} = k_{\text{radiative}} \cdot \tau. \tag{13.37}$$

Phosphorescence
In some cases, the triplet state of molecules may possess an intersecting potential curve of lower energy than the excited singlet state (Fig. 13.19). In such cases, it is possible that molecule populates the triplet state after having reached the excited singlet state. This conversion is called intersystem crossing (Fig. 13.21) and, despite

being a forbidden process since it involves inversion of the spin of an electron (i.e. $\Delta S \neq 0$), this selection rule may be weakened if there is a strong spin-orbit interaction such as e.g. in the presence of a heavy atom. Once in the triplet state, the molecule relaxes to the vibrational ground state of the triplet state ($T_1 v_0$) from where it can return to the electronic ground state by emission of a photon. This process is known as phosphorescence and also requires a spin inversion, possible only because of the weakening of the spin selection rule. The occurrence of two 'forbidden' processes leads to a substantial time delay, therefore the photon is emitted at a much later time than the initial absorption.

13.4.3 Stimulated Emission from Electronically Excited States

In addition to the spontaneous emission of light by fluorescence and phosphorescence, molecules can also emit light in a non-spontaneous fashion when returning from an electronic excited to the ground state. Since this process requires stimulation as well as amplification in order to operate, the process is called light amplification by stimulated emission of radiation (laser).

In contrast to the spontaneous emission of light such as in fluorescence and phosphorescence which happens in a statistical fashion, stimulated emission requires the presence of non-participating but stimulating photon of the same energy. Photons resulting from spontaneous emission have a defined energy equal to the difference between the two energy levels between which the transition occurs. The polarisation and direction of travel, however, are random. In the case of stimulated emission, the energy, as well as other properties of the stimulating and emitted photons are the same.

In the cases of spontaneous emission of light, an intrinsic characteristic is that the electronic transitions take place between two different levels, and excited and a ground state. Light amplification in such systems would only be possible if the population of the excited state was larger than that of the ground state. Such a population ratio is called inversion. Even with sustained excitation, the population of the two states in a two-level system can only achieve a balanced distribution; it is not possible to achieve inversion.

However, in three- or four-level systems, inversion may be possible by a process called optical pumping (Fig. 13.22). Absorption of light with appropriate energy can excite molecules from the ground state to level 1, from which they transit into level 2. If the life time of level 2 is much larger than that of level 1, it is possible to build up a large population of level 2. From this state, the molecule can transition to level 3 by stimulated emission and thus release a photon. If the life time of level 3 is much shorter than that of level 2, the molecule quickly relaxes into the ground state from which it can be excited again into level 1. Importantly, the population of level 3 needs to be fairly small compared to level 2. If this process is repeated many times by means of a constant feedback, an oscillating system is generated that can produce an enormous amplification. This is technically achieved in an optical resonator that sends the photons multiple times through a tube; oscillating properties

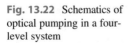

Fig. 13.22 Schematics of optical pumping in a four-level system

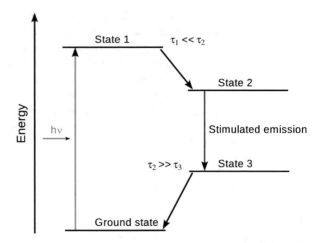

are possible if the resonator length is an integer multiple of half of the wavelength of the emitted light.

The laser medium can be a gas, liquid, or solid semi-conducting material. The excitation of the laser medium may be achieved electrically, with light flashes or even the reaction enthalpy of an exothermic chemical reaction. After excitation, a photon may be emitted spontaneously in the long direction of the resonator. This photon can serve as the stimulating photon that elicits release of another photon by a neighbouring molecule. As this process continues, more and more photons are emitted which then reach a mirror at one end of the resonator. The light is reflected back into the laser medium and leads to further amplification as it travels into the opposite direction. At the other end of the resonator is a partially transparent mirror that reflects most of the light back into the medium. Owing to the relationship between the resonator length and the wavelength of the emitted light, an oscillation results since forth- and back-running light waves are always in phase. Some of the light can leave the resonator through the partially transparent mirror (output coupler) such as to be used for particular purposes.

13.5 Spectroscopic Methods Involving X-rays

13.5.1 Photoelectron Spectroscopy

In Sect. 8.1.3 we discussed the photoelectric effect that described the release of electrons from a metal plate that is exposed to electromagnetic radiation (Fig. 8.3). From Einstein's frequency law (Eq. 8.8) it became clear that the energy of the incident light ($h \cdot \nu$) equals the work required to release electrons from the material ($h \cdot \nu_{cath}$) and the kinetic energy of the released electrons ($\frac{1}{2} \cdot m_e \cdot v^2$):

$$h \cdot \nu = h \cdot \nu_{\text{cath}} + \frac{1}{2} \cdot m_e \cdot v^2. \tag{8.8}$$

This photoelectric effect can also be observed with non-metals, and the frequency law is then formulated more generally as

$$h \cdot \nu = I_i + \frac{1}{2} \cdot m_e \cdot v^2 \tag{13.38}$$

where, according to Koopman's theorem, I_i is the ionisation energy, i.e. the energy required to expel an electron from a particular orbital i.

This effect forms the basis of photoelectron spectroscopy where light of a particular wavelength (monochromatic light) is directed onto a sample and the kinetic energy of released photoelectrons is determined. Energy measurements of electrons require conditions of very high vacuum; therefore, fairly sophisticated equipment as well as careful sample preparation is required for such experiments. The choice of light source depends on whether photoelectrons originating from valence orbitals (with low ionisation energies) or such from deeper orbitals (with high ionisation energies) are investigated. For the former, UV light available e.g. from helium gas discharge lamps is used; this is known as UV photoelectron spectroscopy (UPS). Spectra obtained from UPS show peaks that correspond to the ionisation energies of the valence orbitals, but also a fine structure due to vibrational levels of the molecular ion, which facilitates the assignment of peaks to bonding, non-bonding or anti-bonding molecular orbitals.

Photoelectrons originating from core orbitals require higher energies in the X-ray region. These types of experiments are known as X-ray photoelectron spectroscopy (XPS). In contrast to valence electrons, the electrons in core orbitals of an atom should not be affected by neighbouring atoms. At a first approximation, XP spectra of a particular element therefore show peaks at the same energy irrespective of whether it exists as an isolated atom in the gas phase, covalently bonded in a molecule or embedded in a solid. XP spectra are therefore ideal to determine the elemental composition of samples (a technique known as electron spectroscopy for chemical analysis, ESCA).

Closer inspection shows that the type of bonding does have a small effect on the position of the core photoelectrons in XP spectra and small shifts are seen when comparing the same element in varying bonding environments.

13.5.2 X-ray Diffraction

When discussing the scattering of light by atoms and molecules in previous sections (e.g. such as in the case of Raman spectroscopy), we very much focussed on the scattering by single isolated molecules. When atoms and molecules are arranged in a periodic array with long-range order, the superposition of light scattering from individual atoms gives rise to the phenomenon of diffraction, due to constructive interference of waves at particular angles.

The scattering of light arises from the interaction of electromagnetic radiation with matter which causes the electrons in the exposed sample to oscillate. The accelerated electrons, in turn, will emit radiation of the same frequency as the incident radiation (secondary waves). In a crystalline sample, the individual atoms can be considered point scatterers which all emit secondary waves. The strength with which an atom scatters light is proportional to the number of electrons around that atom. Upon superposition of individual secondary waves, the phenomenon of interference occurs. Depending on the displacement (phase difference) between two waves, their amplitudes either reinforce or cancel each other out. The maximum reinforcement is called constructive interference, the cancelling is called destructive interference. The interference gives rise to dark and bright rings, lines, or spots, depending on the geometry of the object causing the diffraction. Diffraction effects increase as the physical dimension of the diffracting object (aperture) approaches the wavelength of the radiation. When the aperture has a periodic structure, for example in a diffraction grating, repetitive layers or crystal lattices, the features generally become sharper.

Constructive interference of secondary waves from point scatterers occurs only when the diffraction condition is met. Using a geometric approach (Fig. 13.23), the father-son team of William Henry and William Lawrence Bragg found that the diffraction condition is met for re-radiated waves that enclose an angle of $2 \cdot \theta$ with the incident beam of monochromatic radiation (i.e. single wavelength), if the path difference is equal to an integer multiple of the wavelength (Bragg and Bragg 1913). The diffraction condition is known as Bragg's law:

$$n \cdot \lambda = 2 \cdot d \cdot \sin \theta \qquad (13.39)$$

From the observed diffraction rings (powder/multi-crystalline samples) or spots (mono-crystalline samples) that appear at a particular angle $2 \cdot \theta$ with respect to the

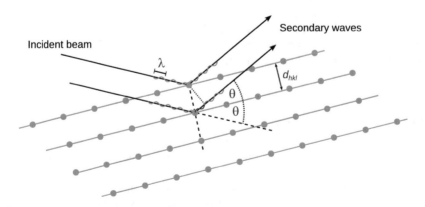

Fig. 13.23 Geometric construction to explain the Bragg diffraction

incident beam, one can therefore calculate distances of lattice planes (d) via the geometric relationship with the integer multiples (n) of the wavelength λ.

Since the distances between atoms or ions are at the order of 10^{-10} m (1 Å), diffraction methods used to determine structures at the atomic level require radiation in the X-ray region of the electromagnetic spectrum, or beams of electrons or neutrons with a similar wavelength. We recall from Sect. 8.1.5, that electrons (as well as neutrons) are particles, but also possess wave properties with the wavelength depending on their energy (DeBroglie relationship). Accordingly, diffraction can not only be observed X-rays but also electron and neutron beams.

Crystalline materials are characterised by the long-range order/periodic arrangement of atoms. The unit cell of a crystal is the basic repeating unit that defines the crystal structure. It repeats in all three dimensions and thus defines the lattice parameters of the crystal. Parallel planes of atoms intersecting the coordinate system of the unit cell define directions and distances in the crystal. The different sets of parallel planes are characterised by their intercepts with the axes of the coordinate system (defined by the unit cell). The reciprocal fractional intercepts are called the Miller indices (h with x-axis, k with y-axis and l with the z-axis), which together with the distance d_{hkl} between the parallel planes in the set, describe a particular set of planes that elicits a diffraction spot at the angle 2·θ.

Notably, in contrast to the definition of resolution in optical microscopy (see Sect. 8.1.5), the resolution of structures derived by diffraction methods is the smallest distance d_{hkl} of lattice planes (corresponding to the maximum reflection angle θ) that can be resolved on the diffraction image.

13.5.3 Mößbauer Spectroscopy

Similar to the resonance phenomena discussed in earlier sections of this chapter, there is a phenomenon of nuclear resonance of γ-rays. If a particular nucleus with Z protons (= order number) and N neutrons exists in an excited state of the energy level E_{exc}, it can transition into the ground state with the energy level E_{ground} and emit the energy difference as a photon; due to the magnitude of the energy difference

$$E_0 = E_{exc} - E_{ground},$$

the photon will be a γ-ray. If this photon hits another nucleus of the same element, that second nucleus can transition from the ground into the excited state. As we have seen earlier, such resonance phenomena are possible only if the incident photon possesses exactly the same energy that corresponds to the difference between the energy levels of the excited and the ground states.

However, similar to a gun, the excited nucleus experiences a recoil effect when emitting a γ-ray photon, owing to the conservation of the total momentum. The recoil effect results in the nucleus to be kicked back, in the direction opposite to the emitted photon. Since the principle of conservation of energy still needs to be considered, too, the emitted photon possesses an energy that is slightly less than the energy difference between the ground and excited states of the transitioning nucleus; the energy of the recoil effect needs to be subtracted from the energy difference between the two nuclear states:

$$E_{\text{photon}} = E_0 - E_{\text{recoil}}. \tag{13.40}$$

The recoil energy of a free atom or molecule can be calculated as per

$$E_{\text{recoil}} = \frac{E_0^2}{2 \cdot m \cdot c^2}. \tag{13.41}$$

Since the energy E_0 takes substantial values (due to the photons being γ-rays), the recoil energy E_{recoil} is orders of magnitudes larger than the natural line width of a transition. The emitted photon thus carries less energy than the resonance difference E_0 (Eq. 13.40) and this difference cannot be ignored; this is different with electromagnetic radiation of lesser energies where the recoil energies are less than the natural line widths, and the recoil effect thus does not play a significant role.

As the absorbing nucleus also experiences a recoil effect as it receives the incoming photon, for a successful resonance transition to occur the energy of the incoming photon needs to carry not only the resonance energy E_0, but also the energy to compensate for the recoil effect:

$$E_{\text{photon}} = E_0 + E_{\text{recoil}}. \tag{13.42}$$

The energies of emitted and received photons are therefore separated by $2 \cdot E_{\text{recoil}}$ (Eqs. 13.40, 13.41) and, due to the large value of E_{recoil} with respect to the natural line width, do not overlap (Fig. 13.24). It is thus virtually impossible to observe this resonance phenomenon in the gas or liquid phase.

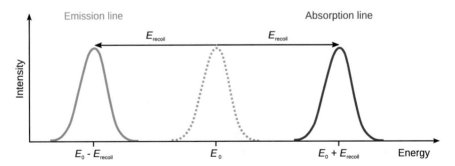

Fig. 13.24 The γ-ray emission and absorption lines observed with free atoms do not appear at the theoretical energy E_0 but are shifted to higher/lower energies due to the recoil effect. In order for the resonance condition to be met in the absorption case, the incoming γ-ray needs to possess a higher energy than the difference between the two atomic energy levels in order to account for the additional energy when transferring momentum (recoil). For emitted photons, the energy of the γ-ray is decreased by the amount of the recoil energy

However, Rudolf Mößbauer discovered in 1958 that in the crystalline phase the crystal lattice is able to absorb the recoil energy and thus allows the observation of nuclear resonance (Mößbauer 1958). This is called the Mößbauer effect, which led to award of a Nobel prize to his discoverer in 1961; the criterion for it to occur is generally formulated as

$$\frac{E_0^2}{2 \cdot m \cdot c^2} < k_B \cdot \Theta \tag{13.43}$$

whereby Θ is a measure for the strength of the crystal lattice (the so-called Debye temperature). For a given temperature the probability of a successful nuclear resonance to occur is higher the stronger atoms are embedded in a crystal. At lower temperatures, the probability of recoil-free emission and absorption of photons increases.

In a Mößbauer spectrometer, the resonance absorption of emitted photons due to nuclear transitions is disturbed in a controlled fashion by varying the energies of the incoming photons. This is achieved by mounting the emitter of the photons (source) on a drive that moves the source with constant velocity either towards or away from the sample (absorber). Due to the externally controlled velocity, the emitted photons experience an additional momentum (Doppler effect) and their energy is thus higher or lower than the transition energy E_0.

If the velocity of the moving source is varied from zero to a maximum value, then the resonance condition is only met at zero velocity and the overlap of the emission and the absorption lines decreases as the photons emitted by the source are subject to significant Doppler shifts and their energies thus deviate from E_0. Since the detector in a Mößbauer spectrometer acquires the transmittance of photons through the sample (absorber), the maximum signal is registered in the regions outside the resonance condition and the lowest transmittance is registered when the resonance condition is met exactly (see Fig. 13.25).

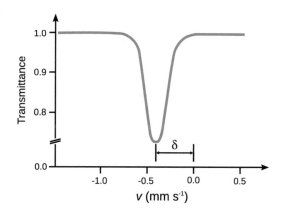

Fig. 13.25 Mößbauer spectra are recorded as relative transmission which is the ratio of γ-quant counts in the resonance range and the number of counts in the non-resonance range. This results in a resonance curve with lowest transmission at the velocity of the source that exactly meets the resonance condition. The isomeric shift δ is defined as the offset to zero velocity

Table 13.7 Summary of the main parameters observed in Mößbauer spectra of substances in solid phase

Parameter	Type of interaction	Description
Isomeric shift	Interaction between nucleus and electrons	Reminiscent of the chemical shift observed in NMR. Information about the absorber: • oxidation state • binding properties of complexes • electronegativity of ligands
Quadrupole splitting	Interaction of the nuclear quadrupole momentum and the electric field at the nucleus	Only possible for nuclei with a spin $I > \frac{1}{2}$; results in $(I + \frac{1}{2})$ sub-levels. Information about the absorber: • spin state • oxidation state • molecular symmetry • binding properties, magnetic behaviour
Magnetic splitting	Interaction between the nuclear magnetic momentum and the magnetic field at the nucleus	Only possible for nuclei with a spin $I > 0$; results in $(2 \cdot I + 1)$ sub-levels. Reminiscent of the hyperfine splitting (Zeeman effect) observed in EPR. Information about the absorber: • spin state • oxidation state • molecular symmetry • binding properties, magnetic behaviour

The number and position of resonance lines in Mößbauer spectra depend on the interactions between nuclei as well as the electric and magnetic fields experienced by the nuclei. These fields are dominated by electrons in the valence shell which define many chemical and physical properties of substances. Three parameters of Mößbauer spectra can be used to assess the interactions between nuclei (see Table 13.7):

- isomeric shift
- electric quadrupole splitting
- magnetic splitting.

Suitable nuclei for Mößbauer spectrometry include Fe, Ni, Zn, Ru, Ag as well as all elements of the third transition metal row, and most lanthanide and actinide metals.

13.6 Atomic Spectroscopy

The interactions of matter with radiation discussed in this chapter so far mainly focussed on molecules; the resulting spectroscopic methods are therefore invaluable for the investigation of structure and properties of molecular matter. However, there also particular interactions between electromagnetic radiation and single atoms. These interactions and the spectroscopic methods arising will be the subject of this section, and also close the loop to some observations and concepts we have discussed earlier. At few previous instances we have made reference to atomic spectra. In Sect. 8.2.1, we used atomic spectroscopy to learn about the fabric of atoms, and in Sect. 10.3.2, we discussed the ionisation energies of elements (derived from atomic spectroscopy) and the concept of shells. Furthermore, the importance of atomic spectroscopy for the study of surfaces and surface processes was in Sect. 7.3.

Clearly, there will be two types of electrons that can reveal information about the particular atoms studied. If the radiation used is of relatively low energy (such as in the optical spectra), the electrons investigated will be those of the outer (valence) shell, i.e. those that define the chemical behaviour of an atom. Radiation of high energy, in contrast, will be probing the tightly bound electrons in the inner (core) shells.

13.6.1 Optical Spectroscopy

When we considered the line spectrum of hydrogen in Sect. 8.2.1, we learned that the different spectral series observed with hydrogen in either atomic absorption or emission spectroscopy followed a particular relationship

$$\frac{1}{\lambda} = R_\infty \cdot \left(\frac{1}{n_0^2} - \frac{1}{n^2} \right) \tag{8.14}$$

whereby R_∞ is the Rydberg constant, and n_0 and n are the principal quantum numbers of the lower and higher energy states, respectively, between which an electronic transition occurs.

If we now consider heavier atoms that also possess only one electron like hydrogen, such as

$$He^+, Li^{2+}, Be^{3+}, B^{4+} \text{ and } C^{5+},$$

we appreciate that their fabric is very similar to that of hydrogen. The difference is that these ions possess a heavier nucleus and multiple nuclear charges (order number $Z > 1$). Their spectra can be observed under extreme conditions, for example when studying the light of stars. The spectra of these one-electron atoms are very similar to that of hydrogen, but the individual lines are shifted to higher frequencies. Importantly, the values of the Rydberg constant for those heavier ions differs from that of hydrogen. The reason for this discrepancy is that when deriving R_∞ it is assumed that an electron of mass m_e orbits around a nucleus of indefinitely high mass; therefore, the nucleus is assumed non-moving. However, nuclei with a real mass $m_{nucleus}$ rotate together with the orbiting electron around the centre of gravity. Instead of the mass of an orbiting electron (m_e), one needs to use the reduced mass (see Sect. 9.2.1), which in this case is given as

$$\mu = \frac{m_e \cdot m_{nucleus}}{m_e + m_{nucleus}}$$

The Rydberg constant for atoms with real mass $m_{nucleus}$ is therefore obtained as

$$R_{m_{nucleus}} = \frac{R_\infty}{1 + \frac{m_e}{m_{nucleus}}}. \tag{13.44}$$

For optical transitions in heavier atoms (than hydrogen) with one electron, one also needs to consider that the nuclear charge is greater than +1e. The field experienced by the orbiting electron is thus different than in the case of hydrogen and depends on the nuclear charge, represented by the order number Z. Equation 8.13 therefore needs to be adjusted for the heavier one-electron atoms and then becomes

$$\frac{1}{\lambda} = Z^2 \cdot R_{m_{nucleus}} \cdot \left(\frac{1}{n_0^2} - \frac{1}{n^2} \right). \tag{13.45}$$

The spectra of atoms with more than one electron are more complicated than those of atoms with just one electron. The wavelengths at which individual lines appear in atomic emission or absorption spectroscopy are, of course, again given by the difference between the energies of the two states between which the transition occurs. However, compared to the fairly simple spectrum of hydrogen-like atoms (see above), the number of lines observed with multi-electron atoms is much larger and can be explained by superposition of different series. Historically, the individual terms (i.e. energy levels) have been called S (sharp), P (principal), D (diffuse) and F (fundamental); the explanation of these different series in terms of the principal quantum number n is illustrated in Table 13.8 (see also Fig. 13.26).

Table 13.8 Term series in optical spectroscopy, illustrated using sodium as an example

Term series	Na ($1s^2\, 2s^2\, 2p^6\, 3s^1$)		General	
Principal series	$3S \rightarrow nP$	$n = 3, 4, 5, \ldots$	$n_0 S \rightarrow n_1 P$	$n_1 = n_0, n_0+1, n_0+2, \ldots$
Sharp series	$3P \rightarrow nS$	$n = 4, 5, 6, \ldots$	$n_0 P \rightarrow n_1 S$	$n_1 = n_0+1, n_0+2, n_0+3, \ldots$
Diffuse series	$3P \rightarrow nD$	$n = 3, 4, 5, \ldots$	$n_0 P \rightarrow n_1 D$	$n_1 = n_0, n_0+1, n_0+2, \ldots$
Fundamental series (Bergmann series)	$3D \rightarrow nF$	$n = 4, 5, 6, \ldots$	$n_0 D \rightarrow n_1 F$	$n_1 = n_0+1, n_0+2, n_0+3, \ldots$

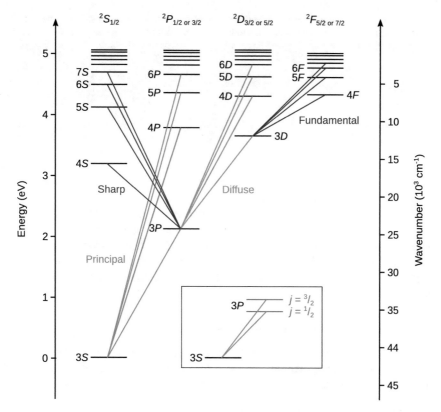

Fig. 13.26 The term scheme of sodium. The inset shows the dublet splitting of lines in spectra of alkali metals due to spin-orbit coupling

Optical Spectra of Alkali Metals

The optical spectra of alkali metals are quite similar to that of the hydrogen atom, owing to the fact that only one electron is responsible for the electronic transitions giving rise to these spectra. In this context, the nucleus and the non-valence electrons are often seen as an entity called the atomic core. The core electrons shield the nuclear charge up to the effective nuclear charge (1 e in case of the alkali metals). The effective nuclear charge is compensated by the valence electron. However, in contrast to the hydrogen atom, the effective nuclear charge experienced by the valence electron is not a point charge. The potential depends on the relative location of the valence electron with respect to the core and is therefore no longer of spherical symmetry. As a consequence, the n^2 different energy levels (see Eq. 10.8) are no longer degenerate, and assume different values depending on the orbital angular momentum l.

Therefore, the spectroscopic terms S, P, D and F are linked to orbital quantum number l, in a similar fashion like the atomic orbitals (see Table 10.5). In order to

Table 13.9 The spectroscopic terms and the orbital quantum number. Possible values of the quantum number for the total angular momentum for atoms with one valence electron are also given

Spectroscopic term L	Orbital quantum number l	Quantum number of the total angular momentum (j)
S	0	0
P	1	$\frac{1}{2}$, $\frac{3}{2}$
D	2	$\frac{3}{2}$, $\frac{5}{2}$
F	3	$\frac{5}{2}$, $\frac{7}{2}$

avoid confusion, atomic orbitals are denoted in lower case and spectroscopic terms in upper case letters (Table 13.9):

In Sect. 10.2.3, we discussed the phenomenon of a magnetic field arising from the motion of a charged particle (electron) that spins around its own axis. This magnetic field couples with the magnetic field arising from the orbiting motion of the electron around the atomic nucleus. This spin-orbit coupling led us to introduce the total angular momentum $\bar{j}$, characterised by the quantum number j that can assume the values

$$j = (l + s), (l + s - 1), \ldots |l - s|. \tag{10.22}$$

Obviously, we need to consider this phenomenon when assessing the possible transitions of a valence electron. The selection rules for the optical spectra of the alkali metals are thus:

$\Delta l = \pm 1$ i.e. transitions are only possible between neighbouring terms, and
$\Delta j = 0, \pm 1$ thereby allowing two transitions between S and P terms, and three transitions between the other terms.

$$\tag{13.46}$$

The different possible transitions are illustrated in the term scheme of sodium (Fig. 13.26). In spectroscopy, the individual quantum states are denoted with symbols such as

$$^2S_{1/2}, {}^2P_{1/2}, {}^2P_{3/2}, {}^2D_{3/2}, {}^2D_{5/2}, \text{etc.; general formula} : {}^{2 \cdot S + 1}L_j$$

In this nomenclature, the spin multiplicity $M = 2 \cdot \Sigma s_i + 1$ (Eq. 13.8) is shown as superscript. Then, the spectroscopic term (dependent on the orbital quantum number l) is given and the quantum number of the total angular momentum (j) is appended as subscript. Since alkali metals have just one valence electron, the total spin is thus $\Sigma s_i = \frac{1}{2}$ and the spin multiplicity is $M = 2$; such terms are called dublet terms.

Optical Spectra of Multi-electron Atoms

In the context of optical spectroscopy, multi-electron atoms are those that possess more than one valence electron. The optical spectra of such atoms are thus much more complicated than those of single-electron atoms, since there are multiple term systems.

In the case of lighter atoms, the orbital momenti $\vec{l}$ of individual electrons couple to form a total orbital momentum $\vec{L}$. Similarly, the spin momenti $\vec{s}$ of individual electrons couple to yield a total spin momentum $\vec{S}$. Both of those total momenti then form a total angular momentum $\vec{J} = \vec{L} + \vec{S}$. This coupling is called Russel-Saunders coupling or L-S-coupling.

Heavier atoms possess a larger nuclear charge and therefore the spin-orbit become as strong as the interactions between individual spins or orbital angular momenti. In such cases, the orbital momenti $\vec{l}$ and spin momenti $\vec{s}$ of individual electrons tend to couple to form individual total angular momenti $\vec{j}$. These individual total momenti j then form a total angular momentum $\vec{j}$, hence this phenomenon is called j–j-coupling.

The selection rule for allowed transitions in multi-electron atoms is

$$\Delta J = 0, \pm 1. \tag{13.47}$$

13.6.2 X-ray Spectroscopy

The energy required to elicit emission of optical spectra from atoms may be provided in the form of thermal energy such as in the flame of a Bunsen burner, a plasma or by a discharge lamp. As discussed in the previous section, the energy difference in the quantum states of valence electrons as of the same order as that of the thermal energy and therefore the spectra arising from transitions of the valence electrons appear in the range of optical and UV light.

If atoms are exposed to electron beams of much higher energy (10 keV–100 keV), emission of light at much shorter wavelengths (X-rays) is observed. Notably, this emission of X-ray light consists of two components, light that contains a continuous distribution of wavelengths (also known as Bremsstrahlung) as well as characteristic X-ray emission at particular wavelengths (see Fig. 13.27).

When accelerated electrons impact on matter they are slowed down by the electric fields of individual atoms. This deceleration results in a loss of kinetic energy which is released in the form of photons. The different wavelengths of the released photons thus result from the different kinetic energies of impacting electrons. Therefore, a minimum wavelength is observed in the Bremsstrahlung; this wavelength corresponds to the highest kinetic energy in the energy distribution of the impacting electrons.

A second process arising from the impact of high energy electrons onto matter is the displacement of an electron from the atomic core (i.e. from an inner shell). The vacated position is subsequently filled by an electron from an outer shell. This transition between two energy levels is accompanied by emission of a photon whose energy equals the energy difference between the two levels of this transition.

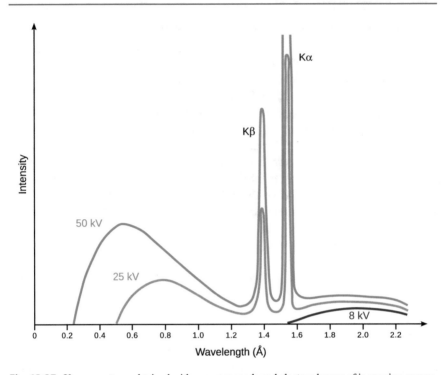

Fig. 13.27 X-ray spectrum obtained with a copper anode and electron beams of increasing energy. The shortest observed wavelength of the continuous X-ray spectrum shifts to higher energies as the kinetic energy of the incoming electrons is increased. At the same time, the intensity at individual wavelengths increases. Note that the wavelengths of characteristic X-ray emission remain constant

The light emitted due to this latter process constitutes the characteristic X-ray emission of an element.

The individual lines of the characteristic X-ray emission can be classified into different series that depend on the principal quantum numbers (shells) of the two energy levels involved in the transition of the electron that fills the gap arising from the impact of high energy electrons (see Fig. 13.28). As with the optical spectra, the orbital angular momentum (represented by the quantum number l) and the total angular momentum (represented by the quantum number j) give rise to a fine structure, hence the selection rules for possible transitions are the same as for the optical spectra:

$$\Delta l = \pm 1 \quad \text{and} \quad \Delta j = 0, \pm 1. \tag{13.46}$$

The emission of characteristic X-ray photons can also be elicited by directing X-ray light onto matter. In this case, the process of X-ray emission is called X-ray fluorescence.

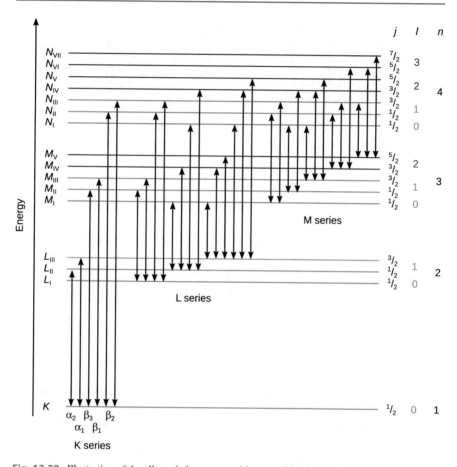

Fig. 13.28 Illustration of the allowed electron transitions resulting in the X-ray spectrum of atoms

The naming of the lines in X-ray spectra frequently uses the traditional Siegbahn notation which denotes the shell of the high energy level as an upper case character (e.g. K) and appends Greek letters (α, β, . . .) as well as numerical indices. Since this notation is non-systematic, it often appears confusing. A more systematic nomenclature has thus been recommended by IUPAC (see Table 13.10).

The Mass Attenuation Coefficient

In Sect. 13.1.1 we saw that there is a loss of intensity as light travels through a sample of interest. The intensity of the exiting beam can be calculated from Eq. 13.2 and is given as per

$$I = I_0 \cdot e^{-\mu \cdot l}, \tag{13.48}$$

where l is the thickness of the material (in cm) and μ the (linear) attenuation coefficient (in cm^{-1}) which comprises of two components: the absorption coefficient

Table 13.10 Naming of characteristic lines in X-ray spectra

Siegbahn notation	Transition between					IUPAC notation
	High energy level		Low energy level			
	Shell	Electron state	Shell	Electron state		
$K\alpha_1$	K	$1s$	L_{III}	$2p_{3/2}$		K-L$_3$
$K\alpha_2$			L_{II}	$2p_{1/2}$		K-L$_2$
$K\beta_1$			M_{III}	$3p_{3/2}$		K-M$_3$
$K\beta_3$			M_{II}	$3p_{1/2}$		K-M$_2$
$L\alpha_1$	L_{III}	$2p_{3/2}$	M_V	$3d_{5/2}$		L$_3$-M$_5$
$L\beta_1$	L_{II}	$2p_{1/2}$	M_{IV}	$3d_{3/2}$		L$_2$-M$_4$
$M\alpha_1$	M_V	$3d_{5/2}$	N_{VII}	$4f_{7/2}$		M$_5$-N$_7$

τ and the loss off intensity based on scattering (σ). For X-rays, the scattering attenuation is negligible, therefore $\mu \approx \tau$.

Since the attenuation is dependent on the mass of matter that is penetrated by the incident beam, the attenuation coefficient is often normalised with respect to the density ρ, thus yielding the so-called mass attenuation coefficient:

$$\left[\frac{\mu}{\rho}\right] = 1\frac{cm^{-1}}{\frac{g}{cm^3}} = 1\frac{cm^2}{g} \tag{13.49}$$

The mass attenuation coefficient is characteristic for a particular element and independent of the chemical and physical state of the sample. In mixtures and compounds, the mass attenuation coefficient is an additive property and therefore be calculated based on the molar ratio of the individual elements in the sample.

Notably, the mass attenuation coefficient depends on the wavelength of the incident radiation (Fig. 13.29). Moving from longer to shorter wavelengths (i.e. from lower to higher energy), the value of the mass attenuation coefficient continuously decreases, until the energy of the incident radiation is sufficient to remove an electron from the next inner shell; at this point, a strong increase in the mass attenuation coefficient is observed, giving rise to a so-called absorption edge. Further increase in the energy of the incident light leads to a renewed decrease of the mass attenuation coefficient until the energy is sufficient for displacement of an electron from the next inner shell.

Moseley's Law

The concepts discussed in the above sections have been concerned with X-ray spectra observed with a particular atom. An important discovery was made by the British physicist Henry Moseley in 1913, when he compared the energies of particular lines, e.g. the $K\alpha_2$ transition, of different elements (Moseley 1913). Expressing the energies of these transitions as either the frequency or wavenumber, a linear correlation between the square root of either ν or $\tilde{\nu}$ with the order number Z is found:

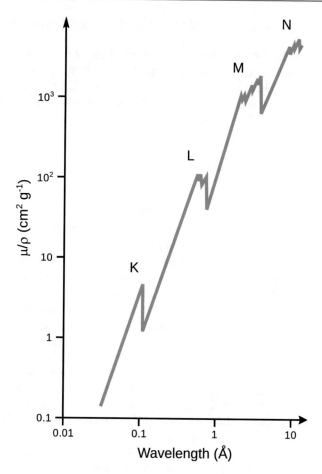

Fig. 13.29 The X-ray absorption spectrum of molybdenum

$$\sqrt{\frac{\tilde{\nu}}{R_\infty}} = \sqrt{\frac{3}{4}} \cdot Z + b. \tag{13.50}$$

For a particular transition between two shells with the principal quantum numbers n_{high} (the inner shell) and n_{low} (the outer shell), the law can be formulated as:

$$\tilde{\nu} = R_\infty \cdot (Z - a)^2 \cdot \left(\frac{1}{n_{high}} - \frac{1}{n_{low}}\right), \quad \text{with} \quad a = -\sqrt{\frac{4}{3}} \cdot b. \tag{13.51}$$

Interestingly, the term $(Z\text{-}a)$ suggests that there is a correction with respect to the positive charge (given by the number of protons = order number). This correction makes sense when we consider that an electron jumping from the outer to the inner shell experiences only the effective nuclear charge, since electrons in inner shells shield the nuclear charge to some degree. This shielding effect is captured by the constant a.

For example, for the Kα transition (principal quantum numbers $n_{high} = 1$ and $n_{low} = 2$), an electron jumps from the L- to the K-shell to fill a gap caused by a displaced electron. There is one remaining electron in the K-shell and provides a shielding of the nuclear charge by about $1 \cdot e$. Consequently, the constant a is found to be approximately 1 for this transition.

13.6.3 Auger Electron Spectroscopy

In Sect. 7.3.2, we have introduced the technique of Auger electron spectroscopy as a method to investigate surface processes. The Auger phenomenon arises as an alternative pathway when core electrons are displaced by impact of high energy electrons or photons. As discussed above, the displacement of an inner shell electron leads to the transition of an electron from an outer shell to fill the gap closer to the nucleus. The energy released during this transition can be emitted in form of a photon (characteristic X-ray lines), but also be transferred onto another electron which is then departing from the atom due to its high energy. Although this process was discovered independently by Lise Meitner (Meitner 1922) and Pierre Auger in the 1920s, it has historically been credited to Auger; the emitted electron is thus called the Auger electron. Auger electron emission and X-ray photon emission are two competing processes. The Auger electron emission typically dominates in lighter atoms, whereas X-ray emission is the preferred process in heavier atoms.

As a consequence of the outlined Auger process, the kinetic energy of the emitted electron equals the energy difference between the two levels involved in the transition of the electron that closes the gap in the inner shell. Notably, this energy is not identical with the energy of a characteristic X-ray photon in the X-ray process. Whereas the latter results in an atom with one positive charge (due to the electron displaced by the impacting high energy electrons), the Auger process results in a doubly charged ion (due to one displaced electron and one emitted Auger electron).

In order to characterise individual Auger process, three electron states need to be described:

- the first state denotes the state from which the primary electron is dislodged
- the second state describes where the electron originates from that fills the inner shell gap
- the third state identifies the level from which the Auger electron originates.

For example, the the Auger spectrum of palladium shows three distinct peaks (Fig. 13.30). For all three types of Auger electrons, a hole is generated in the K-shell by impact of high energy electrons. This core hole can be filled by transition of an electron from either the L_I, L_{II} or L_{III} levels. The energy released during that transition is then transferred to an electron in one of the L levels which is then emitted as an Auger electron from the atom.

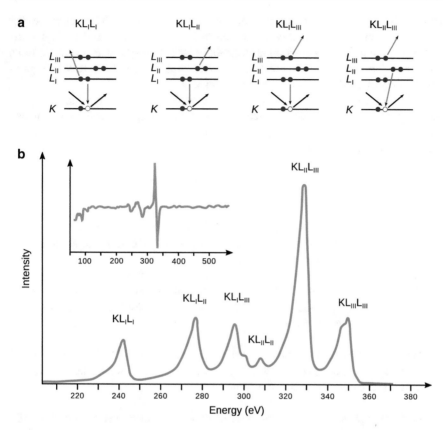

Fig. 13.30 (a) Schematics of particular Auger processes. (b) Auger spectrum of palladium based on data by Babenkov1982. The inset shows the differentiated form of the Auger spectrum

The kinetic energy of Auger electrons is measured by electron energy analysers which are typically based on retardation or deflection of the emitted electrons as they pass through a variable electric or magnetic field. The degree of retardation or deflection is proportional to the kinetic energy. Electrons detected at a particular energy are then directed into an electron multiplier for analysis. Frequently, Auger spectra are measured in a differentiated form (Fig. 13.30b, inset) which allows for a more sensitive detection of peaks.

Importantly, the emission of Auger electrons is not dipole radiation and, therefore, selection rules such as those observed with X-ray emission do not apply. The energy of a particular Auger electron is given by

$$E_{\text{Auger}} = (E_{\text{hole}} - E_{\text{second}}) - E'_{\text{binding}}, \tag{13.52}$$

where E_{hole} is the energy of the electron being displaced, E_{second} is the energy of the electron replacing the dislodged electron and E'_{binding} is the binding energy of the Auger electron in the atom, corrected for the doubly charged state of the atom. Note that the energy of the impacting electron is not featured in Eq. 13.52, since it does not affect the energy of the emitted Auger electrons. The only requirement is that the energy of the impacting electron is sufficiently large to displace an electron from a core shell. The energy of the impacting electron thus needs to be at least as high as the binding energy of the electron to be displaced.

Equation 13.52 shows further that the energies of Auger electrons arise as characteristic for particular elements. Auger electron spectra are thus valuable tools for identification of atoms (see also Sect. 7.3.2), but also for measuring energy levels of the different shells.

13.7   Exercises

1. The infrared spectrum of carbon monoxide shows a vibration ($v = 0 \rightarrow v = 1$) band at 2176 cm^{-1}. What is the force constant of the C-O bond, assuming the molecule behaves as a harmonic oscillator?

2. Determine the ratio of populations of the vibrational levels $v = 1$ and $v = 0$ for carbon monoxide at 300 K and 1000 K. The wavenumber for the transition from $v = 0$ to $v = 1$ is 2176 cm^{-1}.

3. Which spectroscopic symbol denotes the ground state of Li? What is the symbol of the lowest excited state? What feature can be deduced for the spectral line in the optical spectrum for this transition?

4. In the free electron molecular orbital method, the π electrons in an alkene with conjugated double bonds are assumed to be freely moving within extent of the conjugation system. The conjugation system can be treated as a one-dimensional box.
   (a) Calculate the box length for the conjugation system in hexatrien, assuming that the C-C-bond length is 1.4 Å and considering that the electrons are free to move half a bond length beyond each outermost carbon.
   (b) Calculate the wavelength of the lowest energy peak in the absorption spectrum of hexatriene using the free electron molecular orbital method.

5. How many lines/clusters are to be expected in the ^{1}H-NMR spectra of (a) benzene, (b) toluene, (c) o-xylene, (d) p-xylene, (e) m-xylene?

6. Predict the number of hyperfine lines in the proton coupling observed in the ESR spectrum of the benzene anion radical $C_6H_6{}^-$.

7. Determine the Raman and infrared activity of one symmetric and two different anti-symmetric vibrational modes of the square planar molecule XeF_4.

Appendix A Mathematical Appendix

A.1 Basic Algebra and Operations

A.1.1 Exponentials

Definitions

$a^n = a \cdot a \cdot a \cdot a \ldots \cdot a$	$a^{\frac{1}{n}} = \sqrt[n]{a}$
$a^1 = a$	$a^{\frac{m}{n}} = \sqrt[n]{a^m}$
$a^0 = 1$	$a^{-\frac{m}{n}} = \frac{1}{\sqrt[n]{a^m}}$
$a^{-n} = 1/a^n$	

Computation rules

$a^x a^z = a^{x+z}$	$a^x b^x = (ab)^x$
$\frac{a^x}{a^z} = a^{x-z}$	$\frac{a^x}{b^x} = \left(\frac{a}{b}\right)^x$
$(a^x)^z = a^{x \cdot z}$	

A.1.2 Logarithm

Definition

$$\log_b a = x \Leftrightarrow b^x = a$$

Special cases

$$\log_{10} a = \lg a \quad a = 10^x \Leftrightarrow x = \lg a$$
$$\log_e a = \ln a \quad a = e^x \Leftrightarrow x = \ln a \quad e = 2.71828\ldots$$

© Springer International Publishing AG, part of Springer Nature 2018
A. Hofmann, *Physical Chemistry Essentials*,
https://doi.org/10.1007/978-3-319-74167-3

Computation rules

$\log_b(u \cdot v) = \log_b u + \log_b v$	$\log_b \left(\frac{u}{v}\right) = \log_b u - \log_b v$
$\log_b(u^z) = z \cdot \log_b u$	$\log_b \sqrt[n]{u} = \frac{1}{n}\log_b u$

Change of base

$$\log_c a = \frac{\log_b a}{\log_b c}$$

A.1.3 Nonlinear Equations

The quadratic equation $a \cdot x^2 + b \cdot x = -c \;\Leftrightarrow\; a \cdot x^2 + b \cdot x + c = 0$
has two solutions $x_{1,2} = \frac{-b \pm \sqrt{b^2 - 4 \cdot a \cdot c}}{2 \cdot a}$, if $\left(b^2 - 4 \cdot a \cdot c\right) \geq 0$

A.2   Differentials

A.2.1 Functions of One Variable

We consider a quantity U which changes over time t; U is then a function of t and denoted as $U(t)$.

At time t_1 (*start*): $U = U(t_1)$

At time t_2 (*end*): $U = U(t_2)$

The difference in U at times t_1 and t_2 (Fig. A.1a) is indicated as ΔU (differences are always calculated as *end* minus *start*):

$$\Delta U = U(t_2) - U(t_1)$$

The slope of the line joining $U(t_2)$ and $U(t_1)$ is calculated as

$$\text{slope} = \frac{\Delta U}{\Delta t} = \frac{U(t_2) - U(t_1)}{t_2 - t_1}$$

The use of Δ to indicate a difference does not imply anything about the size of the difference. If we want to imply a very small change, we write δU instead of ΔU (Fig. A.1b).

We now consider U to be a continuous function of t as opposed to the discrete function with two individual points as above (Fig. A.1c).

Now consider two time points t_1 and t_2 very close to each other, such that the difference between the two is infinitesimal (infinitely small). The line passing through the two points $U(t_2)$ and $U(t_1)$ is now called the tangent of U at point t_1 (Fig. A.1d).

Fig. A.1 Graphical
illustration of differences and
differentials. For explanation
see text

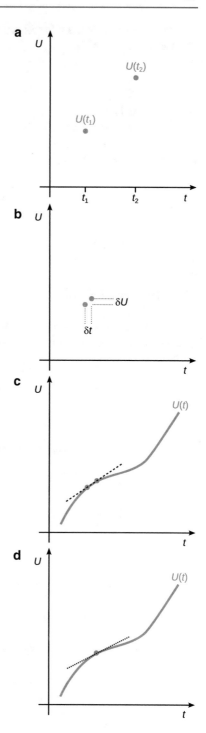

The slope of the tangent is still $\Delta U/\Delta t$, but because Δt is now infinitesimal small, we use 'd' instead of 'Δ':

$$\frac{dU}{dt} = \lim_{t \to 0} \frac{\Delta U}{\Delta t}$$

dU/dt tells us how U is varying with t at a particular point.

The lower case 'd' indicates that the changes in t (and normally U) are infinitesimal. There are various functions f for which the derivative is known analytically, e.g. $f(x) = \ln x \Rightarrow df(x)/dx = x^{-1}$.

If we are told that

$$\frac{dU}{dt} = f(t)$$

then a small change in t will result in a small change in U:

$$dU = f(t)dt$$

dU is called the differential of U. $\left(\frac{dU}{dt}\right)$ is called the Leibniz notation; another notation may be $U'(t)$.

You can integrate both sides of this equation to get the overall change in U as the system changes from its initial to its final state:

$$\int_{U(t_1)}^{U(t_2)} dU = \int_{t_1}^{t_2} f(t)dt$$

$$U(t_2) - U(t_1) = \int_{t_1}^{t_2} f(t)dt$$

$$\Delta U = \int_{t_1}^{t_2} f(t)dt$$

A.2.2 Functions of More Than One Variable

Consider we have a function that varies with two variables, z and t (see Fig. A.2); we denote this as $U(z,t)$.

How does U change with t when z is fixed? This may be denoted as:

$$\left(\frac{dU}{dt}\right)_z = f(t), \quad \text{but} \ldots$$

Fig. A.2 Variation of a
function U in dependence of
two parameters, t and z

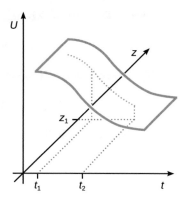

we replace $\frac{d}{dt}$ with $\frac{\delta}{\delta t}$ (which is called the partial derivative) to indicate that U is a function of more than one variable:

$$\left(\frac{\delta U}{\delta t}\right)_z = f(t)$$

So if z is fixed:

$$d(U)_z = f(t)dt = \left(\frac{\delta U}{\delta t}\right)_z dt$$

Similarly, we can now look at how U changes with z when t is fixed:

$$\left(\frac{\delta U}{\delta z}\right)_t = g(z)$$

$$d(U)_t = g(z)dz = \left(\frac{\delta U}{\delta z}\right)_t dz$$

Therefore, the total differential of U when both variables, t and z, are allowed to vary is given by:

$$dU = f(t)dt + g(z)dz = \left(\frac{\delta U}{\delta t}\right)_z dt + \left(\frac{\delta U}{\delta z}\right)_t dz$$

A.2.3 Product Rule

The product rule is a formula used to find the derivatives of products of two or more functions.

Consider the function h being the product of two functions, f and g:

$$h(x) = f(x) \cdot g(x)$$

The derivative of h is then calculated according to the product rule:

$$h'(x) = f'(x) \cdot g(x) + f(x) \cdot g'(x)$$

which can be written in the Leibniz notation:

$$\frac{dh}{dx} = f \cdot \frac{dg}{dx} + g \cdot \frac{df}{dx}$$

or, after multiplying with dx on both sides:

$$dh = f \cdot dg + g \cdot df$$

A.2.4 Functions with Known Derivatives

Table A.1 Functions with known derivatives

Function	Derivative
$f(x) = a$; $a = $ const.	$f'(x) = df(x)/dx = 0$
$f(x) = x$	$f'(x) = df(x)/dx = 1$
$f(x) = x^n$	$f'(x) = df(x)/dx = n \cdot x^{n-1}$
$f(x) = 1/x = x^{-1}$	$f'(x) = df(x)/dx = -1/x^2 = -x^{-2}$
$f(x) = 1/x^n = x^{-n}$	$f'(x) = df(x)/dx = -n/x^{n+1} = -n \cdot x^{-n-1}$
$f(x) = \sqrt[n]{x} = x^{1/n}$	$f'(x) = df(x)/dx = 1/n \cdot x^{n-1}$
$f(x) = e^x$	$f'(x) = df(x)/dx = e^x$
$f(x) = a^x$; $a = $ const.	$f'(x) = df(x)/dx = a^x \cdot \ln a$
$f(x) = \ln x$	$f'(x) = df(x)/dx = 1/x = x^{-1}$
$f(x) = \sin x$	$f'(x) = df(x)/dx = \cos x$
$f(x) = \cos x$	$f'(x) = df(x)/dx = -\sin x$

A.3   Integration

Integrating is similar to summing a property as is it gradually changes from one state to another. Therefore, U is simply the sum of all the infinitesimal changes dU.

There are various functions f for which the integral is known analytically. You may remember a number of these (e.g. $\int x^{-1} \, dx = \ln x$).

A.3.1 Functions with Known Integrals

Table A.2 Functions with known integrals

Function	Integral
$f(x) = a; a = \text{const.}$	$\int f(x)\,dx = \int a\,dx = a \cdot \int dx = a \cdot x$
$f(x) = x$	$\int f(x)\,dx = \int x\,dx = \frac{1}{2}x^2$
$f(x) = x^n$	$\int f(x)\,dx = \int x^n\,dx = 1/(n+1)\,x^{n+1}$
$f(x) = 1/x = x^{-1}$	$\int f(x)\,dx = \int 1/x\,dx = \ln|x|; x \neq 0$
$f(x) = 1/x^n = x^{-n}$	$\int f(x)\,dx = \int x^{-n}\,dx = 1/(-n+1)\,x^{-n+1}; n \neq 1$
$f(x) = \sqrt[n]{x} = x^{1/n}$	$\int f(x)\,dx = \int x^{1/n}\,dx = 1/(1/n+1)\,x^{1/n+1}$
$f(x) = e^x$	$\int f(x)\,dx = \int e^x\,dx = e^x$
$f(x) = e^{ax}$	$\int f(x)\,dx = \int e^{ax}\,dx = 1/a \cdot e^x$
$f(x) = a^x; a = \text{const.}$	$\int f(x)\,dx = \int a^x\,dx = a^x/(\ln a)$
$f(x) = \ln x$	$\int (\ln x)\,dx = x \cdot \ln x - x$
$f(x) = \sin x$	$\int (\sin x)\,dx = -\cos x$
$f(x) = \cos x$	$\int (\cos x)\,dx = \sin x$
$f(x) = \sin^2(a \cdot x)$	$\int \sin^2(a \cdot x)\,dx = \dfrac{x}{2} - \dfrac{\sin(2 \cdot a \cdot x)}{4 \cdot a}$

A.3.2 Rule for Partial Integration

The rule for partial integration is useful for integration of more complicated functions. In particular, the rule relates the integral of a product of two functions, $\int u(x)v'(x)$, to the integral of their derivative and anti-derivative, $\int u'(x)v(x)dx$. It is frequently used to transform the anti-derivative of a product of functions into an anti-derivative for which a solution can be more easily found.

$$\int u(x)v'(x)dx = u(x) \cdot v(x) - \int u'(x)v(x)dx$$

Example: Integrate the function $\ln x$.

$$\int \ln x\,dx$$

This can be solved by using partial integration. Define two appropriate functions u and v:

$$u = \ln x \quad \text{and} \quad dv = dx, \quad \text{i.e. } v = x.$$

Therefore:

$$\int \ln x \mathrm{d}x = \int u \mathrm{d}v$$

We know the derivative of u:

$$u = \ln x \Rightarrow \mathrm{d}u = \frac{1}{x} \mathrm{d}x$$

The anti-derivative of $\mathrm{d}v$ is v, therefore:

$$\mathrm{d}v = \mathrm{d}x \Rightarrow v = x$$

So we can now solve the substituted integral:

$$\int \ln x \, \mathrm{d}x = \int u \mathrm{d}v = u \cdot v - \int v \mathrm{d}u$$
$$\int \ln x \, \mathrm{d}x = \ln x \cdot x - \int x \cdot \frac{1}{x} \mathrm{d}x$$
$$\int \ln x \, \mathrm{d}x = x \cdot \ln x - \int \mathrm{d}x$$

With $\int \mathrm{d}x = x$ this yields:

$$\int \ln x \, \mathrm{d}x = x \cdot \ln x - x$$

A.4   Data Visualisation and Fitting

A.4.1   Software

Very commonly, basic data visualisation and analysis (such as averaging, error estimation and fitting with linear functions) can be carried out with generic spreadsheet programs such as *MS Excel* or the spreadsheet programs from the open-source suites *LibreOffice* or *OpenOffice*. More sophisticated software dedicated to data analysis and visualisation include the commercial products *SigmaPlot*, *Origin*, *Prism*, *IGOR* and others. There are also free and open-source programs (see Table A.3).

A.4.2   Independent and Technical Repeats

Measurements are repeated in order to obtain estimates of the precision of the experimental method. Repeating the measurement of an individual sample several times does not constitute an independent repeat; rather, it is a technical repeat. In order to obtain independent repeats, the same condition needs to be reproduced multiple times. A commonly used parameter to assess precision is the estimated standard deviation σ_{exp}:

Table A.3 Some free and open-source software for data analysis and visualisation

Software	Web page	Reference
R	http://www.R-project.org/	
SDAR	http://www.structuralchemistry.org/pcsb/sdar.php	Weeratunga et al. (2012)
Grace	http://plasma-gate.weizmann.ac.il/Grace/	
gnu-plot	http://www.gnuplot.info/	
Fityk		Wojdr (2010)
peak-o-mat	http://lorentz.sourceforge.net/	
HippoDraw	http://www.slac.stanford.edu/grp/ek/hippodraw/	
Veusz	http://home.gna.org/veusz/	
ParaView	http://www.paraview.org/	

$$\sigma_{exp} = \sqrt{\frac{1}{n} \sum_{i=1}^{n} \left(y_{exp} - \mu_{exp} \right)^2}$$

where $\mu_{exp} = \frac{1}{n} \sum_{i=1}^{n} y_{exp}$ is the mean of the experimental values.

A.4.3 Data Visualisation

The way data are presented can make a substantial difference to the perception of experimental results by the reader. A typical example is the scale chosen; the plot of a baseline can be presented as a steady straight line or as noisy data, depending on the scale of the y-axis. Other aspects to consider are the type of plot used (line graphs, bar graphs, pie charts, etc).

Here, we will focus on some basic considerations for simple two-dimensional plotting of data. In general, the following should be followed (Fig. A.3):

- Plots are presented with respect to two dimensions, one on the x-, the other on the y-axis. Always label the axes with the parameter that is plotted in that dimension; include the units. For example: x-axis: $c(NaOH)$ in mM; y-axis: A(280 nm)—no units, since absorbance is a scalar.
- The scale needs to be available to the reader. This means that in most cases one needs to indicate the origin of each axis ('0') and at least one data point (e.g. '1 mM').
- Choose appropriate scales on the axes, and avoid overly long numbers. For example, instead of '0.0001 M, 0.0002 M, . . .' label ticks with '0.1 mM, 0.2 mM, . . .'.
- If an experiment comprised of acquisition of individual data points (e.g. absorbance of a sample at select concentrations of a reactant), then the data are discrete data points and should be plotted as individual points, without connecting them with lines; this is called a scatter plot. Only if (quasi-)continuous data have been acquired

time	V(KMnO4)	ln(V/cm^3)
min	cm^3	
5	37.1	1.56937
10	29.8	1.47422
20	19.6	1.29226
30	12.3	1.08991
50	5.0	0.69897

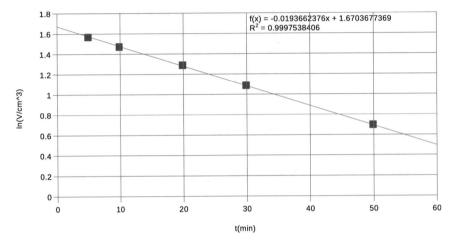

Fig. A.3 Example plot of experimental data with superimposed linear fit using the free spreadsheet software *LibreOffice*

(e.g. by using sensor readings at a reasonably high sampling rate), the data can be shown as line graphs. In a line plot, individual data points may or may not be visible and subsequent data points are connected by a line.

- If independently repeated measurements for individual conditions are available, the values for each condition are averaged and the estimated standard deviation is calculated. Error bars are constructed by adding and subtracting the estimated standard deviation to/from the averaged value.
- Discrete data (scatter plots) are never smoothed; only (quasi-)continuous data may be subjected to smoothing procedures.
- Data fitting may be undertaken as a means to show agreement with a theoretical model. Data fits are superimposed on the experimental data as line plots. Since data fits arise from a theoretical model with a numerical equation, they are per definition continuous data.

A.4.4 Data Fitting

When fitting a theoretical model to experimental data, statistical parameters need to be calculated to assess the goodness of fit between the theoretical and experimental data. Commonly used statistical parameters include:

Table A.4 Commonly used goodness-of-fit parameters

Parameter	Definition	Value for perfect fit
Chi-square	$\chi^2 = \sum \left(\frac{y_{exp}-y_{fit}}{\sigma_{exp}}\right)^2$	0
R-square	$R^2 = \frac{\sum(y_{fit}-\mu_{exp})^2}{\sum(y_{exp}-\mu_{exp})^2}$ with $\mu_{exp} = \frac{1}{n}\sum y_{exp}$	1
Summed square error	$SSE = \sum(y_{exp} - y_{fit})^2$	0
R-factor	$R = \frac{\sum|y_{exp}|-|y_{fit}|}{\sum|y_{exp}|}$	0

In Table A.4, μ_{exp} is the mean and σ_{exp} the estimated standard deviation of the experimental data. The parameter χ^2 is a weighted measure of error, since it is divided by the estimated standard deviation (i.e. the error appearing in independent repeats); the non-weighted χ^2 is identical to the summed square error.

A.4.5 Correlation

In order to evaluate the correlation of two different quantities (e.g. the variables x and y), correlation coefficients can be determined.

The Pearson product-moment correlation coefficient r is a measure of the linear correlation between two variables x and y, giving a value between $+1$ and -1:

- $+1$: total positive correlation
- 0: no correlation
- -1: total negative correlation

$$r = \frac{\sum_{i=1}^{n}(x_i - \bar{x})\cdot(y_i - \bar{y})}{\sqrt{\sum_{i=1}^{n}(x_i - \bar{x})^2}\cdot\sqrt{\sum_{i=1}^{n}(y_i - \bar{y})^2}}$$

Here, $\bar{x}$ and $\bar{y}$ denote the arithmetic mean over all x- and y-values, respectively.

If the two variables x and y are related in a monotonous fashion, but a linear relationship cannot be expected, then Spearman's rank correlation coefficient instead of the Pearson correlation needs to be applied.

Appendix B Solutions to Exercises

Chapter 2

2.1. Calculate the work done per mole when an ideal gas is expanded reversibly by a factor of 3 at 120 °C.

$$\text{Expansion work}: \quad dW = p \cdot dV$$

The work is done by the system (hence we count it as negative) in a reversible process from state '1' at the volume $V_{\text{start}} = V_1$ to state '2' at volume $V_{\text{end}} = 3 \cdot V_1$. In order to obtain the amount work during the entire process, we need to integrate above equation:

$$\int_{W_1}^{W_2} dW = -\int_{V_{\text{start}}}^{V_{\text{end}}} p \cdot dV$$

During the *reversible* process, the outside pressure is constantly adjusted to match the pressure in the system. Hence, p is a function V and calculated as per the ideal gas equation:

$$p = \frac{n \cdot R \cdot T}{V}$$

Therefore:

$$\int_{W_1}^{W_2} dW = -\int_{V_{\text{start}}}^{V_{\text{end}}} \frac{n \cdot R \cdot T}{V} \cdot dV = -n \cdot R \cdot T \int_{V_1}^{3 \cdot V_1} \frac{dV}{V}$$

$$[W]_{W_1}^{W_2} = -n \cdot R \cdot T \cdot [\ln V]_{V_1}^{3 \cdot V_1}$$

$$(W_2 - W_1) = -n \cdot R \cdot T \cdot [\ln (3 \cdot V_1) - \ln V_1]$$

$$\Delta W = -n \cdot R \cdot T \cdot \ln \frac{3 \cdot V_1}{V_1}$$

$$\Delta W = -1 \text{ mol} \cdot 8.3144 \frac{J}{K \cdot mol} \cdot (120 + 273) \text{ K} \cdot \ln 3$$

$$\Delta W = -3.6 \text{ kJ}$$

The work done when 1 mol of an ideal gas is expanded reversibly by a factor of 3 at 120 °C is −3.6 kJ.

2.2. A friend thinking of investing in a company that makes engines shows you the company's prospectus. Analysing the machine, you realise that it works by harnessing the expansion of a gas. The operating temperature is approx. constant at 120 °C, and the gas doubles in volume during the power extraction phase of operation. It is claimed that the machine produces 5.5 kJ mol^{-1} of work during expansion. What advice would you give your friend? Briefly explain the thermodynamic rationale.

No process can produce more work than an equivalent process (see Exercise 2.1) running under reversible conditions. The claim that the machine produced 5.5 kJ mol^{-1} is thus in error. Your friend should not invest.

2.3. Which of the following equations embodies the first law of thermodynamics:

(a) $dS = \frac{dQ_{rev}}{T}$
(b) $U = Q + W$
(c) $\Delta S_{universe} > 0$
(d) $S_{system} > 0$
(e) $C_p = \left(\frac{\delta H}{\delta T}\right)_p$.

In the integrated form, the first law of thermodynamics states that the internal energy of a system (U) comprises of heat (Q) and work (W); the correct answer is therefore (b). (a) is the definition of entropy change of a system for a reversible process; (c) describes the second, and (d) the third law of thermodynamics; (e) is the definition of the heat capacity for processes under constant pressure.

2.4. If the pressure is constant, the system is in mechanical equilibrium with its surroundings and no work is done other than work due to expansion and compression, which of the following are true:

(a) $\Delta H = \Delta Q$ and $\Delta W = -p \cdot \Delta V$
(b) $\Delta U = \Delta Q$ and $\Delta W = -p \cdot \Delta V$
(c) $\Delta U = \Delta Q - p \cdot \Delta V$ and $\Delta W = p \cdot \Delta V$
(d) $\Delta H = \Delta Q - p \cdot \Delta V$ and $\Delta W = -p \cdot \Delta V$
(e) $\Delta U = \Delta Q$ and $\Delta W = p \cdot \Delta V$

The correct answer is (a). Under constant external pressure the enthalpy change equals the transferred heat if only work due to expansion and compression is done.

In solution (c), '$\Delta U = \Delta Q - p \cdot \Delta V$' is correct, but '$\Delta W = p \cdot \Delta V$' is not correct!

2.5. Which of the following equations define fugacity?

(a) $\mu(p) = \mu^{\varnothing}(p) + R \cdot T \cdot \ln \frac{f}{p^{\varnothing}}$
(b) $f = \frac{p}{p^{\varnothing}}$
(c) $\mu_A = \mu_A^* + R \cdot T \cdot \ln a_A$

The correct answer is (a). The fugacity replaces the pressure in the equation for the chemical potential to account for non-ideal behaviour of gases. Fugacity measures how compressible a gas is compared to the ideal gas; it is measured in units of pressure. Therefore, (b) is not correct, as this would yield a scalar. (c) describes the chemical equation for a solution containing a solute A of activity a.

2.6. The chemical potential is defined as:

(a) $\mu = G$
(b) $\mu = G_m$
(c) $\mu = G_m^{\varnothing}$
(d) $\mu = n \cdot G_m$
(e) $\mu = G^{\varnothing} + n \cdot R \cdot T \cdot \ln \frac{p}{p^{\varnothing}}$
(f) $\mu = G_m^{\varnothing} + R \cdot T \cdot \ln \frac{p}{p^{\varnothing}}$

The correct answers are (b) and (f). The chemical potential is defined as the molar free energy, i.e. the free energy of 1 mol of substance. Note that (e) is not correct, as these are not molar free energies.

Chapter 3

3.1. Is it possible for a one-component system to exhibit a quadruple point?

For a one-component system ($C = 1$) to exhibit a quadruple point ($P = 4$), the Gibbs phase rule yields $F = C - P + 2 = 1 - 4 + 2 = -1$. Since the number of degrees of freedom cannot be negative, a quadruple point cannot exist in a one-component system.

3.2. Henry's law is valid for dilute solutions. Using the Henry's law constant for oxygen (solute) and water (solvent) of $K(O_2) = 781 \cdot 10^5$ Pa M^{-1}, calculate the molar concentration of oxygen in water at sea level with an atmospheric pressure of $p_{atm} = p^{\varnothing}$.

Since the fraction of O_2 in the atmosphere is 21%, its partial pressure is $p(O_2) = 0.21 \cdot p_{atm} = 0.21 \cdot p^{\emptyset} = 0.21 \cdot 10^5$ Pa. From Henry's law, this yields $c(O_2) = 0.27$ μM.

3.3. Below is the T-x phase diagram of the benzene/toluene system acquired at a constant pressure of two bar. A mixture that contains 40% benzene is heated steadily to 122 °C. How many phases are present at this point and what are their compositions? If the total amount of 1 mol of substances was in the initial mixture with 40% benzene, how many moles of substances are in the phase(s) at 122 °C?

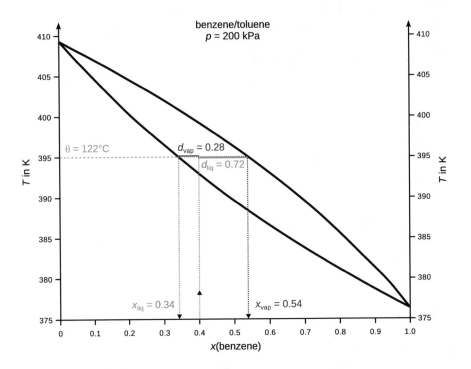

At 122 °C (395 K), there are two phases: x_{liq}(benzene) $= 0.34$, x_{vap}(benzene) $= 0.54$. According to the lever rule, the ratio of molar amounts in the liquid and vapour phase is $\frac{n_{liq}}{n_{vap}} = \frac{0.72}{0.28}$. Since the total amount of substances in the mixture was set at $n_0 = 1$ mol, there is $n_{liq} = 0.72$ mol and $n_{vap} = 0.28$ mol.

3.4. A mixture of benzene and toluene with x(benzene) $= 0.4$ is subjected to fractional distillation at two bar (see Exercise 3.3 above). What is the boiling temperature of this mixture? How many theoretical plates are required as a minimum to obtain pure benzene in the distillate?

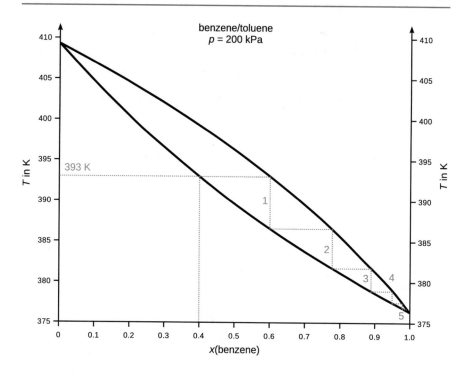

The mixture boils at 393 K (120 °C). A minimum of five theoretical plates are required to obtain pure benzene.

Chapter 4

4.1. Nickel-cadmium batteries are based on the following half-cell reactions:

$$NiO(OH)_{(s)} + H_2O_{(l)} + e^- \rightarrow Ni(OH)_{2(s)} + OH^-_{(aq)}$$
$$Cd_{(s)} + 2\,OH^-_{(aq)} \rightarrow Cd(OH)_{2(s)} + 2\,e^-$$

(a) The *e.m.f.* of a nickel-cadmium cell is 1.4 V, and the standard Redox potential of $Cd(OH)_{2(s)}$ is -0.809 V. What is the standard Redox potential of $NiO(OH)_{(s)}$?

The half-reactions for the nickel cadmium cell are:

$$NiO(OH)_{(s)} + H_2O_{(l)} + e^- \rightarrow Ni(OH)_{2(s)} + OH^-_{(aq)} \quad E^\varnothing\left(NiO(OH)_{(s)}\right) = ?$$
$$Cd(OH)_{2(s)} + 2\,e^- \rightarrow Cd_{(s)} + 2\,OH^-_{(aq)} \quad E^\varnothing\left(Cd(OH)_{2(s)}\right) = -0.809\ V$$

In order for the nickel-cadmium cell to operate, the second reaction needs to run in reverse, i.e. it constitutes the oxidation; the first reaction constitutes the reduction.

$$e.m.f. = E^{\emptyset}(\text{reduction cell}) - E^{\emptyset}(\text{oxidation cell})$$

$$e.m.f. = \left(E^{\emptyset}\text{NiO(OH)}_{(s)} \mid \text{Ni(OH)}_{2(s)}\right) - E^{\emptyset}\left(\text{Cd(OH)}_{2(s)} \mid \text{Cd}_{(s)}\right)$$

$$E^{\emptyset}\left(\text{NiO(OH)}_{(s)} \mid \text{Ni(OH)}_{2(s)}\right) = e.m.f. + E^{\emptyset}\left(\text{Cd(OH)}_{2(s)} \mid \text{Cd}_{(s)}\right)$$

$$E^{\emptyset}\left(\text{NiO(OH)}_{(s)} \mid \text{Ni(OH)}_{2(s)}\right) = 1.4 \text{ V} - 0.809 \text{ V} = 0.6 \text{ V}$$

$$E^{\emptyset}\left(\text{NiO(OH)}_{(s)} \mid \text{Ni(OH)}_{2(s)}\right) = 0.6 \text{ V}.$$

(b) What is the standard free energy of formation of $\text{NiO(OH)}_{(s)}$?
The standard Gibbs free energy change for the overall nickel half-cell reaction can be calculated as per:

$$\Delta G_r^{\emptyset} = -z \cdot F \cdot E^{\emptyset}\left(\text{NiO(OH)}_{(s)} \mid \text{Ni(OH)}_{2(s)}\right)$$

$$\Delta G_r^{\emptyset} = -2 \cdot 96485 \text{ C mol}^{-1} \cdot 0.6 \text{ V} = -58 \text{ kJ mol}^{-1}$$

The standard Gibbs free energy of formation is then obtained from the reaction free energy:

$$\Delta G_r^{\emptyset} = \Delta G_f^{\emptyset}\left(\text{Ni(OH)}_{2(s)}\right) + \Delta G_f^{\emptyset}(\text{OH}^-_{(aq)}) - \Delta G_f^{\emptyset}\left(\text{NiO(OH)}_{(s)}\right) - \Delta G_f^{\emptyset}(\text{H}_2\text{O}_{(l)})$$

$$\Delta G_f^{\emptyset}\left(\text{NiO(OH)}_{(s)}\right) = \Delta G_f^{\emptyset}\left(\text{Ni(OH)}_{2(s)}\right) + \Delta G_f^{\emptyset}(\text{OH}^-_{(aq)}) - \Delta G_f^{\emptyset}(\text{H}_2\text{O}_{(l)}) - \Delta G_r^{\emptyset}$$

$$\Delta G_f^{\emptyset}\left(\text{NiO(OH)}_{(s)}\right) = -444 \text{ kJ mol}^{-1} - 157 \text{ kJ mol}^{-1} - (-237 \text{ kJ mol}^{-1}) - (-58 \text{ kJ mol}^{-1})$$

$$\Delta G_f^{\emptyset}\left(\text{NiO(OH)}_{(s)}\right) = -306 \text{ kJ mol}^{-1}.$$

4.2. Calculate the standard *e.m.f.* and the equilibrium constant under standard conditions for the following cell:

$$\text{Pt}_{(s)}, \text{H}_{2(g)} \mid \text{HCl}_{(aq)} \mid \text{AgCl}_{(s)}, \text{Ag}_{(s)}$$

This cell consists of the following two half-cells:

$$2 \text{ H}^+_{(aq)} + 2 \text{ e}^- \rightarrow \text{H}_{2(g)} \quad E^{\emptyset}(\text{H}^+ \mid \text{H}_2) = 0$$

$$\text{AgCl}_{(s)} + \text{e}^- \rightarrow \text{Ag(s)} + \text{Cl}^-_{(aq)} \quad E^{\emptyset}(\text{AgCl} \mid \text{Ag}, \text{Cl}^-) = 0.223 \text{ V}$$

According to the reduction potentials, the first reaction needs to run as oxidation, and the second one as reduction, just like the electrochemical notation suggests. The standard *e.m.f.* is then calculated as per:

$$e.m.f. = E^{\emptyset}(\text{AgCl} \mid \text{Ag}, \text{Cl}^-) - E^{\emptyset}(\text{H}^+ \mid \text{H}_2)$$
$$e.m.f. = 0.223 \text{ V}$$

The net reaction in the electrochemical cell is thus:

$$2 \, AgCl_{(s)} + H_{2(g)} \rightleftharpoons 2 \, Ag(s) + 2 \, Cl^-_{(aq)} + 2 \, H^+_{(aq)}$$

When this reaction is in equilibrium, there is no potential difference between the two half cells:

$$\Delta E(Pt, H_2|HCl|AgCl, Ag) = 0 = \Delta E^{\emptyset}(Pt, H_2|HCl|AgCl, Ag) - \frac{R \cdot T}{z \cdot F} \ln K$$

$$\frac{R \cdot T}{z \cdot F} \ln K = \Delta E^{\emptyset}(Pt, H_2|HCl|AgCl, Ag)$$

$$\ln K = \frac{z \cdot F \cdot \Delta E^{\emptyset}(Pt, H_2|HCl|AgCl, Ag)}{R \cdot T}$$

$$\ln K = \frac{2 \cdot 96485 \cdot 0.223 \cdot C \cdot V \cdot mol \cdot K}{8.3144 \cdot 298 \cdot J \cdot K \cdot mol}$$

$$\ln K = 17.37$$

$$K = 35 \cdot 10^6$$

4.3. Calculate the pH of a solution with a formal concentration of $5 \cdot 10^{-7}$ M of the strong acid HI at 25 °C.

Strong acids fully dissociate in aqueous solution:

$$HI_{(aq)} \rightarrow H^+_{(aq)} + I^-_{(aq)}$$

Therefore, the concentration of protons and base anion is:

$$c(H^+) = c(I^-) = c(HI) = 5 \cdot 10^{-7} \, M$$

The pH can thus be calculated as:

$$pH = -\lg \frac{c(H^+)}{1M} = -\lg 5 \cdot 10^{-7} = 6.3.$$

4.4. What e.m.f. would be generated by the following cell at 25 °C, assuming ideal behaviour:

$$Pt_{(s)}, H_{2(g)}(1 \text{ bar}) \mid H^+_{(aq)}(0.03 \text{ M}) \mid Cl^-_{(aq)}(0.004 \text{ M}) \mid AgCl_{(s)}, Ag_{(s)}$$

This cell consists of the following two half-cells:

$$2H^+_{(aq)} + 2e^- \rightarrow H_{2(g)} \quad E^{\emptyset}(H^+|H_2) = 0$$
$$AgCl_{(s)} + e^- \rightarrow Ag_{(s)} + Cl^-_{(aq)} \quad E^{\emptyset}(AgCl|Ag) = 0.223 \, V$$

According to the reduction potentials, the first reaction needs to run as oxidation, and the second one as reduction, just like the electrochemical notation suggests. The net reaction in the electrochemical cell is thus:

$$2\,AgCl_{(s)} + H_{2(g)} \rightleftharpoons 2\,Ag_{(s)} + 2\,Cl^-_{(aq)} + 2\,H^+_{(aq)}$$

The concentration dependence of an electrochemical cell is given by the Nernst equation:

$$\Delta E(Pt, H_2|H^+|Cl^-|AgCl, Ag) = \Delta E^{\varnothing}(Pt, H_2|H^+|Cl^-|AgCl, Ag) - \frac{R \cdot T}{z \cdot F} \cdot \ln \frac{\prod \frac{c(\mathrm{Red})}{c^{\varnothing}}}{\prod \frac{c(\mathrm{Ox})}{c^{\varnothing}}}$$

$$\Delta E(Pt, H_2|H^+|Cl^-|AgCl, Ag) = E^{\varnothing}(AgCl|Ag) - E^{\varnothing}(H_2|H^+) - \frac{R \cdot T}{z \cdot F} \ln \frac{\prod \frac{c(\mathrm{Red})}{c^{\varnothing}}}{\prod \frac{p(\mathrm{Ox})}{p^{\varnothing}}}$$

$$\Delta E = E^{\varnothing}(AgCl|Ag) - E^{\varnothing}(H_2|H^+) - \frac{R \cdot T}{z \cdot F} \cdot \ln \frac{\left[\frac{c(Cl^-)}{c^{\varnothing}}\right]^2 \cdot \left[\frac{c(H^+)}{c^{\varnothing}}\right]^2}{\left[\frac{p(H_2)}{p^{\varnothing}}\right]}$$

$$\Delta E = 0.22\,\mathrm{V} - 0\,\mathrm{V} - \frac{8.3144 \cdot 298\,\mathrm{J \cdot K \cdot mol}}{296485\,\mathrm{C \cdot K \cdot mol}} \ln \frac{0.004^2 \cdot 0.03^2}{1}$$

$$\Delta E = 0.22\,\mathrm{V} - 0.013\,\mathrm{V} \cdot \ln\left(14.4 \cdot 10^{-9}\right)$$

$$\Delta E = 0.22\,\mathrm{V} - 0.013\,\mathrm{V} \cdot (-18.1)$$

$$\Delta E = 0.46\,\mathrm{V}.$$

The e.m.f. generated is 0.46 V.

4.5. The lactate/pyruvate Redox system can be described as per:

$$\text{pyruvate} + 2\,H^+ + 2\,e^- \rightleftharpoons \text{lactate}$$

The standard reduction potential is measured at $c(\text{lactate}) = c(\text{pyruvate}) = c(H^+) = 1$ M and has a value of $E^{\varnothing} = 0.21$ V. Calculate and plot the pH dependency of the reduction potential for this system at 298 K, assuming $c(\text{lactate}) = c(\text{pyruvate})$.

The concentration dependence is given by the Nernst equation:

$$E = E^{\varnothing} - \frac{R \cdot T}{z \cdot F} \cdot \ln \frac{\prod \frac{c(\mathrm{Red})}{c^{\varnothing}}}{\prod \frac{c(\mathrm{Ox})}{c^{\varnothing}}}$$

where $c(\text{Ox})$ refers to the concentration on the oxidised side and $c(\text{Red})$ to the concentrations on the reduced side of the equilibrium reaction. Therefore:

$$E = E^{\varnothing} - \frac{R \cdot T}{z \cdot F} \cdot \ln \frac{\left[\frac{c(\text{lactate})}{c^{\varnothing}}\right]}{\left[\frac{c(\text{pyruvate})}{c^{\varnothing}}\right] \cdot \left[\frac{c(H^+)}{c^{\varnothing}}\right]^2}$$

$$E = 0.21 \text{ V} - \frac{8.314 \cdot 298 \cdot \text{J} \cdot \text{K}}{2 \cdot 96485 \cdot \text{C} \cdot \text{K} \cdot \text{mol}} \cdot \ln \frac{\left[\frac{c(\text{lactate})}{c^{\varnothing}}\right]}{\left[\frac{c(\text{pyruvate})}{c^{\varnothing}}\right] \cdot \left[\frac{c(\text{H}^+)}{c^{\varnothing}}\right]^2}$$

pH	c(H+) M	E = EØ - (RT)/(zF) ln [(cL)/(cP c^2(H+)] V
0	1	0.21
1	1E-01	0.1844
2	1E-02	0.1588
3	1E-03	0.1332
4	1E-04	0.1076
5	1E-05	0.082
6	1E-06	0.0564
7	1E-07	0.0308
8	1E-08	0.0052
9	1E-09	-0.0204
10	1E-10	-0.046
11	1E-11	-0.0716
12	1E-12	-0.0972
13	1E-13	-0.1228
14	1E-14	-0.1484

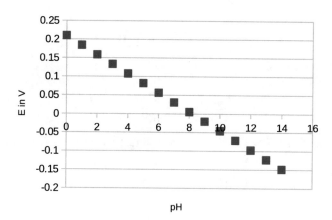

The relationship between the reduction potential and the pH is linear with $E(\text{pH} = 0) = 0.21$ V and $E(\text{pH} = 14) = -0.15$ V, i.e. a slope of -0.026 V per pH unit.

Chapter 5

5.1. At 25 °C, the molar ionic conductivities Λ_m of alkali ions are

Table B.1 Molar
conductivities for select
alkali ions

Ion	Λ_m (mS m^2 mol^{-1})
Li$^+$	3.87
Na$^+$	5.01
K$^+$	7.35

What are their mobilities?

The mobility of ions provides a link between their charge and the conductivity in solution:

$$\lambda = z \cdot u \cdot F$$

The molar conductivity Λ_m is linked to the conductivity λ of the counter-ions by:

$$\Lambda_m = \nu_+ \cdot \lambda_+ + \nu_- \cdot \lambda_-$$

Since we are interested in the isolated alkali ions, we set $\Lambda_m = \nu_+ \cdot \lambda_+$ with $\nu_+ = 1$. Therefore:

$$u = \frac{\lambda}{z \cdot F} = \frac{\Lambda_m}{\nu \cdot z \cdot F}$$

For Li$^+$, this yields:

$$u = \frac{\Lambda_m(Li^+)}{\nu \cdot z \cdot F} = \frac{3.87 \cdot 10^{-3}}{9.649 \cdot 10^4} \frac{S \cdot m^2}{C}$$

Remember that $1\ S = 1\ \Omega^{-1} = 1\ A/V$, and that $1\ C = 1\ A\ s$. Therefore:

$$u = 0.401 \cdot 10^{-7} \frac{A \cdot m^2}{V \cdot A \cdot s} = 4.01 \cdot 10^{-8} m^2 V^{-1} s^{-1}$$

In summary:

Table B.2 Ion
mobilities for select
alkali ions

Ion	Λ_m (mS m^2 mol^{-1})	u (m^2 V^{-1} s^{-1})
Li$^+$	3.87	$4.01 \cdot 10^{-8}$
Na$^+$	5.01	$5.19 \cdot 10^{-8}$
K$^+$	7.35	$7.62 \cdot 10^{-8}$

5.2. Fullerene (C$_{60}$) has a diameter of 10.2 Å. Estimate the diffusion coefficient of fullerene in benzonitrile at 25 °C. The viscosity of benzonitrile at that temperature is 12.4 mP. How large is the error of your estimate when comparing the result to the experimental value of $4.1 \cdot 10^{-10}$ m^2 s^{-1}?

The diffusion coefficient can be calculate by the Stokes-Einstein relationship:

$$D = \frac{k_B \cdot T}{6 \cdot \pi \cdot \eta \cdot r}$$

The viscosity is $\eta(\text{benzonitrile}) = 12.4 \text{ mP} = 12.4 \cdot 10^{-3} \text{ P} = 1.24 \cdot 10^{-3} \text{ N s m}^{-2}$
The radius is $r(\text{C}_{60}) = \frac{1}{2} \cdot 10.2 \text{ Å} = 5.1 \text{ Å} = 5.1 \cdot 10^{-10} \text{ m}$
Therefore:

$$D = \frac{1.381 \cdot 10^{-23} \text{JK}^{-1} \cdot 298\text{K}}{6 \cdot \pi \cdot 1.24 \cdot 10^{-3} \text{Nsm}^{-2} \cdot 5.1 \cdot 10^{-10}\text{m}}$$

$$D = \frac{298 \cdot 1.381 \cdot 10^{-23} \cdot \text{J} \cdot \text{K} \cdot \text{m}^2}{6 \cdot \pi \cdot 1.24 \cdot 5.1 \cdot 10^{-13} \cdot \text{N} \cdot \text{K} \cdot \text{s} \cdot \text{m}} \qquad 1\text{N} = 1\text{Jm}^{-1}$$

$$D = \frac{298 \cdot 1.381 \cdot 10^{-10} \cdot \text{J} \cdot \text{K} \cdot \text{m}^2 \cdot \text{m}}{6 \cdot \pi \cdot 1.24 \cdot 5.1 \cdot \text{J} \cdot \text{K} \cdot \text{s} \cdot \text{m}}$$

$$D = \frac{298 \cdot 1.381 \cdot 10^{-10} \cdot \text{J} \cdot \text{K} \cdot \text{m}^2 \cdot \text{m}}{6 \cdot \pi \cdot 1.24 \cdot 5.1 \cdot \text{J} \cdot \text{K} \cdot \text{s} \cdot \text{m}}$$

$$D = 3.5 \cdot 10^{-10} \text{ m}^2 \text{ s}^{-1}.$$

The calculated diffusion coefficient thus has a value of $3.5 \cdot 10^{-10}$ m^2 s^{-1}. The difference to the experimental value is:

$$\Delta D = 4.1 \cdot 10^{-10} \text{m}^2\text{s}^{-1} - 3.5 \cdot 10^{-10}\text{m}^2\text{s}^{-1} = 0.6 \cdot 10^{-10}\text{m}^2\text{s}^{-1}$$

This is an error of approximately

$$\frac{\Delta D}{D_{\text{exp}}} = \frac{0.6}{4.1} = 15\%$$

The relatively large error is due to the very nature of the Stokes-Einstein relationship which applies thermodynamic quantities (temperature, the average energy of particles $k_B \cdot T$) that relate to large numbers of particles to single molecules.

5.3. The mean free path length of gas molecules can be calculated according to Maxwell as $\lambda = \frac{1}{\sqrt{2} \cdot \mathbb{N} \cdot \sigma}$, where the particle density is $\mathbb{N} = \frac{p}{k_B \cdot T}$. Calculate the diffusion coefficient of argon at 298 K and a pressure of 1 bar. The collisional cross-section of argon is $\sigma = 0.41$ nm^2.
The diffusion coefficient for an ideal gas is calculated as per

$$D = \frac{1}{3} \cdot \lambda \cdot c$$

The mean speed of molecules, c, is calculated as per

$$c = \sqrt{\frac{3 \cdot R \cdot T}{M}}.$$

The mean free path length is accessible from above formulae as per

$$\lambda = \frac{1}{\sqrt{2} \cdot \mathbb{N} \cdot \sigma} = \frac{k_B \cdot T}{\sqrt{2} \cdot p \cdot \sigma}.$$

This yields for the diffusion coefficient:

$$D = \frac{1}{3} \cdot \frac{k_B \cdot T}{\sqrt{2} \cdot p \cdot \sigma} \cdot \sqrt{\frac{3 \cdot R \cdot T}{M}}$$

$$D = \frac{1}{3} \cdot \frac{k_B}{p \cdot \sigma} \cdot \sqrt{\frac{3 \cdot R \cdot T^3}{2 \cdot M}}$$

$$D = \frac{1}{3} \cdot \frac{1.381 \cdot 10^{-23} \text{ JK}^{-1}}{1\text{bar} \cdot 0.41 \text{ nm}^2} \cdot \sqrt{\frac{3 \cdot 8.3144 \text{ JK}^{-1} \text{ mol}^{-1} \cdot 298^3 \text{ K}^3}{2 \cdot 39.948 \text{ gmol}^{-1}}}$$

$$D = \frac{1}{3} \cdot \frac{1.381 \cdot 10^{-23} \text{ JK}^{-1}}{10^5\text{Pa} \cdot 0.41 \cdot 10^{-18} \text{ m}^2} \cdot \sqrt{\frac{3 \cdot 8.3144 \text{ J} \cdot 298^3 \text{ K}^2}{2 \cdot 39.948 \cdot 10^{-3} \text{ kg}}}$$

$$D = \frac{1}{3} \cdot \frac{1.381 \cdot 10^{-5} \text{ kg m}^2 \text{ s}^{-2} \text{ K}^{-1}}{10^5 \text{ kg m}^{-1} \text{ s}^{-2} \cdot 0.41 \text{ m}^2} \cdot \sqrt{\frac{3 \cdot 8.3144 \text{ kg m}^2 \text{ s}^{-2} \cdot 298^3 \text{ K}^2}{2 \cdot 39.948 \cdot 10^{-3} \text{ kg}}}$$

$$D = \frac{1}{3} \cdot \frac{1.381 \cdot 10^{-10}}{0.41} \cdot \frac{\text{kgm}^2\text{s}^{-2}\text{K}^{-1}}{\text{kgm}^2\text{s}^{-2}\text{m}^{-1}} \cdot \sqrt{\frac{3 \cdot 8.3144 \cdot 298^3}{2 \cdot 39.948 \cdot 10^{-3}} \cdot \frac{\text{kg m}^2 \text{ s}^{-2} \text{ K}^2}{\text{kg}}}$$

$$D = 1.123 \cdot 10^{-10} \cdot \frac{\text{m}}{\text{K}} \cdot \sqrt{8.261 \cdot 10^9 \cdot \frac{\text{m}^2 \text{ K}^2}{\text{s}^2}}$$

$$D = 1.123 \cdot 10^{-10} \cdot \frac{\text{m}}{\text{K}} \cdot 9.089 \cdot 10^4 \cdot \frac{\text{mK}}{\text{s}}$$

$$D = 10.21 \cdot 10^{-6}\frac{\text{m}^2}{\text{s}} = 1.021 \cdot 10^{-5}\frac{\text{m}^2}{\text{s}}.$$

The diffusion coefficient of argon thus has a value of $1.021 \cdot 10^{-5} \text{ m}^2 \text{ s}^{-1}$.

5.4. The diffusion coefficient of sucrose in water is $5.2 \cdot 10^{-6} \text{ cm}^2 \text{ s}^{-1}$ at room temperature. Estimate the effective radius of a sucrose molecule, if water has a viscosity of 10 mP.

The diffusion coefficient can be calculate by the Stokes-Einstein relationship:

$$D = \frac{k_B \cdot T}{6 \cdot \pi \cdot \eta \cdot r}$$

The effective radius of the molecule is then:

$$r = \frac{k_B \cdot T}{6 \cdot \pi \cdot \eta \cdot D}$$

The viscosity is $\eta(\text{water}) = 10 \text{ mP} = 1.0 \cdot 10^{-2} \text{ P} = 1.0 \cdot 10^{-3} \text{ N s m}^{-2}$
The diffusion coefficient is $D(\text{sucrose}) = 5.2 \cdot 10^{-6} \text{ cm}^2 \text{ s}^{-1} = 5.2 \cdot 10^{-10} \text{ m}^2 \text{ s}^{-1}$.
Therefore:

$$r = \frac{1.381 \cdot 10^{-23} \text{ JK}^{-1} \cdot 298 \text{ K}}{6 \cdot \pi \cdot 1.0 \cdot 10^{-3} \text{ N s m}^{-2} \cdot 5.2 \cdot 10^{-10} \text{ m}^2 \text{ s}^{-1}}$$

$$r = \frac{1.381 \cdot 298 \cdot 10^{-10}}{6 \cdot \pi \cdot 1.0 \cdot 5.2} \cdot \frac{\text{J}}{\text{N}}$$

$$r = 4.2 \cdot 10^{-10} \cdot \frac{\text{J m}}{\text{J}}.$$

The effective radius of a sucrose molecule is thus estimated at 4.2 Å.

Chapter 6

6.1. Consider the gas-phase reaction

$$H_2 + I_2 \quad \rightarrow \quad 2\,HI$$

(a) Assume that the reaction order is as suggested by the chemical equation.
Calculate the rate constant at 681 K, assuming that from an initial pressure of
iodine of 823 N m^{-2}, the rate of loss of iodine was 0.192 N m^{-2} s^{-1}. The initial
pressure of hydrogen was 10500 N m^{-2}.
If the reaction order is as suggested by the chemical equation, we are dealing
with first order kinetics:

$$v = -\frac{dc(H_2)}{dt} = -\frac{dc(I_2)}{dt} = k_r \cdot c(H_2) \cdot c(I_2)$$

Since this is a gas phase reaction, and all parameters are given in the units of
pressure, we can substitute the molar concentration against the pressure:

$$v = -\frac{dp(H_2)}{dt} = -\frac{dp(I_2)}{dt} = k_r \cdot p(H_2) \cdot p(I_2)$$

The data given are:

$$-\frac{dp(I_2)}{dt} = v = 0.192 \text{ N m}^{-2}\text{ s}^{-1}$$

$$p(I_2) = 823 \text{ N m}^{-2}$$

$$p(H_2) = 10500 \text{ N m}^{-2}.$$

Therefore

$$k_r = \frac{v}{p(H_2) \cdot p(I_2)}$$

$$k_r = \frac{0.192}{10500 \cdot 823} \cdot \frac{N \cdot m^2 \cdot m^2}{m^2 \cdot s \cdot N \cdot N}$$

$$k_r = 22.2 \cdot 10^{-9} \text{ m}^2 \text{ s}^{-1} \text{ N}^{-1}.$$

To obtain the rate constant in molar units, we assume ideal gas conditions, and thus
$$\frac{p}{R \cdot T} = \frac{n}{V} = c.$$

$$k_r = \frac{v}{c(H_2) \cdot c(I_2)} = \frac{-\dfrac{dc(I_2)}{dt}}{c(H_2) \cdot c(I_2)} = \frac{-\dfrac{dp(I_2)}{dt \cdot R \cdot T}}{\dfrac{p(H_2)}{R \cdot T} \cdot \dfrac{p(I_2)}{R \cdot T}} = \frac{v}{p(H_2) \cdot p(I_2)} \cdot R \cdot T$$

$$k_r = 22.2 \cdot 10^{-9} \text{ m}^2 \text{ s}^{-1} \text{ N}^{-1} \cdot 8.3144 \text{ J K}^{-1} \text{ mol}^{-1} \cdot 681 \text{ K}$$
$$k_r = 0.126 \cdot 10^{-3} \text{ m}^3 \text{ s}^{-1} \text{ mol}^{-1}$$
$$k_r = 0.126 \text{ M}^{-1} \text{ s}^{-1}.$$

(b) What is the rate of the reaction if the iodine pressure was unchanged and the initial hydrogen pressure was 39500 N m^{-2}?
The rate of the reaction for the new hydrogen pressure $p(H_2) = 39500$ N m^{-2} is then

$$v = k_r \cdot p(H_2) \cdot p(I_2)$$
$$v = 22.2 \cdot 10^{-9} \text{ m}^2 \text{ s}^{-1} \text{ N}^{-1} \cdot 39500 \text{ N m}^{-2} \cdot 823 \text{ N m}^{-2}$$
$$v = 0.72 \text{ N m}^{-2} \text{ s}^{-1}.$$

Alternatively, if we calculate the rate in molar units instead of pressure:

$$v = k_r \cdot c(H_2) \cdot c(I_2)$$

$$v = k_r \cdot \frac{p(H_2)}{R \cdot T} \cdot \frac{p(I_2)}{R \cdot T} = \frac{k_r \cdot p(H_2) \cdot p(I_2)}{R^2 \cdot T^2}$$

$$v = \frac{0.126 \cdot 39500 \cdot 823 \cdot dm^3 \cdot N \cdot N \cdot K \cdot K \cdot mol \cdot mol}{8.3144^2 \cdot 681^2 \cdot mol \cdot s \cdot m^2 \cdot m^2 \cdot Nm \cdot Nm \cdot K \cdot K}$$

$$v = 0.128 \cdot 10^{-6} \, M \, s^{-1}$$

6.2. The rate constant for the decomposition of a particular substance is $2.80 \cdot 10^{-3}$ $dm^3 \, mol^{-1} \, s^{-1}$ at 30 °C, and $1.38 \cdot 10^{-2} \, dm^3 \, mol^{-1} \, s^{-1}$ at 50 °C. Evaluate the Arrhenius parameters of the reaction.

Data compilation:

$$\theta_1 = 30\,°C \Rightarrow T_1 = 303\,K \quad k_{r1} = 2.80 \cdot 10^{-3} \, dm^3 \, mol^{-1} \, s^{-1}$$
$$\theta_2 = 50\,°C \Rightarrow T_2 = 323\,K \quad k_{r2} = 1.38 \cdot 10^{-2} \, dm^3 \, mol^{-1} \, s^{-1}$$

As above, we build the ratio of the two Arrhenius equations for the two conditions:

$$\frac{k_{r1}}{k_{r2}} = \frac{A \cdot e^{-\frac{E_a}{R \cdot T_1}}}{A \cdot e^{-\frac{E_a}{R \cdot T_2}}}$$

and obtain after cancellation of the pre-exponential parameter A and some re-arrangements:

$$E_a = \frac{R \cdot \ln \frac{k_{r1}}{k_{r2}}}{\frac{1}{T_2} - \frac{1}{T_1}}$$

This yields for the data given:

$$E_a = \frac{8.3144 \cdot \ln \left(\frac{2.80 \cdot 10^{-3}}{1.38 \cdot 10^{-2}} \right)}{\frac{1}{323} - \frac{1}{303}} \cdot \frac{J}{K \cdot mol \cdot \frac{1}{K}}$$

$$E_a = \frac{8.3144 \cdot \ln (0.203)}{0.003096 - 0.003300} \cdot \frac{J}{mol} = 65 \, kJ \, mol^{-1}$$

This allows calculation of the pre-exponential factor:

$$A = k_{r1} \cdot e^{-\frac{E_a}{R \cdot T_1}}$$
$$A = 2.8 \cdot 10^{-3} \, dm^3 \, mol^{-1} \, s^{-1} \cdot e^{-\frac{64989 \, J \cdot K \cdot mol}{8.3144 \cdot 303 \cdot K \cdot J \cdot mol}} = 4.5 \cdot 10^8 \, dm^3 \, mol^{-1} \, s^{-1}.$$

6.3. The reaction mechanism for the reaction of A_2 and B to product P involves the intermediate A:

$$
\begin{array}{lll}
A_2 & \rightleftharpoons & A + A & \text{(fast)} \\
A + B & \rightarrow & P & \text{(slow)}
\end{array}
$$

Deduce the rate law for the reaction assuming a pre-equilibrium.

If we assume an equilibrium for the first reaction, we know that the equilibrium constant K is the ration of the rate of the forward (k_1) and reverse reaction (k_{-1}). We also know that this equals the ratio of the equilibrium concentrations of A_2 and A:

$$
K = \frac{k_1}{k_{-1}} = \frac{c(A)^2}{c(A_2)}
$$

This yields for the equilibrium concentration of A:

$$
c(A)^2 = c(A_2) \cdot K \quad => \quad c(A) = \sqrt{c(A_2) \cdot K}
$$

For the second reaction, the rate equation is:

$$
v_2 = -\frac{dc(A)}{dt} = \frac{dc(P)}{dt} = k_2 \cdot c(A) \cdot c(B)
$$

Substituting the expression for the equilibrium concentration of A yields:

$$
v_2 = k_2 \cdot K^{1/2} \cdot c(A_2)^{1/2} \cdot c(B).
$$

6.4. The rate constant of a first-order reaction was measured as $1.11 \cdot 10^{-3}\ \text{s}^{-1}$.

(a) What is the half-life of the reaction?

The first order rate law is

$$
v = -\frac{dc(A)}{dt} = k \cdot c(A)
$$

and the half-life is

$$
t_{1/2} = \frac{\ln 2}{k}
$$

Therefore:

$$t_{1/2} = \frac{\ln 2}{1.11 \cdot 10^{-3} \, \text{s}^{-1}} = \frac{\ln 2}{1.11} \cdot 10^3 \, \text{s} = 624 \, \text{s}.$$

(b) What time is needed for the concentration of the reactant to fall to $^1/_8$ of its initial value?

In order to calculate a different life time, the differential rate law needs to be integrated:

$$-\frac{dc(A)}{dt} = k \cdot c(A)$$

$$\frac{dc(A)}{c(A)} = -k \cdot dt$$

$$d\ln c(A) = -k \cdot dt$$

$$\int_{c_0}^{1/8c_0} d\ln c(A) = -k \cdot \int_0^t dt$$

$$[\ln c(A)]_{c_0}^{1/8c_0} = -k \cdot [t]_0^t$$

$$\ln \left(\frac{1}{8} c_0 \right) - \ln c_0 = -k \cdot (t - 0)$$

$$\ln \frac{c_0}{8 \cdot c_0} = -k \cdot t$$

$$\ln 8 = k \cdot t$$

$$t = \frac{\ln 8}{k}$$

$$t = \frac{\ln 8}{1.11 \cdot 10^{-3} \, \text{s}^{-1}} = \frac{\ln 8}{1.11} \cdot 10^3 \, \text{s} = 1873 \, \text{s} = 31.2 \, \text{min}.$$

(c) What time is needed for the concentration of the reactant to fall to $^3/_4$ of its initial value?

In analogy to the above, one derives:

$$t = \frac{\ln \frac{4}{3}}{k}$$

$$t = \frac{\ln \frac{4}{3}}{1.11 \cdot 10^{-3} \, \text{s}^{-1}} = \frac{\ln \frac{4}{3}}{1.11} \cdot 10^3 \, \text{s} = 259 \, \text{s}.$$

Chapter 7

7.1. The initial rates of the myosin-catalysed hydrolysis of ATP were measured in the presence of varying starting concentrations of ATP:

Table B.3 Initial rates for myosin-catalysed ATP hydrolysis

c_0(ATP)	mmol dm^{-3}	0.005	0.010	0.020	0.030	0.050	0.100	0.200	0.300
v_0	μmol dm^{-3} s^{-1}	0.051	0.083	0.118	0.138	0.158	0.178	0.190	0.194

Assume that the enzymatic reaction follows a Michaelis-Menten mechanism and determine the maximum rate and the Michaelis constant.

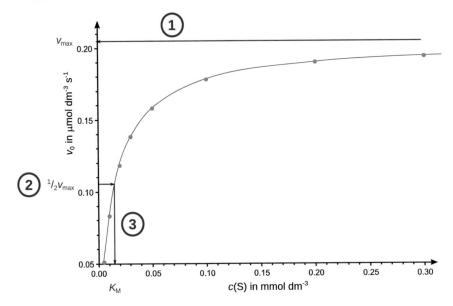

For an accurate determination of v_{max}, the original data either need to be fitted with by non-linear regression (above) or transformed into a reciprocal relationship that linearises the Michaelis-Menten equation, e.g. by using the Lineweaver-Burk approach:

$$\frac{1}{v} = \frac{1}{v_{max}} + \frac{K_M}{v_{max}} \cdot \frac{1}{c_0}$$

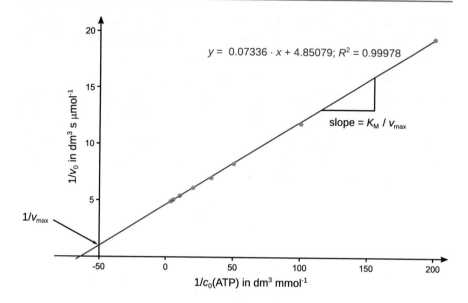

Table B.4 Lineweaver-Burk transformation of the initial data

$1/c_0(\text{ATP})$	$\text{dm}^3\,\text{mmol}^{-1}$	200	100	50.0	33.3	20.0	10.0	5.00	3.33
$1/v_0$	$\text{dm}^3\,\text{s}\,\mu\text{mol}^{-1}$	19.6	12.1	8.47	7.25	6.33	5.62	5.26	5.15

The fit with a linear equation yields a function:

$$\frac{1}{v} = 4.85079 \frac{\text{sdm}^3}{\mu\text{mol}} + 0.07336 \frac{\text{mmol}\cdot\text{dm}^3\cdot\text{s}}{\text{dm}^3\cdot\mu\text{mol}}\cdot\frac{1}{c_0}$$

with the goodness of fit $R^2 = 0.99978$

From the line fit one obtains:

$$\frac{1}{v_{\text{max}}} = \frac{1}{4.85079}\frac{\mu\text{mol}}{\text{dm}^3\text{s}} = 0.206\frac{\mu\text{mol}}{\text{dm}^3\text{s}}$$

$$\frac{K_{\text{M}}}{v_{\text{max}}} = 0.07336\frac{\text{mmol}\cdot\text{dm}^3\cdot\text{s}}{\text{dm}^3\cdot\mu\text{mol}}$$

$$K_{\text{M}} = 0.07336\frac{\text{mmol}\cdot\text{dm}^3\cdot\text{s}}{\text{dm}^3\cdot\mu\text{mol}}\cdot v_{\text{max}}$$

$$K_{\text{M}} = 0.07336\frac{\text{mmol}\cdot\text{dm}^3\cdot\text{s}}{\text{dm}^3\cdot\mu\text{mol}}\cdot 0.206\frac{\mu\text{mol}}{\text{dm}^3\text{s}}$$

$$K_{\text{M}} = 0.015\frac{\text{mmol}}{\text{dm}^3}.$$

7.2. The adsorption behaviour of 1 g activated carbon at $0\,^{\circ}$C has been quantitatively assessed using different pressures of N_2. The following molar amounts of N_2 are adsorbed:

Table B.5 Data for adsorption of N_2 by activated carbon

$p(N_2)$	kPa	0.524	1.73	3.06	4.13	7.50	10.3
$n_{ads}(N_2)$	10^{-4} mol	0.440	1.35	2.27	3.14	4.60	5.82

Assuming Langmuir adsorption behaviour, determine (a) the maximum amount of N_2 that can be adsorbed by 1 g activated carbon at $0\,^{\circ}$C, and (b) the Langmuir constant.

The Langmuir adsorption isotherm is

$$\Gamma = \frac{n_{ads}}{n_{max}} = \frac{K \cdot p}{K \cdot p + 1}$$

Taking the reciprocal value on both side of above equation yields:

$$\frac{n_{max}}{n_{ads}} = \frac{K \cdot p + 1}{K \cdot p} = 1 + \frac{1}{K} \cdot \frac{1}{p}$$

and thus

$$\frac{1}{n_{ads}} = \frac{K \cdot p + 1}{K \cdot p} = \frac{1}{n_{max}} + \frac{1}{K \cdot n_{max}} \cdot \frac{1}{p}$$

From this equation, it is obvious that a plot of $1/n_{ads}$ versus $1/p$ yields a line with y-intercept $1/n_{max}$ and slope $1/(K \cdot n_{max})$.

The reciprocal values of the original data yield:

Table B.6 Transformation of initial data for Langmuir adsorption isotherm

$1/p(N_2)$	kPa^{-1}	1.91	0.578	0.327	0.242	0.133	0.0968
$1/n_{ads}(N_2)$	10^4 mol^{-1}	2.27	0.741	0.441	0.318	0.217	0.172

p(N2) kPa	n(N2)ads 10^-4 mol	1/p(N2) kPa^-1	1/n(N2)ads 10^4 mol^-1
0.524	0.44	1.908396947	2.272727273
1.73	1.35	0.578034682	0.740740741
3.06	2.27	0.326797386	0.440528634
4.13	3.14	0.242130751	0.318471338
7.5	4.6	0.133333333	0.217391304
10.33	5.82	0.096805421	0.171821306

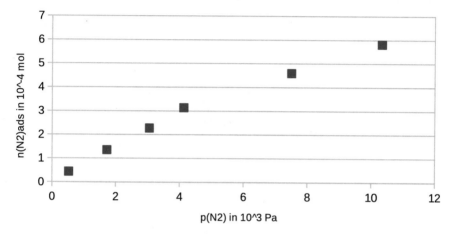

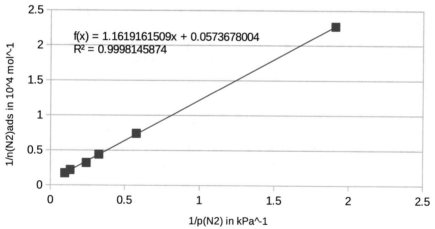

Linear fit of these data result in the following equation:

$$1/n_{ads} = 1.162 \cdot 10^4\,\text{kPa}\,\text{mol}^{-1}\,1/p + 0.0574 \cdot 10^4\,\text{mol}^{-1}$$
$$\text{with } R^2 = 0.9998$$

(a) From the linear fit, the y-intercept yields the $1/n_{max}$:

$$\frac{1}{n_{max}} = 0.0574 \cdot 10^4 \frac{1}{mol}$$
$$n_{max} = 1.74 \cdot 10^{-3}\ mol$$

(b) The slope reveals the product of the Langmuir constant K and n_{max}:

$$\frac{1}{K \cdot n_{max}} = 1.162 \cdot 10^4 \frac{kPa}{mol}$$
$$K = \frac{1}{1.162 \cdot 10^4\ kPa} \frac{mol}{} \cdot \frac{1}{1.74 \cdot 10^{-3}\ mol}$$
$$K = \frac{1}{1.162 \cdot 1.74 \cdot 10\ kPa} \frac{1}{}$$
$$K = 0.049\ kPa^{-1}.$$

Chapter 8

8.1. Calculate the wavelength of an electron that is accelerated by a potential difference of 10.0 kV.

The energy gained by an electron accelerated in an electric field (its kinetic energy) equals the potential energy provided by the electric field:

$$E_{kin} = E_{pot}$$
$$\frac{1}{2} \cdot m_e \cdot v^2 = e \cdot U$$

The DeBroglie relationship links the wave and corpuscular properties of a particle in form of the wavelength λ and the momentum p. Specifically for the electron, this yields:

$$\lambda = \frac{h}{p} = \frac{h}{m_e \cdot v}$$

The squared equation yields:

$$\lambda^2 = \frac{h^2}{(m_e \cdot v)^2} \text{ or } (m_e \cdot v)^2 = \frac{h^2}{\lambda^2}$$

We now multiply me into both sides of above energy equation and obtain:

$$\frac{1}{2} \cdot m_e^2 \cdot v^2 = m_e \cdot e \cdot U$$
$$m_e^2 \cdot v^2 = 2 \cdot m_e \cdot e \cdot U$$

For the squared momentum we can substitute the wavelength expression from the DeBroglie relationship:

$$\frac{h^2}{\lambda^2} = 2 \cdot m_e \cdot e \cdot U$$

Resolving for the wavelength yields:

$$\lambda^2 = \frac{h^2}{2 \cdot m_e \cdot e \cdot U}$$
$$\lambda = \frac{h}{\sqrt{2 \cdot m_e \cdot e \cdot U}}$$
$$\lambda = \frac{6.626 \cdot 10^{-34} \text{ J s}}{\sqrt{2 \cdot 9.108 \cdot 10^{-31} \text{ kg} \cdot 1.602 \cdot 10^{-19} \text{C} \cdot 10 \cdot 10^3 \text{ V}}}$$
$$\lambda = \frac{6.626 \cdot 10^{-34} \text{ J s}}{\sqrt{2 \cdot 9.108 \cdot 1.602 \cdot 10^{-46} \text{ kg C V}}}$$
$$\lambda = \frac{6.626 \cdot 10^{-34} \text{ J s}}{\sqrt{29.18 \cdot 10^{-46} \text{ kgJ}}}$$
$$\lambda = \frac{6.626 \cdot 10^{-34} \text{ kg m}^2 \text{ s}^{-2} \text{ s}}{\sqrt{29.18 \cdot 10^{-46} \text{ kgkg m}^2 \text{ s}^{-2}}}$$
$$\lambda = \frac{6.626 \cdot 10^{-34} \text{ kg m}^2 \text{ s}^{-1}}{5.402 \cdot 10^{-23} \text{ kg m s}^{-1}}$$
$$\lambda = \frac{6.626}{5.402} \cdot 10^{-11} \text{ m}$$
$$\lambda = 1.23 \cdot 10^{-11} \text{ m} = 0.0123 \text{ nm} = 0.123 \text{ Å}.$$

8.2. The atomic model suggested by Niels Bohr in 1913 depicts atoms as systems very much like a solar system, where electrons travel in circular orbits around the positively charged nucleus. If one wanted to determine the location of an electron at a particular point in time with a certainty of ± 0.05 Å, what is the uncertainty with respect to the speed of the electron?

The uncertainty relationship by Heisenberg informs of the minimum value of the product ($\Delta x \cdot \Delta p$):

$$\Delta x \cdot \Delta p \geq \frac{h}{2\pi}$$
$$\Delta x \cdot \Delta(m \cdot v) \geq \frac{h}{2\pi}$$

With $m = m_e$ and $\Delta x = 0.1$ Å, this yields:

$$\Delta v \geq \frac{h}{2\pi \cdot \Delta x \cdot m_e}$$

$$\Delta v \geq \frac{6.626 \cdot 10^{-34} \text{ J s}}{2\pi \cdot 0.1 \cdot 10^{-10} \text{ m} \cdot 9.109 \cdot 10^{-31} \text{ kg}}$$

$$\Delta v \geq \frac{6.626}{2\pi \cdot 9.109} \cdot \frac{10^{-34}}{10^{-11} \cdot 10^{-31}} \cdot \frac{\text{kg m}^2 \text{ s}^{-2} \text{ s}}{\text{m kg}}$$

$$\Delta v \geq \frac{6.626}{2\pi \cdot 9.109} \cdot 10^8 \cdot \frac{\text{m}}{\text{s}}$$

$$\Delta v \geq 0.12 \cdot 10^8 \cdot \frac{\text{m}}{\text{s}}.$$

The uncertainty about the speed of the orbiting electron is approx. $0.12 \cdot 10^8$ m s^{-1}. This dramatically high uncertainty shows that no statement as to the exact location of the electron can be made at a particular time.

8.3. The wavelength of macroscopic objects. What is the wavelength of a person of 65 kg walking at a speed of 0.8 m s^{-1}?

According to the deBroglie relationship, the wavelength λ is related to the momentum of an object as per:

$$\lambda = \frac{h}{p} = \frac{h}{m \cdot v}$$

This yields:

$$\lambda = \frac{6.626 \cdot 10^{-34} \text{ J s}}{65 \text{kg} \cdot 0.8 \text{ m s}^{-1}}$$

$$\lambda = 0.127 \cdot 10^{-34} \frac{\text{kg m}^2 \text{ s}^{-2} \text{ s}}{\text{kg m s}^{-1}}$$

$$\lambda = 1.27 \cdot 10^{-35} \text{ m}.$$

This is a very tiny wavelength; for comparison, the proton has a radius of about 0.8 fm $= 0.8 \cdot 10^{-15}$ m. This explains why the waves of macroscopic objects cannot be measured.

8.4. Two consecutive lines in the atomic spectrum of hydrogen have the wave numbers $\tilde{v}_i = 2.057 \cdot 10^6$ m^{-1} and $\tilde{v}_{i+1} = 2.304 \cdot 10^6$ m^{-1}. Calculate to which series these two transitions belong and which transitions they describe.

According to Bohr's term scheme, the wave numbers for the two transitions are related to the states n_0, n_i and n_{i+1} as per:

$$\text{h} \cdot c \cdot \tilde{\nu}_i = \frac{R_\infty \cdot c \cdot \text{h}}{n_0^2} - \frac{R_\infty \cdot c \cdot \text{h}}{n_i^2} \Rightarrow \tilde{\nu}_i = \frac{R_\infty}{n_0^2} - \frac{R_\infty}{n_i^2}$$

$$\text{h} \cdot c \cdot \tilde{\nu}_{i+1} = \frac{R_\infty \cdot c \cdot \text{h}}{n_0^2} - \frac{R_\infty \cdot c \cdot \text{h}}{n_{i+1}^2} \Rightarrow \tilde{\nu}_{i+1} = \frac{R_\infty}{n_0^2} - \frac{R_\infty}{n_{i+1}^2}$$

Subtraction of the first equation from the second yields:

$$\tilde{\nu}_{i+1} - \tilde{\nu}_i = \frac{R_\infty}{n_0^2} - \frac{R_\infty}{n_{i+1}^2} - \frac{R_\infty}{n_0^2} + \frac{R_\infty}{n_i^2}$$

$$\tilde{\nu}_{i+1} - \tilde{\nu}_i = \frac{R_\infty}{n_i^2} - \frac{R_\infty}{n_{i+1}^2}$$

$$\frac{\tilde{\nu}_{i+1} - \tilde{\nu}_i}{R_\infty} = \frac{1}{n_i^2} - \frac{1}{n_{i+1}^2}$$

The left hand side of this equation computes as per:

$$\frac{\tilde{\nu}_{i+1} - \tilde{\nu}_i}{R_\infty} = \frac{2.304 \cdot 10^6 \, \text{m}^{-1} - 2.057 \cdot 10^6 \, \text{m}^{-1}}{1.097 \cdot 10^7 \, \text{m}^{-1}} = 0.0225$$

From comparison of this value with the following transitions between n_i and n_{i+1} we find a suitable pair n_i, n_{i+1}:

Table B.7 Term scheme differences

n_i	n_{i+1}	$\dfrac{1}{n_i}$	$\dfrac{1}{n_{i+1}}$	$\left(\dfrac{1}{n_i}\right)^2$	$\left(\dfrac{1}{n_{i+1}}\right)^2$	$\left(\dfrac{1}{n_i}\right)^2 - \left(\dfrac{1}{n_{i+1}}\right)^2$
1	2	1	0.5	1	0.25	0.75
2	3	0.5	0.3333	0.25	0.1111	0.1389
3	4	0.3333	0.25	0.1111	0.0625	0.0486
4	5	0.25	0.2	0.0625	0.04	0.0225
5	6	0.2	0.1667	0.04	0.0278	0.0122

$$n_i = 4, n_{i+1} = 5.$$

With this knowledge, we can calculate n_0:

$$\tilde{v}_i = \frac{R_\infty}{n_0^2} - \frac{R_\infty}{n_i^2}$$

$$\frac{\tilde{v}_i}{R_\infty} = \frac{1}{n_0^2} - \frac{1}{n_i^2}$$

$$\frac{1}{n_0^2} = \frac{\tilde{v}_i}{R_\infty} + \frac{1}{n_i^2}$$

$$\frac{1}{n_0^2} = \frac{2.057 \cdot 10^6 \text{ m}^{-1}}{1.097 \cdot 10^7 \text{ m}^{-1}} + \frac{1}{16} = 0.25$$

$$n_0^2 = \frac{1}{0.25} = 4$$

$$n_0 = 2.$$

The two transitions thus are: $n_0 = 2 \rightarrow n = 4$ and $n_0 = 2 \rightarrow n = 5$. Since $n_0 = 2$, the transitions belong to the Balmer series.

8.5. Assuming that (a) the sun ($T = 5780$ K) and (b) the earth ($T = 298$ K) behave as black-bodies, calculate and plot the spectral flux densities for the sun and the earth. What are the similarities and differences between both radiation curves? Use Planck's law to calculate $E(\lambda)$ for the wavelength range 100 nm–8 μm. Calculation and plotting might be best done with a spreadsheet software.

Planck's law describes radiation (spectral flux density) $E(\lambda)$ of a black-body at varying wavelengths for a a particular temperature:

$$E(\lambda) = \frac{h \cdot c^2}{\lambda^5 \cdot \left(e^{\frac{h \cdot c}{\lambda \cdot k_B \cdot T}} - 1\right)}.$$

We set out to calculate $E(\lambda)$ for the following wavelengths in a spreadsheet program:

Table B.8 Wavelength values in μm

0.1	0.2	0.3	0.4	0.5	1.0	1.5	2.0	2.5	3.0	3.5	4.0	4.5	5.0	5.5	6.0	6.5	7.0	7.5	8.0

Since above equation involves several factors with large exponential values, we separate the problem into different individual steps and deal with the exponential factors manually, since such software use floating point numbers, which is a finite sized representation. Therefore, approximated results or error conditions might occur when using very large numbers.

(1) $h \cdot c^2 = 6.626 \cdot 10^{-34} \text{ J s} \cdot \left(3 \cdot 10^8 \frac{\text{m}}{\text{s}}\right)^2 = 59.634 \cdot 10^{-18} \frac{\text{J m}^2}{\text{s}}$

(2) $\frac{h \cdot c}{k_B \cdot T} = \frac{6.626 \cdot 10^{-34} \text{ J s} \cdot 3 \cdot 10^8 \text{ ms}^{-1}}{1.381 \cdot 10^{-23} \text{ JK}^{-1} \cdot 5780 \text{ K}} = 0.002491 \cdot 10^{-3} \text{ m} = 2.491 \cdot 10^{-6} \text{ m}$

(3) The exponential factor $exp\_factor = e^{\frac{h \cdot c}{\lambda \cdot k_B \cdot T}}$ can now be calculated in a spreadsheet program as a function of the wavelength λ using the factor of 2.491 from

above; note that "10^{-6} m" cancels when using the wavelength values from Table B.9: $exp\_factor = e^{\frac{2.491 \cdot 10^{-6} \text{ m}}{\lambda}} = e^{\frac{2.491}{[\lambda]}}$.

(4) When using the wavelength values from Table B.8, λ^5 values are obtained as multiples of 10^{-30} m^5:

$$(0.1 \ \mu m)^5 = (0.1 \cdot 10^{-6} \text{ m})^5 = 0.1^5 \cdot 10^{-30} \text{ m}^5$$

(5) We then combine (3) and (4) to obtain $\lambda^5 \cdot \left(e^{\frac{h \cdot c}{\lambda \cdot k_B \cdot T}} - 1\right)$. For this purpose we multiple the fifth power of the wavelength values from Table B.8 $[\lambda]^5$ with (exp_factor-1). The resulting values are multiples of 10^{-30} m.

(6) Now, we can combine (1) with (5). In the spreadsheet, this will be calculated as $\frac{59.634}{[\text{result from (5)}]}$. Taking care of the exponential factors and units, this calculation yields:

$$\frac{59.634}{[\text{result from (5)}]} \cdot \frac{10^{-18} \text{ J m}^2 \text{ s}^{-1}}{10^{-30} \text{ m}^5} = \frac{59.634}{[\text{result from (5)}]} \cdot 10^{12} \cdot \frac{\text{J}}{\text{m}^3 \text{s}}$$

Table B.9 The four quantum numbers of the electron

Name	Symbol	Values	Description
Principal quantum number	n	0, 1, 2, ...	Size of the wave function, energy level.
Angular momentum quantum number	l	0, 1, ..., $(n-1)$	Shape of the wave function.
Magnetic quantum number	m_l	$-l, (-l+1), ...,$ 0, ..., $(l-1), l$	Orientation of the wave function in space.
Magnetic spin quantum number	m_s	$-\frac{1}{2}, +\frac{1}{2}$	Orientation of the magnetic moment of the electron.

(a) The sun

Sun

T [K]	5780
hc/(kT) [10⁶ m]	2.490297137

λ [10^6 m]	exp_factor	λ^5 [10^{30} m⁵]	λ^5 (exp_factor - 1) [10^{30} m⁵]	E [10^{12} J m⁻³ s⁻¹]
0.1	65346607881	0.00001	653466.0788	0.000091258
0.2	255629.8259	0.00032	81.80122428	0.7290110939
0.3	4027.859838	0.00243	9.785269406	6.09426246
0.4	505.5984829	0.01024	5.167088465	11.5411223184
0.5	145.560859	0.03125	4.517526842	13.2005856484
1	12.0648605	1	11.0648605	5.3894940641
1.5	5.260352771	7.59375	32.35205386	1.8432832817
2	3.473450806	32	79.1504258	0.7534261427
2.5	2.707752229	97.65625	166.7726786	0.3575765558
3	2.293545895	243	314.3316524	0.1897168152
3.5	2.037071969	525.21875	544.6896434	0.1094825296
4	1.863719616	1024	884.4488865	0.0674250382
4.5	1.739155001	1845.28125	1363.948864	0.0437215804
5	1.64552491	3125	2017.265343	0.0295618027
5.5	1.5726802	5032.84375	2882.209965	0.0206903733
6	1.514445738	7776	4000.330062	0.0149072699
6.5	1.466857909	11602.90625	5416.908553	0.0110088622
7	1.427260302	16807	7180.963894	0.0083044562
7.5	1.393808068	23730.46875	9345.250043	0.0063812097
8	1.365181166	32768	11966.25644	0.0049835135

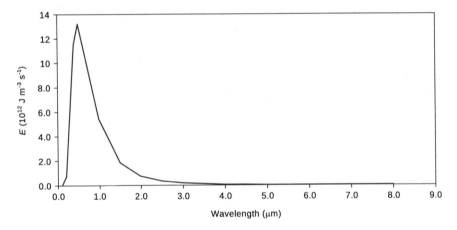

(b) The Earth

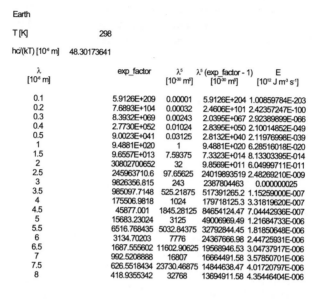

Earth

T [K]	298
hc/(kT) [10^{-6} m]	48.30173641

λ [10^{-6} m]	exp_factor	λ^5 [10^{-30} m^5]	λ^5 (exp_factor - 1) [10^{-30} m^5]	E [10^{12} J m^{-3} s^{-1}]
0.1	5.9126E+209	0.00001	5.9126E+204	1.00859784E-203
0.2	7.6893E+104	0.00032	2.4606E+101	2.42357247E-100
0.3	8.3932E+069	0.00243	2.0395E+067	2.92389899E-066
0.4	2.7730E+052	0.01024	2.8395E+050	2.10014852E-049
0.5	9.0023E+041	0.03125	2.8132E+040	2.11976998E-039
1	9.4881E+020	1	9.4881E+020	6.28516018E-020
1.5	9.6557E+013	7.59375	7.3323E+014	8.13303395E-014
2	30802700652	32	9.8569E+011	6.04999711E-011
2.5	245963710.6	97.65625	24019893519	2.48269210E-009
3	9826356.815	243	2387804463	0.000000025
3.5	985097.7148	525.21875	517391265.2	1.15259000E-007
4	175506.9818	1024	179718125.3	3.31819620E-007
4.5	45877.001	1845.28125	84654124.47	7.04442936E-007
5	15683.23024	3125	49006969.49	1.21684733E-006
5.5	6516.768435	5032.84375	32792844.45	1.81850648E-006
6	3134.70203	7776	24367666.98	2.44725931E-006
6.5	1687.555602	11602.90625	19568946.53	3.04737917E-006
7	992.5208888	16807	16664491.58	3.57850701E-006
7.5	626.5518434	23730.46875	14844638.47	4.01720797E-006
8	418.9355342	32768	13694911.58	4.35446404E-006

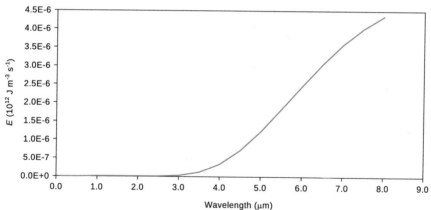

From comparison of both radiation curves, we see that the earth emits less radiation than the sun (note the difference of orders of magnitude on the *y*-axes). Also, the interesting wavelength range is 0.4–0.7 μm (or 400–700 nm), as this is the visual range. Whereas the earth as a comparably cold planet emits no radiation in the visual range, this is different for hot stars such as the sun. The maximum of the spectral flux density for the sun is exactly in the range of visual light.

Chapter 9

9.1. The radius of the first orbit in the Bohr model is $r_1 = 5.3 \cdot 10^{-9}$ cm. As a rough estimate, calculate the energy of the electron in the hydrogen atom, assuming it was a particle in a cubic box of the same volume as that of a sphere with radius r_1. Compare the result with the energy that is predicted by the Bohr model.

(1) Energy in the Bohr model

The energy of the electron in different orbits is given by

$$E_n = -\frac{m_e \cdot e^4}{8 \cdot \varepsilon_0^2 \cdot n^2 \cdot h^2}$$

for $n = 1$ this yields:

$$E_1 = -\frac{9.11 \cdot 10^{-31} \text{ kg} \cdot \left(1.602 \cdot 10^{-19} \text{ C}\right)^4}{8 \cdot \left(8.854 \cdot 10^{-12} \text{ A}^2 \text{ s}^4 \text{ m}^{-3} \text{ kg}^{-1}\right)^2 \cdot 1 \cdot \left(6.626 \cdot 10^{-34} \text{ J s}\right)^2}$$

$$E_1 = -\frac{9.11 \cdot 1.602^4 \cdot 10^{-31} \cdot 10^{-76} \cdot \text{kg A}^4 \text{ s}^4}{8 \cdot 8.854^2 \cdot 6.626^2 \cdot 10^{-24} \cdot 10^{-68} \cdot \text{A}^4 \text{ s}^8 \text{ m}^{-6} \text{ kg}^{-2} \text{ J}^2 \text{ s}^2}$$

$$E_1 = -\frac{9.11 \cdot 1.602^4}{8 \cdot 8.854^2 \cdot 6.626^2} \cdot 10^{-15} \cdot \frac{\text{kg A}^4 \text{ s}^4}{\text{A}^4 \text{ s}^8 \text{ m}^{-6} \text{ kg}^{-2} \text{ kg}^2 \text{ m}^4 \text{ s}^{-4} \text{ s}^2}$$

$$E_1 = -0.00218 \cdot 10^{-15} \cdot \frac{\text{kg}}{\text{m}^{-2} \text{ s}^2}$$

$$E_1 = -2.18 \cdot 10^{-18} \text{ J}.$$

(2) Energy of the particle in a box

The energy of a particle in a three-dimensional box is given by

$$E_n = \frac{h^2}{8 \cdot m \cdot a^2} \cdot n^2$$

Since we assume a cubic box where the volume equals the volume of a sphere with the radius of the first orbit in the Bohr model, we can calculate a as follows:

$$V_{sphere} = \frac{4}{3} \cdot \pi \cdot r_{Bohr}^3 = a^3 = V_{cube}$$

$$a = \left(\frac{4}{3} \cdot \pi\right)^{\frac{1}{3}} \cdot r_{Bohr}$$

$$a = \left(\frac{4}{3} \cdot \pi\right)^{\frac{1}{3}} \cdot 5.3 \cdot 10^{-9} \text{ cm}$$

$$a = 8.54 \cdot 10^{-11} \text{ m}.$$

We can now calculate the energy of the particle in a cubic box of width a with $n = 1$:

$$E_1 = \frac{\left(6.626 \cdot 10^{-34} \text{ J s}\right)^2}{8 \cdot 9.11 \cdot 10^{-32} \text{ kg} \cdot \left(8.54 \cdot 10^{-11} \text{ m}\right)^2} \cdot 1$$

$$E_1 = \frac{6.626^2 \cdot 10^{-68} \text{ J}^2 \text{ s}^2}{8 \cdot 9.11 \cdot 8.54^2 \cdot 10^{-32} \cdot 10^{-22} \cdot \text{kg m}^2}$$

$$E_1 = \frac{6.626^2}{8 \cdot 9.11 \cdot 8.54^2} \cdot 10^{-14} \cdot \frac{\text{kg}^2 \text{ m}^4 \text{ s}^{-4} \text{ s}^2}{\text{kg m}^2}$$

$$E_1 = 0.00826 \cdot 10^{-14} \cdot \frac{\text{kg m}^2 \text{ s}^{-2}}{1}$$

$$E_1 = 8.26 \cdot 10^{-17} \text{ J}.$$

Comparison of the numerical values of the energies for the first orbit in the Bohr model with the lowest energy state in the quantum mechanical model shows that the energy predicted by the Bohr model is by an order of magnitude lower.

9.2. Calculate the zero-point energy of $^1\text{H}^{35}\text{Cl}$ (a) for one molecule, and (b) for 1 mol, assuming a force constant of 480.6 Nm^{-1}.

The zero-point energy is given by

$$E_0 = \frac{1}{2} \cdot h \cdot \nu_0$$

where ν_0 is the frequency of vibrational oscillation of a di-atomic molecule defined by

$$\nu_0 = \frac{1}{2\pi} \sqrt{\frac{D}{\mu}}, \text{ with } \mu = \frac{m_1 \cdot m_2}{m_1 + m_2}.$$

Therefore:

$$v_0 = \frac{1}{2\pi} \cdot \sqrt{\frac{D}{\frac{m_1 \cdot m_2}{m_1 + m_2}}} = \frac{1}{2\pi} \cdot \sqrt{\frac{D \cdot (m_1 + m_2)}{m_1 \cdot m_2}}, \text{ and thus,}$$

$$E_0 = \frac{1}{2} \cdot h \cdot v_0 = \frac{1}{2} \cdot h \cdot \frac{1}{2\pi} \cdot \sqrt{\frac{D \cdot (m_1 + m_2)}{m_1 \cdot m_2}} = \frac{h}{4\pi} \cdot \sqrt{\frac{D \cdot (m_1 + m_2)}{m_1 \cdot m_2}}$$

$$E_0 = \frac{6.626 \cdot 10^{-34} \text{ J s}}{4\pi} \cdot \sqrt{\frac{480.6 \text{ Nm}^{-1} \cdot (1.008 \text{ u} + 34.969 \text{ u})}{1.008 \text{ u} \cdot 34.969 \text{ u}}}$$

$$E_0 = \frac{6.626 \cdot 10^{-34} \text{ J s}}{4\pi} \cdot \sqrt{\frac{480.6 \text{ Nm}^{-1} \cdot 35.977 \text{ u}}{35.249 \text{ u}^2}}$$

$$E_0 = \frac{6.626 \cdot 10^{-34} \text{ J s}}{4\pi} \cdot \sqrt{\frac{480.6 \text{ kg ms}^{-2} \text{ m}^{-1} \cdot 35.977}{35.249 \cdot 1.661 \cdot 10^{-27} \text{ kg}}}$$

$$E_0 = \frac{6.626 \cdot 10^{-34} \text{ J s}}{4\pi} \cdot \sqrt{\frac{480.6 \cdot 35.977}{35.249 \cdot 1.661} \cdot 10^{27} \frac{1}{\text{s}^2}}$$

$$E_0 = \frac{6.626 \cdot 10^{-34} \text{ J s}}{4\pi} \cdot \sqrt{29.532 \cdot 10^{28} \frac{1}{\text{s}^2}}$$

$$E_0 = \frac{6.626 \cdot 10^{-34} \text{ J s}}{4\pi} \cdot 5.434 \cdot 10^{14} \frac{1}{\text{s}}$$

$$E_0 = \frac{6.626 \cdot 5.434}{4\pi} \cdot 10^{-20} \text{ J}$$

$$E_0 = 2.87 \cdot 10^{-20} \text{ J}.$$

For 1 mol of HCl, one needs to multiply with Avogadro's constant:

$$E_0(1\text{mol}) = E_0 \cdot N_A = 2.87 \cdot 10^{-20} \text{ J} \cdot 6.022 \cdot 10^{23} \frac{1}{\text{mol}}$$

$$E_0(1\text{mol}) = 17.3 \cdot 10^3 \frac{\text{J}}{\text{mol}} = 17.3 \text{ kJ mol}^{-1}.$$

9.3. What is the value of the transmission coefficient for an electron with an energy of 1 eV that moves against a potential barrier of 5 eV and 2 nm thickness?

The transmission coefficient is given by

$$T = \frac{16 \cdot E \cdot (V_0 - E)}{V_0^2} \cdot e^{-2 \cdot \sqrt{\frac{8\pi^2 \cdot m \cdot a^2 \cdot (V_0 - E)}{h^2}}}$$

With $m = m_e = 9.110 \cdot 10^{-31}$ kg, one obtains:

$$T = \frac{16 \cdot 1\,\text{eV} \cdot (5\,\text{eV} - 1\,\text{eV})}{(5\,\text{eV})^2} \cdot e^{-2 \cdot \sqrt{\frac{8\pi^2 \cdot 9.110 \cdot 10^{-31}\ \text{kg} \cdot (2\text{nm})^2 \cdot (5\ \text{eV} - 1\ \text{eV})}{(6.626 \cdot 10^{-34}\ \text{J s})^2}}}$$

$$T = \frac{16 \cdot 4\,\text{eV}^2}{25\,\text{eV}^2} \cdot e^{-2 \cdot \sqrt{\frac{8\pi^2 \cdot 9.110 \cdot 10^{-31}\ \text{kg} \cdot (2 \cdot 10^{-9}\ \text{m})^2 \cdot 4\ \text{eV}}{(6.626 \cdot 10^{-34}\ \text{J s})^2}}}$$

$$T = 2.56 \cdot e^{-2 \cdot \sqrt{\frac{8\pi^2 \cdot 9.110 \cdot 10^{-31}\ \text{kg} \cdot 4 \cdot 10^{-18}\ \text{m}^2 \cdot 4 \cdot 1.602 \cdot 10^{-19}\ \text{J}}{6.626^2 \cdot 10^{-68}\ \text{J}^2\ \text{s}^2}}}$$

$$T = 2.56 \cdot e^{-2 \cdot \sqrt{\frac{8\pi^2 \cdot 9.110 \cdot 4 \cdot 4 \cdot 1.602 \cdot 10^{-31 - 18} \cdot 10^{-18} \cdot 10^{-19}\ \text{J kg m}^2}{6.626^2 \cdot 10^{-68}\ \text{J}^2\ \text{s}^2}}}$$

$$T = 2.56 \cdot e^{-2 \cdot \sqrt{\frac{8\pi^2 \cdot 9.110 \cdot 4 \cdot 4 \cdot 1.602}{6.626^2} \cdot 10^{-31 - 18 - 19 + 68}\ \text{J kg m}^2\ \text{J}^{-2}\ \text{s}^{-2}}}$$

$$T = 2.56 \cdot e^{-2 \cdot \sqrt{419.5 \cdot 10^0\ \text{J J J}^{-2}}}$$

$$T = 2.56 \cdot e^{-2 \cdot 20.48} = 2.56 \cdot 1.63 \cdot 10^{-18}$$

$$T = 4.2 \cdot 10^{-18}.$$

The transition coefficient is $4.2 \cdot 10^{-18}$.

9.4. Calculate the probability of locating a particle in a potential-free one-dimensional box of length a between $\frac{1}{4}\,a$ and $\frac{3}{4}\,a$, assuming the particle being in its lowest energy state.

The probability of finding the particle is described the squared wave function $|\Psi|^2$. The normalised wave function solving the Schrödinger equation for a one-dimensional box without a potential is given by

$$\Psi_n(x) = \sqrt{\frac{2}{a}} \cdot \sin\left(\frac{n \cdot \pi}{a} \cdot x\right),$$

which yields for the probability:

$$|\Psi_n(x)|^2 = \frac{2}{a} \cdot \sin^2\left(\frac{n \cdot \pi}{a} \cdot x\right).$$

In order to calculate the probability density, the accumulated probability to find the particle between $\frac{1}{4}a$ and $\frac{3}{4}a$, the probability needs to be integrated in this region:

$$\int_{\frac{1}{4}a}^{\frac{3}{4}a} |\Psi_n(x)|^2 dx = \int_{\frac{1}{4}a}^{\frac{3}{4}a} \frac{2}{a} \cdot \sin^2\left(\frac{n \cdot \pi}{a} \cdot x\right) dx = \frac{2}{a} \cdot \int_{\frac{1}{4}a}^{\frac{3}{4}a} \sin^2\left(\frac{n \cdot \pi}{a} \cdot x\right) dx.$$

We find that we need to integrate a function of the type $\sin^2(C \cdot x)$, with $C = \frac{n \cdot \pi}{a}$ for which the following integral is known:

$$\int \sin^2(C \cdot x) dx = \frac{x}{2} - \frac{\sin(2 \cdot C \cdot x)}{4 \cdot C} + \text{const.}$$

With $n = 1$ (lowest energy state), this yields:

$$\int_{\frac{1}{4}\cdot a}^{\frac{3}{4}\cdot a} |\Psi_1(x)|^2 dx = \frac{2}{a} \cdot \int_{\frac{1}{4}\cdot a}^{\frac{3}{4}\cdot a} \sin^2\left(\frac{\pi}{a}\cdot x\right) dx$$

$$\int_{\frac{1}{4}\cdot a}^{\frac{3}{4}\cdot a} |\Psi_1(x)|^2 dx = \frac{2}{a} \cdot \left[\frac{x}{2} - \frac{\sin\left(2\cdot\frac{\pi}{a}\cdot x\right)}{4\cdot\frac{\pi}{a}}\right]_{\frac{1}{4}\cdot a}^{\frac{3}{4}\cdot a}$$

$$\int_{\frac{1}{4}\cdot a}^{\frac{3}{4}\cdot a} |\Psi_1(x)|^2 dx = \frac{2}{a} \cdot \left[\frac{\frac{3}{4}\cdot a}{2} - \frac{\sin\left(2\cdot\frac{\pi}{a}\cdot\frac{3}{4}\cdot a\right)}{4\cdot\frac{\pi}{a}} - \frac{\frac{1}{4}\cdot a}{2} + \frac{\sin\left(2\cdot\frac{\pi}{a}\cdot\frac{1}{4}\cdot a\right)}{4\cdot\frac{\pi}{a}}\right]$$

$$\int_{\frac{1}{4}\cdot a}^{\frac{3}{4}\cdot a} |\Psi_1(x)|^2 dx = \frac{2}{a} \cdot \left[\frac{3\cdot a}{8} - \frac{\sin\left(\frac{3}{2}\cdot\pi\right)}{4\cdot\frac{\pi}{a}} - \frac{a}{8} + \frac{\sin\left(\frac{1}{2}\cdot\pi\right)}{4\cdot\frac{\pi}{a}}\right]$$

$$\int_{\frac{1}{4}\cdot a}^{\frac{3}{4}\cdot a} |\Psi_1(x)|^2 dx = \frac{2}{a} \cdot \left[\frac{3\cdot a - a}{8} + \frac{a}{4\pi}\cdot\sin\left(\frac{1}{2}\cdot\pi\right) - \frac{a}{4\pi}\cdot\sin\left(\frac{3}{2}\cdot\pi\right)\right]$$

$$\int_{\frac{1}{4}\cdot a}^{\frac{3}{4}\cdot a} |\Psi_1(x)|^2 dx = \frac{2}{a} \cdot \left[\frac{2\cdot a}{8} + \frac{a}{4\pi}\cdot\left(\sin\left(\frac{1}{2}\cdot\pi\right) - \sin\left(\frac{3}{2}\cdot\pi\right)\right)\right]$$

$$\int_{\frac{1}{4}\cdot a}^{\frac{3}{4}\cdot a} |\Psi_1(x)|^2 dx = \frac{2\cdot 2\cdot a}{8\cdot a} + \frac{2\cdot a}{4\pi\cdot a}\cdot[1-(-1)]$$

$$\int_{\frac{1}{4}\cdot a}^{\frac{3}{4}\cdot a} |\Psi_1(x)|^2 dx = \frac{1}{2} + \frac{1}{2\pi}\cdot 2 = \frac{1}{2} + \frac{1}{\pi}$$

$$\int_{\frac{1}{4}\cdot a}^{\frac{3}{4}\cdot a} |\Psi_1(x)|^2 dx = 0.82$$

The probability to find the particle between $^1/_4\,a$ and $^3/_4\,a$ is 82%.

Chapter 10

10.1. Calculate at which distances from the nucleus the 3s orbital possesses radial nodes.

The radial eigenfunction for the 3s orbital ($n = 3, l = 0$) is

$$R_{3,0} = \frac{2}{81 \cdot \sqrt{3}} \cdot e^{-\frac{\rho}{3}} \cdot \left(27 - 18 \cdot \rho + 2 \cdot \rho^2\right).$$

Nodes are areas in which the probability to find the electron is zero. For the $3s$ orbital, the number of radial nodes is $(n–l–1) = (3–0–1) = 2$. In the eigenfunction, these are the points where the function has a zero-crossing, i.e. the function assumes a value of zero. Therefore:

$$R_{3,0} = \frac{2}{81 \cdot \sqrt{3}} \cdot e^{-\frac{\rho}{3}} \cdot \left(27 - 18 \cdot \rho + 2 \cdot \rho^2\right) = 0.$$

This function is a product of three factors; neither the scalar factor $\frac{2}{81 \cdot \sqrt{3}}$ nor the exponential factor $e^{-\frac{\rho}{3}}$ can assume a value of zero. The only factor that can assume a value of zero is the expression in brackets:

$$\left(27 - 18 \cdot \rho + 2 \cdot \rho^2\right) = 0$$

This is a quadratic equation; the zero-crossings are thus calculated as per the formula $x_{1,2} = \frac{-b \pm \sqrt{b^2 - 4 \cdot a \cdot c}}{2 \cdot a}$. This yields:

$$\rho_{1,2} = \frac{+18 \pm \sqrt{18^2 - 4 \cdot 2 \cdot 27}}{2 \cdot 2}$$

$$\rho_{1,2} = \frac{18 \pm \sqrt{324 - 216}}{4} = \frac{18 \pm \sqrt{108}}{4} = \frac{18 \pm 10.4}{4}$$

$$\rho_1 = 1.9, \rho_2 = 7.1.$$

Since ρ is the ratio between a a radial distance and the radius of the first orbit r_{Bohr} ($\rho = \frac{r}{r_{Bohr}}$), the metric distances of the two nodes are:

$$r_1 = \rho_1 \cdot r_{Bohr} = 1.9 \cdot 5.3 \cdot 10^{-11} \text{ m} = 1.0 \cdot 10^{-10} \text{ m}$$
$$r_2 = \rho_2 \cdot r_{Bohr} = 7.1 \cdot 5.3 \cdot 10^{-11} \text{ m} = 3.8 \cdot 10^{-10} \text{ m}.$$

10.2. Derive the electron configuration for (a) vanadium, and (b) chromium in their ground states.

(a) The order number of vanadium is $Z = 23$, i.e. there are 23 electrons to be filled into the atomic orbital scheme.

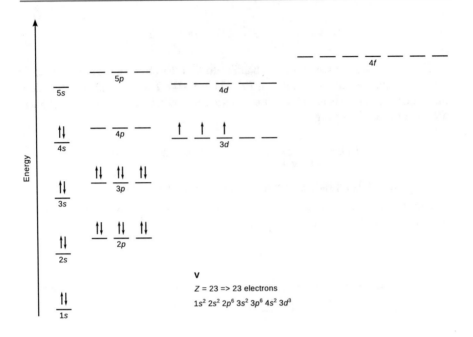

The electronic configuration of vanadium is $1s^2\ 2s^2\ 2p^6\ 3s^2\ 3p^6\ 3d^3\ 4s^2$.

(b) Chromium has an order number of $Z = 24$, i.e. there is one more electron to be filled in as compared to vanadium. However, due to the fact that the $4s$ and $3d$ orbitals are energetically close and a half-filled $3d$ subshell can be achieved, there is only one electron in the $4s$ orbital.

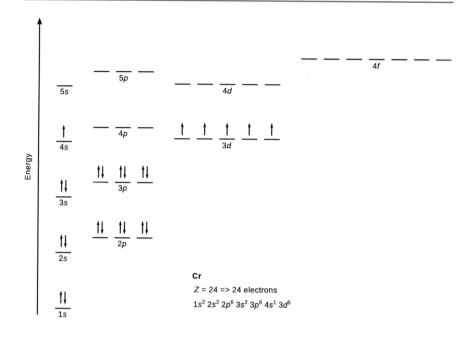

The electronic configuration of chromium is $1s^2\ 2s^2\ 2p^6\ 3s^2\ 3p^6\ 3d^5\ 4s^1$.

10.3. Name the four quantum numbers of the electron and state their relationships/possible values.

10.4. Describe the overall shape of the orbitals characterised by the angular momentum quantum number $l = 0, 1, 2$.

$$l = 0$$

This is an *s*-orbital; it has spherical symmetry.

$$l = 1$$

These are *p*-orbitals and they have a two-lobed structure.

$$l = 2$$

These are *d*-orbitals; four of them adopt a four-lobed structure, one of them has rotational symmetry and consists of a two-lobed structure with a torus.

10.5. Determine the ground-state electronic configuration of the following species: S⁻, Zn^{2+}, Cl⁻, Cu^+.

S^-

17 electrons

$$1s^2 2s^2 2p^6 3s^2 3p^5 \quad \text{or} \quad [\text{Ne}] \, 3s^2 3p^5$$

Zn^{2+}

28 electrons

$$1s^2 2s^2 2p^6 3s^2 3p^6 3d^8 4s^2 \quad \text{or} \quad [\text{Ne}] \, 3s^2 3p^6 3d^8 4s^2 \quad \text{or} \quad [\text{Ar}] \, 3d^8 4s^2$$

Cl^-

18 electrons

$$1s^2 2s^2 2p^6 3s^2 3p^6 \quad \text{or} \quad [\text{Ne}] \, 3s^2 3p^6 \quad \text{or} \quad [\text{Ar}]$$

Cu^+

28 electrons

$$1s^2 2s^2 2p^6 3s^2 3p^6 3d^{10} 4s^0 \quad \text{or} \quad [\text{Ne}] \, 3s^2 3p^6 3d^{10} 4s^0 \quad \text{or} \quad [\text{Ar}] \, 3d^{10} 4s^0 .$$

10.6. a) Using the knowledge about the allowed energy states (eigenvalues), calculate the ionisation potential of the hydrogen atom.

The allowed energy states (eigenvalues) of the hydrogen atom are obtained as

$$E_n = -\frac{m_e \cdot e^4}{8 \cdot \varepsilon_0^2 \cdot h^2 \cdot n^2},$$

with $n = 1$, since the only electron in hydrogen in the ground state occupies the first shell. In order to remove the electron, the energy of $|E_1|$ needs to be provided, hence this is the ionisation energy.

$$| E_1 | = \frac{9.11 \cdot 10^{-31} \text{ kg} \cdot \left(1.602 \cdot 10^{-19} \text{ C}\right)^4}{8 \cdot \left(8.85 \cdot 10^{-12} \text{ A}^2 \text{ s}^4 \text{ m}^{-3} \text{ kg}^{-1}\right)^2 \cdot \left(6.63 \cdot 10^{-34} \text{ J s}\right)^2 \cdot 1^2}$$

$$| E_1 | = \frac{9.11 \cdot 10^{-31} \cdot 1.602^4 \cdot 10^{-76}}{8 \cdot 8.85^2 \cdot 10^{-24} \cdot 6.63^2 \cdot 10^{-68}} \cdot \frac{\text{s}^4 \text{ kg A}^4 \text{ s}^4 \text{ m}^6 \text{ kg}^2}{\text{A}^4 \text{ s}^8 \text{ s}^2 \text{ kg}^2 \text{ m}^4}$$

$$| E_1 | = \frac{9.11 \cdot 1.602^4}{8 \cdot 8.85^2 6.63^2} \cdot 10^{-15} \cdot \frac{\text{kg m}^6}{\text{s}^2 \text{ m}^4}$$

$$| E_1 | = 0.00218 \cdot 10^{-15} \cdot \frac{\text{kg m}^2}{\text{s}^2} = 2.18 \cdot 10^{-18} \text{ J}$$

$$| E_1 | = 2.18 \cdot 10^{-18} \text{J} \cdot 6.24 \cdot 10^{18} \frac{\text{eV}}{\text{J}} = 13.6 \text{ eV}.$$

The ionisation energy of hydrogen is 13.6 eV.

b) The equation yielding the allowed energy states of a multi-electron atom is often modified to $E_n = -\dfrac{m_e \cdot e^4 \cdot (Z - \sigma)^2}{8 \cdot \varepsilon_0^2 \cdot h^2 \cdot n^2}$ to accommodate the shielding of the nuclear charge by electrons in the various orbitals; the parameter σ is thus called the screening constant. Calculate the value of σ for helium, assuming its first ionisation potential is 24.5 eV.

For non-hydrogen atoms ($Z > 1$), the above equation for eigenvalues needs to consider the number of electrons (which equals the order number Z in the periodic system)

$$E_n = -\frac{m_e \cdot e^4 \cdot Z^2}{8 \cdot \varepsilon_0^2 \cdot h^2 \cdot n^2}$$

but then needs to be adjusted for the shielding of the nuclear charge by electrons present. One way to do this is to replace Z with $(Z-\sigma)$ where σ is called the screening constant:

$$E_n = -\frac{m_e \cdot e^4 \cdot (Z - \sigma)^2}{8 \cdot \varepsilon_0^2 \cdot h^2 \cdot n^2}.$$

From the first ionisation energy of He (24.6 eV), the screening constant s can be calculated using the above equation. For convenience, we will first determine $(Z-\sigma)^2$ and then determine σ. The He-specific parameters are:

$n = 1$

$|E_1| = 24.6$ eV

$Z = 2$

$$| E_1 | = \frac{m_e \cdot e^4 \cdot (Z-\sigma)^2}{8 \cdot \varepsilon_0^2 \cdot h^2 \cdot 1^2}$$

$$(Z-\sigma)^2 = \frac{|E_1| \cdot 8 \cdot \varepsilon_0^2 \cdot h^2}{m_e \cdot e^4}$$

$$(Z-\sigma)^2 = \frac{24.6 \text{ eV} \cdot 8 \cdot (8.85 \cdot 10^{-12} \text{ A}^2 \text{s}^4 \text{m}^{-3} \text{kg}^{-1})^2 \cdot (6.63 \cdot 10^{-34} \text{Js})^2}{9.11 \cdot 10^{-31} \text{kg} \cdot (1.602 \cdot 10^{-19} \text{C})^4}$$

$$(Z-\sigma)^2 = \frac{24.6 \cdot 1.602 \cdot 10^{-19} \text{J} \cdot 8 \cdot (8.85 \cdot 10^{-12} \text{ A}^2 \text{s}^4 \text{m}^{-3} \text{kg}^{-1})^2 \cdot (6.63 \cdot 10^{-34} \text{Js})^2}{9.11 \cdot 10^{-31} \text{kg} \cdot (1.602 \cdot 10^{-19} \text{C})^4}$$

$$(Z - \sigma)^2 = \frac{24.6 \, \text{J} \cdot 8 \cdot \left(8.85 \cdot 10^{-12} \, \text{A}^2 \, \text{s}^4 \, \text{m}^{-3} \, \text{kg}^{-1}\right)^2 \cdot \left(6.63 \cdot 10^{-34} \, \text{J s}\right)^2}{9.11 \cdot 10^{-31} \, \text{kg} \cdot \left(1.602 \cdot 10^{-19}\right)^3 \text{C}^4}$$

$$(Z - \sigma)^2 = \frac{24.6 \cdot 8 \cdot 8.85^2 \cdot 10^{-24} \cdot 6.63^2 \cdot 10^{-68} \, \text{J A}^4 \, \text{s}^8 \, \text{m}^{-6} \, \text{kg}^{-2} \, \text{J}^2 \, \text{s}^2}{9.11 \cdot 10^{-31} \cdot 1.602^3 \cdot 10^{-57} \, \text{kg A}^4 \, \text{s}^4}$$

$$(Z - \sigma)^2 = \frac{24.6 \cdot 8 \cdot 8.85^2 \cdot 6.63^2}{9.11 \cdot 1.602^3} \cdot 10^{-4} \cdot \frac{\text{J}^3 \, \text{s}^6}{\text{kg}^3 \, \text{m}^6}$$

$$(Z - \sigma)^2 = \frac{24.6 \cdot 8 \cdot 8.85^2 \cdot 6.63^2}{9.11 \cdot 1.602^3} \cdot 10^{-4} \cdot \frac{\text{J}^3}{\text{J}^3}$$

$$(Z - \sigma)^2 = 18090 \cdot 10^{-4} = 1.809$$

$$(Z - \sigma) = \sqrt{1.809} = 1.345$$

$$\sigma = Z - 1.345 = 2 - 1.345 = 0.66.$$

The screening constant σ for He is thus calculated as 0.66, i.e. 0.33 per electron in the $1s$ orbital. This value agrees qualitatively the semi-empirical screening constants suggested by Slater's rules (Slater 1930). These rules provide numerical values for each electron in an atom. For the $1s$ orbital, Slater's value for σ is 0.30 per electron.

Chapter 11

11.1. NaCl crystallises in the lattice type sodium chloride ($M = 1.747565$) with $r_0 = 2.82$ Å and $\rho = 0.321$ Å. Calculate the lattice energy of NaCl.

The lattice energy of an ionic crystal can be calculated by the following expression:

$$\Delta U_{\text{lattice}} = \frac{z_+ \cdot z_- \cdot e^2 \cdot N_A \cdot M}{4\pi \cdot \varepsilon_0 \cdot r_0} \cdot \left(1 - \frac{\rho}{r_0}\right)$$

We thus require the following parameters:

$$z_+ = +1(\text{Na}^+)$$
$$z_- = -1 \quad (\text{Cl}^-)$$
$$M = 1.747565$$
$$r_0 = 2.82 \text{Å} = 2.82 \cdot 10^{-10} \text{ m}$$
$$\rho = 0.321 \text{Å} = 0.321 \cdot 10^{-10} \text{ m}$$
$$e = 1.602 \cdot 10^{-19} \text{ C}$$
$$\varepsilon_0 = 8.854 \cdot 10^{-12} \text{ A}^2 \, \text{s}^4 \, \text{m}^{-3} \, \text{kg}^{-1}$$

This yields:

$$\Delta U_{\text{lattice}} = \frac{(+1) \cdot (-1) \cdot \left(1.602 \cdot 10^{-19}C\right)^2 \cdot 6.022 \cdot 10^{23} \, \text{mol}^{-1} \cdot 1.747565}{4\pi \cdot 8.854 \cdot 10^{-12} \, \text{A}^2 \, \text{s}^4 \, \text{m}^{-3} \, \text{kg}^{-1} \cdot 2.82 \cdot 10^{-10} \, \text{m}} \cdot$$

$$\left(1 - \frac{0.321 \cdot 10^{-10} \, \text{m}}{2.82 \cdot 10^{-10} \, \text{m}}\right)$$

$$\Delta U_{\text{lattice}} = -\frac{1.602^2 \cdot 10^{-38} \cdot 6.022 \cdot 10^{23} \cdot 1.747565 \cdot \text{C}^2 \, \text{mol}^{-1}}{4\pi \cdot 8.854 \cdot 10^{-12} \cdot 2.82 \cdot 10^{-10} \cdot \text{A}^2 \, \text{s}^4 \, \text{m}^{-3} \, \text{kg}^{-1} \, \text{m}} \cdot \left(1 - \frac{0.321}{2.82}\right)$$

$$\Delta U_{\text{lattice}} = -\frac{1.602^2 \cdot 6.022 \cdot 1.747565 \cdot 10^{-15} \cdot \text{A}^2 \, \text{s}^2 \, \text{mol}^{-1}}{4\pi \cdot 8.854 \cdot 2.82 \cdot 10^{-22} \cdot \text{A}^2 \, \text{s}^4 \, \text{m}^{-2} \, \text{kg}^{-1}} \cdot (0.8862)$$

$$\Delta U_{\text{lattice}} = -\frac{1.602^2 \cdot 6.022 \cdot 1.747565 \cdot 0.8862}{4\pi \cdot 8.854 \cdot 2.82} \cdot 10^7 \cdot \frac{\text{mol}^{-1}}{\text{s}^2 \, \text{m}^{-2} \, \text{kg}^{-1}}$$

$$\Delta U_{\text{lattice}} = -0.0763 \cdot 10^7 \cdot \frac{\text{m}^2 \, \text{kg}}{\text{s}^2 \, \text{mol}}$$

$$\Delta U_{\text{lattice}} = -763 \cdot 10^3 \cdot \frac{\text{J}}{\text{mol}}$$

$$\Delta U_{\text{lattice}} = -763 \, \frac{\text{kJ}}{\text{mol}}.$$

The lattice energy of sodium chloride is -763 kJ mol^{-1}.

11.2. Discuss the bonding of carbon monoxide (CO) using the valence bond theory without and with hybridised atomic orbitals as well as the molecular orbital theory. Compare the bond order derived by the three approaches.
(a) VB theory:

Several resonance structure can be drawn for CO:

(i) a double bond structure without formal charges;
(ii) a triple bond structure with a positive formal charge on oxygen and a negative formal charge on carbon, not in agreement with the atomic electronegativities;
(iii) a single bond structure that has a formal positive charge on carbon and a negative charge on oxygen, in agreement with the atomic electronegativities.

For (i) and (iii), the octet rule is nor satisfied for carbon, making the triple bond structure more favourable.

Experimentally, CO is found to possess a small dipole moment of 0.1 D; this suggests that structures with formal charges are less populated.

Overall, one could conclude for a bond order that is between a double and a triple bond.

(b) Hybrid orbitals:

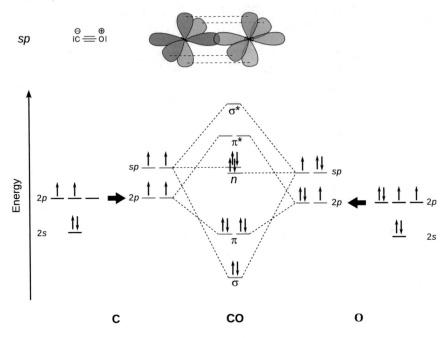

The CO bonding can be explained by *sp* hybridisation which suggests two lone pairs (n) and predicts a bond order of:

$\frac{1}{2} \cdot (8 - 2) = \frac{1}{2} \cdot 6 = 3$. (Note that the MO scheme above does not show the first shell electrons.)

The overlap of one *sp* orbital of C and O each establishes a σ bond; the overlap of p_y and p_z orbitals of C and O established two π bonds: this suggests a C–O triple bond. One *sp* orbital on C and O each is occupied by a lone pair.

(c) MO theory:

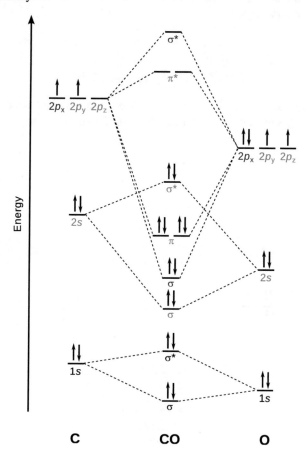

Bond order:

$$\frac{1}{2} \cdot (10 - 4) = \frac{1}{2} \cdot 6 = 3.$$

11.3. Determine for each of the species S^-, Zn^{2+}, Cl^- and Cu^+ whether they are diamagnetic or paramagnetic.

S^-

17 electrons

$$1s^2 2s^2 2p^6 3s^2 3p^5 \quad \text{or} \quad [\text{Ne}]\, 3s^2 3p^5$$

The ground state of S is $[\text{Ne}]\, 3s^2\, 3p^4$. The extra electron is added to the 3p orbitals and renders a species with one unpaired electron. Therefore, S^- is paramagnetic.

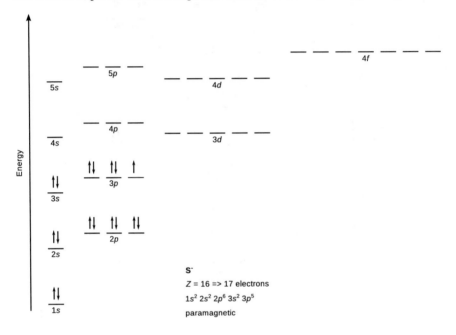

Zn^{2+}

28 electrons

$$1s^2 2s^2 2p^6 3s^2 3p^6 3d^8 4s^2 \quad \text{or} \quad [\text{Ne}]\, 3s^2 3p^6 3d^8 4s^2 \quad \text{or} \quad [\text{Ar}]\, 3d^8 4s^2$$

The ground state of Zn is $[\text{Ar}]\, 3d^{10}\, 4s^2$. In Zn^{2+}, two electrons are removed from the $3d$ orbital, rendering a species with two unpaired electrons, hence Zn^{2+} is paramagnetic.

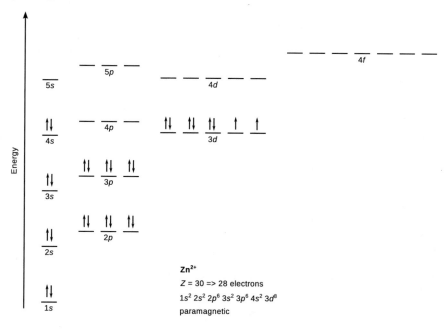

Zn²⁺
Z = 30 => 28 electrons
1s² 2s² 2p⁶ 3s² 3p⁶ 4s² 3d⁸
paramagnetic

Cl⁻
18 electrons

$$1s^2 2s^2 2p^6 3s^2 3p^6 \quad \text{or} \quad [\text{Ne}]\, 3s^2 3p^6 \quad \text{or} \quad [\text{Ar}]$$

With the ground state of Cl being $[\text{Ne}]\, 3s^2\, 3p^5$, the extra electron completes the octet of the third shell and Cl⁻ attains noble gas configuration. There are no unpaired electrons, hence Cl⁻ is diamagnetic.

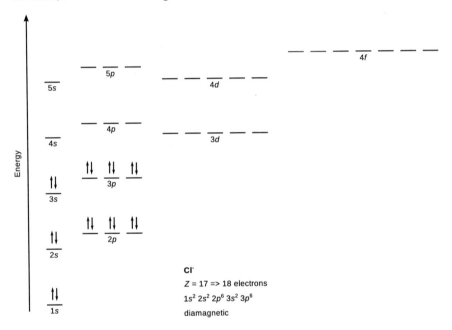

Cu$^+$

28 electrons

$$1s^2 2s^2 2p^6 3s^2 3p^6 3d^{10} 4s^0 \quad \text{or} \quad [\text{Ne}]\, 3s^2 3p^6 3d^{10} 4s^0 \quad \text{or} \quad [\text{Ar}]\, 3d^{10} 4s^0$$

The ground state configuration for Cu is [Ar] $3d^{10}\, 4s^1$. In Cu$^+$, the $4s^1$ electron is removed and the outer shell is now $3d^{10}$, which has no unpaired electrons. Therefore Cu$^+$ is diamagnetic.

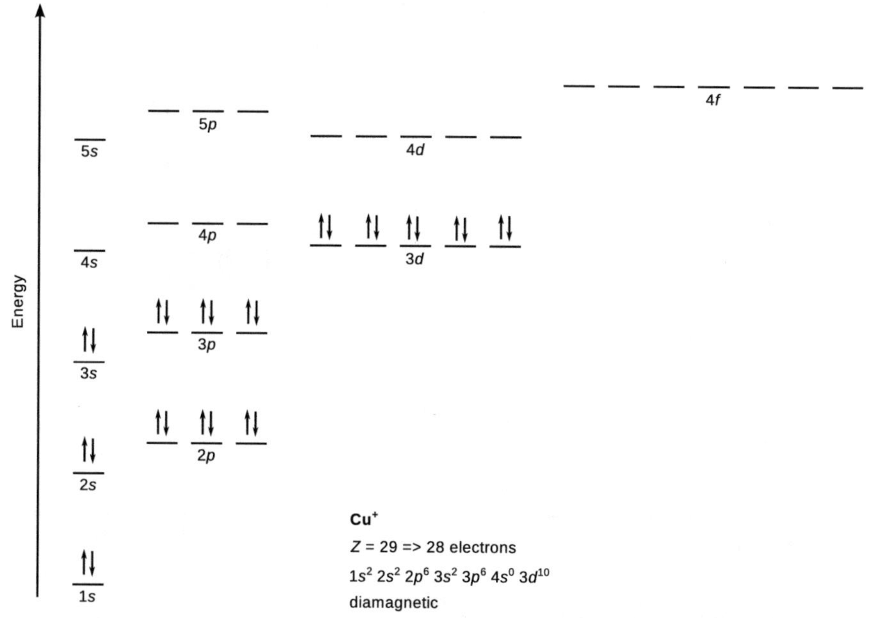

Cu$^+$

$Z = 29 \Rightarrow$ 28 electrons

$1s^2\ 2s^2\ 2p^6\ 3s^2\ 3p^6\ 4s^0\ 3d^{10}$

diamagnetic

11.4. Using an appropriate molecular orbital scheme, explain why [Co(NH$_3$)$_6$]$^{3+}$ is a diamagnetic low-spin complex, whereas [CoF$_6$]$^{3-}$ is a paramagnetic high-spin complex.

The ground state electron configuration of Co is $1s^2\ 2s^2\ 2p^6\ 3s^2\ 3p^6\ 3d^7\ 4s^2$. In both complexes, cobalt takes the oxidation state +3, and has thus lost three electrons: two from the $4s$ and one from the $3d$ orbitals. The resulting electron configuration is thus $3d^6$.

[Co(NH$_3$)$_6$]$^{3+}$

According to the spectrochemical series, NH$_3$ is a ligand that elicits a strong field, hence the crystal field splitting Δ is large. The six electrons are filled into the three low lying d orbitals which results in a low-spin, diamagnetic species.

[CoF$_6$]$^{3-}$

According to the spectrochemical series, F$^-$ is a ligand that elicits a weak field, hence the crystal field splitting Δ is small. The first three electrons populate the low lying d orbitals, then the fourth and fifth electrons occupy the high-lying d orbitals. The sixth electron pairs up with an electron in the low-lying d orbitals. This results in a high-spin, paramagnetic species.

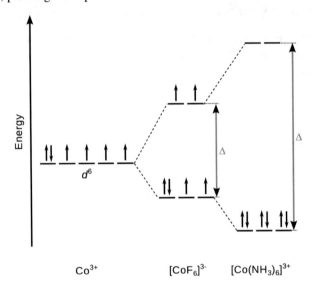

Chapter 12

12.1. The dipole moment of HF is 1.92 D.

(a) Calculate the potential energy of the attractive dipole-dipole interaction between two HF molecules oriented along the x-axis in a plane, separated by 5 Å.

The attractive arrangement of the two permanent HF dipoles in the plane along the x-axis can be illustrated as:

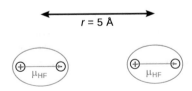

The potential energy for this orientation is given by

$$V = -\frac{2 \cdot \mu^2}{4\pi \cdot \varepsilon_0 \cdot r^3}$$

$$V = -\frac{2 \cdot (1.92\,\text{D})^2}{4\pi \cdot 8.854 \cdot 10^{-12}\,\text{A}^2\,\text{s}^4\,\text{m}^{-3}\,\text{kg}^{-1} \cdot (5 \cdot 10^{-10}\,\text{m})^3}$$

$$V = -\frac{2 \cdot (1.92 \cdot 3.336 \cdot 10^{-30}\,\text{C m})^2}{4\pi \cdot 8.854 \cdot 10^{-12}\,\text{A}^2\,\text{s}^4\,\text{m}^{-3}\,\text{kg}^{-1} \cdot (5 \cdot 10^{-10}\,\text{m})^3}$$

$$V = -\frac{2 \cdot 1.92^2 \cdot 3.336^2 \cdot 10^{-60}\,\text{C}^2\,\text{m}^2}{4\pi \cdot 8.854 \cdot 10^{-12}\,\text{A}^2\,\text{s}^4\,\text{m}^{-3}\,\text{kg}^{-1} \cdot 125 \cdot 10^{-30}\,\text{m}^3}$$

$$V = -\frac{2 \cdot 1.92^2 \cdot 3.336^2 \cdot 10^{(-60+12+30)}\,\text{C}^2\,\text{m}^2}{4\pi \cdot 8.854 \cdot 125\,\text{A}^2\,\text{s}^4\,\text{m}^{-3}\,\text{kg}^{-1}\,\text{m}^3}$$

$$V = -\frac{2 \cdot 1.92^2 \cdot 3.336^2}{4\pi \cdot 8.854 \cdot 125} \cdot 10^{-18} \cdot \frac{\text{A}^2\,\text{s}^2\,\text{m}^2}{\text{A}^2\,\text{s}^4\,\text{m}^{-3}\,\text{kg}^{-1}\,\text{m}^3}$$

$$V = -0.0059 \cdot 10^{-18} \cdot \frac{\text{m}^2\,\text{kg}}{\text{s}^2}$$

$$V = -0.0059 \cdot 10^{-18} \cdot \frac{\text{m}^2\,\text{kg}}{\text{s}^2}$$

$$V = -5.9 \cdot 10^{-21}\,\text{J}.$$

(b) What is the potential energy of the dipole–dipole interaction for 1 mol HF?
For 1 mol HF:

$$E_{pot} = V \cdot N_A = -5.9 \cdot 10^{-21}\,\text{J} \cdot 6.022 \cdot 10^{23}\,\text{mol}^{-1}$$
$$E_{pot} = -3553\,\frac{\text{J}}{\text{mol}} = -3.6\,\frac{\text{kJ}}{\text{mol}}.$$

(c) Calculate the average thermal energy of bulk matter at room temperature. Can the dipole–dipole interaction of HF be sustained at room temperature?

Average thermal energy of bulk matter:

$$E_{thermal} = R \cdot T = 8.3144\,\frac{\text{J}}{\text{Kmol}} \cdot 298\,\text{K} = 2478\,\frac{\text{J}}{\text{mol}} = 2.5\,\frac{\text{kJ}}{\text{mol}}$$

The dipole–dipole interaction is stronger than the thermal motion and can be sustained at 25 °C.

12.2. Explain why the boiling point of the two isomers of butane, n-butane and i-butane, are different. Which isomer has the higher boiling point?

Being isomers, n-butane and i-butane (2-methylpropane) have the same chemical formula C_4H_{10}. The isomers differ in their structures; whereas n-butane is an unbranched alkane, i-butane is branched:

n-butane i-butane

Stronger inter-molecular forces result in a higher boiling point, as they must be overcome to free molecules into the gas phase. Of the different inter-molecular forces, only London dispersion forces apply to butane as both isoforms are alkanes and neutral molecules (no charge). They possess no hetero-atoms (i.e. C and H only) and thus no permanent dipole.

The longer, unbranched n-butane molecule has a larger surface area which provides more possibilities to anneal to neighbouring molecules and thus exert/experience dispersion forces. With i-butane being more compact, there are less dispersion forces between neighbouring molecules.

n-butane therefore has the higher boiling point.

This conclusion agrees with the literature values for the boiling points of n- and i-butane:

Table B.10 Comparison of the boiling points of butane isomers		Boiling point
	n-butane	0 °C
	i-butane	−11 °C

Therefore, $T_b(n\text{-butane}) > T_b(i\text{-butane})$.

12.3. In which of the following substances are molecules held together by hydrogen bonds? Draw the hydrogen bonds where applicable.

(a) XeF_4

Since this does not contain hydrogen, there are no hydrogen bonds possible in XeF_4.

(b) CH_4

Whereas hydrogen is bonded covalently to carbon, the latter is not very electronegative and possesses no lone pairs. CH_4 is a non-polar molecule and does not exhibit hydrogen bonds.

(c) H_2O

There is a considerable electronegativity difference between oxygen and hydrogen and the lone pairs of oxygen engage in intra-molecular hydrogen bonds.

(d) NaH

This is not a covalent molecule, but rather an ionic structure consisting of Na⁺ and H⁻ (hydride) ions. There are no hydrogen bonds possible.

(e) BH₃

Boron is less electronegative than hydrogen, hence hydrogen is not experiencing sufficiently positive partial charge; there are not hydrogen bonds.

(f) NH₃

Compared to hydrogen, nitrogen is considerably more electronegative (polar N–H bond) and also possesses a lone pair which can engage in hydrogen bonding.

(g) HI

There is no hydrogen bonding; iodine is not sufficiently electronegative to generate a polar H–I bond.

H_2O

NH_3

12.4. What is the relationship of the interaction energy between two particles due to Coulomb forces and dispersion forces with the distance? Compare the falloff of these energies with distance by calculating their ratio for the distances 1, 2, 3, 4 and 5 Å. Based on these results, assess the implications for ideal/non-ideal behaviour of solutions.

The Coulomb force depends on the charges on the two particles as well as their distance as per

$$V_C = \frac{Q_1 \cdot Q_2}{\varepsilon \cdot r}$$

and is thus indirect proportional to the distance: $V_C \sim \frac{1}{r}$.

In contrast, the London dispersion force for a mixture of two particles A and B is given by

$$V_L = -\frac{3}{2} \cdot \frac{I_A \cdot I_B}{I_A + I_B} \cdot \frac{\alpha_A \cdot \alpha_B}{r^6}$$

and therefore indirect proportional to the sixth power of the distance: $V_L \sim \frac{1}{r^6}$.

Table B.11 Comparison of Coulomb and dispersion force falloff with distance

Distance r (Å)	Coulomb falloff r^{-1} (Å^{-1})	Dispersion falloff r^{-6} (Å^{-6})	Ratio Coulomb:Dispersion falloff
1	1.00000	1.00000	1
2	0.25000	0.01563	16
3	0.11111	0.00137	81
4	0.06250	0.00024	256
5	0.04000	0.00006	625

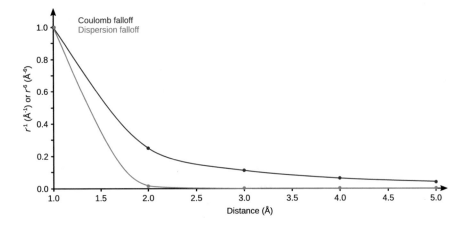

This (simplified) analysis shows that Coulomb forces between charged particles fall off much slower with separation distance than the weak dispersion forces. At a separation distance of 5 Å, Coulomb forces are several 100 times more pronounced than the weak dispersion forces. Therefore, even in dilute solutions, where the average separation distance of particles is large, Coulomb forces are major contributors.

For a solution to be ideal it is required that the interactions of solvent–solvent, solvent–solute, and solute–solute molecules are identical. This is clearly not the case when electrolytes are present, hence ionic solutions typically show non-ideal behaviour (which has implications for colligative properties; concentrations need to be replaced by activities, etc).

Chapter 13

13.1. The infrared spectrum of carbon monoxide shows a vibration ($v = 0 \rightarrow v = 1$) band at 2176 cm^{-1}. What is the force constant of the C–O bond, assuming the molecule behaves as a harmonic oscillator?

The frequency of the harmonic oscillator is given by $\nu = \frac{1}{2\pi} \cdot \sqrt{\frac{k}{\mu}}$, and the wavenumber is related to the frequency as per $\tilde{\nu} = \frac{\nu}{c}$. Therefore:

$$\tilde{\nu} = \frac{1}{2\pi \cdot c} \cdot \sqrt{\frac{k}{\mu}}.$$

The reduced mass μ needs to be calculated using particular isotopes (^{12}C and ^{16}O), yielding

$$\mu = \frac{m(^{12}C) \cdot m(^{16}O)}{m(^{12}C) + m(^{16}O)} = \frac{12.0000 \text{ Da} \cdot 15.9949 \text{ Da}}{12.0000 \text{ Da} + 15.9949 \text{ Da}} = 6.856 \text{ Da}$$

$$= 6.856 \cdot 1.661 \cdot 10^{-24} \text{ g} \quad \mu = 1.139 \cdot 10^{-23} \text{ g} = 1.139 \cdot 10^{-26} \text{ kg}.$$

The force constant can now be calculated as follows:

$k = 4 \cdot \pi^2 \cdot \tilde{\nu}^2 \cdot c^2 \cdot \mu$

$k = 4 \cdot \pi^2 \cdot (2170 \text{ cm}^{-1})^2 \cdot (2.998 \cdot 10^8 \text{ ms}^{-1})^2 \cdot 1.139 \cdot 10^{-26} \text{ kg}$

$k = 4 \cdot \pi^2 \cdot 4.709 \cdot 10^6 \text{ cm}^{-2} \cdot 8.988 \cdot 10^{16} \text{ m}^2 \text{ s}^{-2} \cdot 1.139 \cdot 10^{-26} \text{ kg}$

$k = 4 \cdot \pi^2 \cdot 4.709 \cdot 10^6 \cdot 10^4 \text{ m}^{-2} \cdot 8.988 \cdot 10^{16} \text{ m}^2 \text{ s}^{-2} \cdot 1.139 \cdot 10^{-26} \text{ kg}$

$k = 4 \cdot \pi^2 \cdot 4.709 \cdot 8.998 \cdot 1.139 \cdot 10^{6+4+16-26} \dfrac{\text{m}^2 \cdot \text{kg}}{\text{m}^2 \cdot \text{s}^2}$

$k = 1905.3 \dfrac{\text{kg} \cdot \text{m}}{\text{s}^2} \dfrac{1}{\text{m}}$

$k = 1.905 \cdot 10^3 \dfrac{\text{N}}{\text{m}}.$

13.2. Determine the ratio of populations of the vibrational levels $v = 1$ and $v = 0$ for carbon monoxide at 300 K and 1000 K. The wavenumber for the transition from $v = 0$ to $v = 1$ is 2176 cm^{-1}.

The population ratio is given by the equation based on the Boltzmann statistics:

$$\frac{N_v}{N_{v=0}} = e^{-\frac{E_v - E_{v=0}}{k_B \cdot T}}.$$

The difference ($E_v - E_{v=0}$) is the energy difference between the two vibrational levels and can be accessed via the wavenumber observed for this transition:

$$\Delta E = h \cdot \nu = h \cdot \tilde{\nu} \cdot c$$
$$\Delta E = 6.626 \cdot 10^{-34} \text{ J s} \cdot 2176 \text{ cm}^{-1} \cdot 2.998 \cdot 10^{8} \text{ ms}^{-1}$$
$$\Delta E = 6.626 \cdot 10^{-34} \text{ J s} \cdot 2176 \cdot 10^{2} \text{ m}^{-1} \cdot 2.998 \cdot 10^{8} \text{ ms}^{-1}$$
$$\Delta E = 6.626 \cdot 2176 \cdot 2.998 \cdot 10^{-34+2+8} \frac{\text{J} \cdot \text{s} \cdot \text{m}}{\text{m} \cdot \text{s}}$$
$$\Delta E = 43226 \cdot 10^{-24} \text{ J}$$
$$\Delta E = 4.323 \cdot 10^{-20} \text{ J}.$$

For $T = 300$ K, the population ratio is then:

$$\frac{N_{v=1}}{N_{v=0}} = e^{-\frac{4.323 \cdot 10^{-20} \text{ J}}{1.381 \cdot 10^{-23} \text{ JK}^{-1} \cdot 300 \text{ K}}} = e^{-0.0104 \cdot 10^{3}} = e^{-10.4} = 0.00003.$$

For $T = 1000$ K, the population ratio is:

$$\frac{N_{v=1}}{N_{v=0}} = e^{-\frac{4.323 \cdot 10^{-20} \text{ J}}{1.381 \cdot 10^{-23} \text{ JK}^{-1} \cdot 1000 \text{ K}}} = e^{-0.0031 \cdot 10^{3}} = e^{-3.1} = 0.045.$$

Whereas at 300 K, the first excited vibrational mode is virtually not populated, at 1000 K, about 4.5% of the CO molecules populate the mode $v = 1$.

13.3. Which spectroscopic symbol denotes the ground state of Li? What is the symbol of the lowest excited state? What feature can be deduced for the spectral line in the optical spectrum for this transition?

Li ground state:

Electronic configuration	$1s^{2}\,2s^{1}$
Quantum numbers of the valence electron	$n = 2, l = 0, m_s = {}^{1}/_{2}, j = 0 + {}^{1}/_{2} = {}^{1}/_{2}$
Total spin	$S = |\Sigma\, s_i| = |+ {}^{1}/_{2} - {}^{1}/_{2} + {}^{1}/_{2}| = {}^{1}/_{2}$
Spin multiplicity	$M = 2 \cdot S + 1 = 2 \cdot {}^{1}/_{2} + 1 = 2$
Spectroscopic symbol	${}^{2}S_{1/2}$

Li first excited state:

Electronic configuration	$1s^{2}\,2p^{1}$
Quantum numbers of the valence electron	$n = 2, l = 1, m_s = {}^{1}/_{2}, j = 1 + {}^{1}/_{2} = {}^{3}/_{2}$
	$n = 2, l = 1, m_s = -{}^{1}/_{2}, j = 1 - {}^{1}/_{2} = {}^{1}/_{2}$
Total spin	$S = |\Sigma\, s_i| = |+ {}^{1}/_{2} - {}^{1}/_{2} - {}^{1}/_{2}| = {}^{1}/_{2}$
Spin multiplicity	$M = 2 \cdot S + 1 = 2 \cdot {}^{1}/_{2} + 1 = 2$
Spectroscopic symbol	${}^{2}P_{3/2}$ and ${}^{2}P_{1/2}$

The transition from the ground state to the first excited state can be from ${}^{2}S_{1/2}$ to either ${}^{2}P_{3/2}$ or ${}^{2}P_{1/2}$. The spectral line therefore appears as a doublet with the individual lines having slightly differing energies.

13.4. In the free electron molecular orbital method, the π electrons in an alkene with conjugated double bonds are assumed to be freely moving within extent of the conjugation system. The conjugation system can be treated as a one-dimensional box.

(a) Calculate the box length for the conjugation system in hexatrien, assuming that the C–C-bond length is 1.4 Å and considering that the electrons are free to move half a bond length beyond each outermost carbon.
(b) Calculate the wavelength of the lowest energy peak in the absorption spectrum of hexatriene using the free electron molecular orbital method.

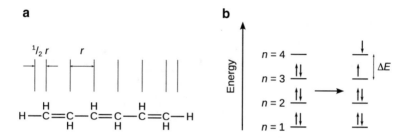

(a) With six consecutive carbon atoms participating in the π conjugation system, the box length a can be calculated as per:

$$a = 5 \cdot r + 2 \cdot \left(\frac{1}{2} \cdot r\right) = 6 \cdot r$$
$$a = 6 \cdot 1.4 \cdot 10^{-10} \text{ m} = 8.4 \cdot 10^{-10} \text{ m}.$$

(b) The lowest energy transition in the absorption spectrum of hexatriene is the one occurring between the highest occupied and the lowest unoccupied orbital. These frontier orbitals will be filled by π electrons, since the σ electrons occupy lower lying orbitals. With six p electrons, we need to consider three energy levels ($n = 1$, $n = 2$, $n = 3$) when applying the model of a particle in a one-dimensional box. The energy level at $n = 4$ constitutes the lowest unoccupied model. The electronic transition giving rise to the lowest energy peak in the absorption spectrum will be between $n = 3$ and $n = 4$.

The energy of a particle in the one-dimensional box is given by

$$E_n = \frac{h^2 \cdot n^2}{8 \cdot m \cdot a^2}, \text{where } m = m_e = 9.11 \cdot 10^{-31} \text{ kg.}$$

This yields for the energy difference of the transition:

$$\Delta E = E_4 - E_3 = \frac{h^2}{8 \cdot m_e \cdot a^2} \cdot \left(4^2 - 3^2\right)$$

The wavelength is obtained as per:

$$\Delta E = h \cdot \nu = \frac{h \cdot c}{\lambda} \quad \Rightarrow \quad \lambda = \frac{h \cdot c}{\Delta E}$$

Therefore:

$$\lambda = \frac{8 \cdot m_e \cdot a^2 \cdot h \cdot c}{h^2 \cdot \left(4^2 - 3^2\right)} = \frac{8 \cdot m_e \cdot a^2 \cdot c}{h \cdot (16 - 9)} = \frac{8 \cdot m_e \cdot a^2 \cdot c}{7 \cdot h}$$

$$\lambda = \frac{8 \cdot m_e \cdot a^2 \cdot c}{7 \cdot h} = \frac{8 \cdot 9.11 \cdot 10^{-31} \text{ kg} \cdot \left(8.4 \cdot 10^{-10} \text{ m}\right)^2 \cdot 2.99 \cdot 10^8 \text{ ms}^{-1}}{7 \cdot 6.63 \cdot 10^{-34} \text{ J s}}$$

$$\lambda = \frac{8 \cdot 9.11 \cdot 8.4^2 \cdot 2.99}{7 \cdot 6.63} \cdot 10^{-31-(2 \cdot 10)+8+34} \cdot \frac{\text{kg m}^2 \text{ m}}{\text{J s s}}$$

$$\lambda = 331 \cdot 10^{-9} \cdot \frac{\text{kg m}^2 \text{ m}}{\text{kg m}^2 \text{ s}^{-2} \text{ s s}}$$

$$\lambda = 331 \cdot 10^{-9} \text{ m} = 331 \text{ nm.}$$

13.5. How many lines/clusters are to be expected in the ^{1}H-NMR spectra of (a) benzene, (b) toluene, (c) o-xylene, (d) p-xylene, (e) m-xylene?

	Protons	Lines/clusters	δ (ppm)
(a) benzene	6 aromatic H	Singlet	7.33
(b) toluene	2+2+1 aromatic H	Multiplet	7.00–7.38
	3 methyl H	Singlet	2.34
(c) o-xylene	2+2 aromatic H	Multiplet	7.22–7.28
	6 methyl H	Singlet	2.40
(d) p-xylene	4 aromatic H	Singlet	7.24
	6 methyl H	Singlet	2.48
(e) m-xylene	1 aromatic H	Doublet of doublet	7.28–7.31
	2+1 aromatic H	Multiplet	7.11–7.14
	6 methyl H	Singlet	2.47

13.6. Predict the number of hyperfine lines in the proton coupling observed in the ESR spectrum of the benzene anion radical $C_6H_6^-$.

The six protons in the benzene anion are chemically equivalent ($N = 6$); the nuclear spin of the proton is $I = {}^1/_2$. Therefore:

$$2 \cdot n \cdot I + 1 = 2 \cdot 6 \cdot \frac{1}{2} + 1 = 7.$$

The benzene anion radical shows seven lines in the proton hyperfine coupling.

13.7. Determine the Raman and infrared activity of one symmetric and two different anti-symmetric vibrational modes of the square planar molecule XeF_4.

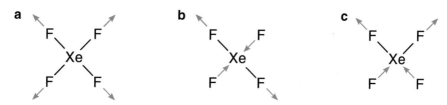

The symmetric stretching mode (a) does not lead to a change in the dipole momentum, but a change in the distribution electrons, i.e. polarisability. This mode is thus IR inactive and Raman active.

The anti-symmetric stretching mode (b) does not lead to a change in the dipole momentum but to a change in the electron distribution (polarisability). This mode is therefore IR inactive and Raman active.

The anti-symmetric stretching mode in (c) does indeed lead to a change in the dipole momentum as the individual dipole momentum changes by bond stretching on opposite sides do no longer cancel. The overall electron distribution does not change, since the compression on one side of the molecule is compensated for by an expansion on the other side; the polarisability does not change. Therefore, this mode is Raman inactive, but IR active.

References

Arrhenius SA (1889a) Über die Dissociationswärme und den Einfluß der Temperatur auf den Dissociationsgrad der Elektrolyte. Z Phys Chem 4:96–116

Arrhenius SA (1889b) Über die Reaktionsgeschwindigkeit bei der Inversion von Rohrzucker durch Säuren. Z Phys Chem 4:226–248

Bodenstein M, Lind S (1907) Geschwindigkeit der Bildung des Bromwasserstoffs aus seinen Elementen. Z Phys Chem 57:168–192

Bragg WH, Bragg WL (1913) The reflexion of X-rays by cystals. Proc Royal Soc Lond A 88:428–438

Bureau international des poids at mesures (2006) SI Brochure. http://www.bipm.org/en/publications/si-brochure/. Accessed 4 Nov 2016

Bureau international des poids at mesures (2013) http://www.bipm.org/en/measurement-units/new-si/, http://www.bipm.org/en/measurement-units/new-si/. Accessed 4 Nov 2016

Carothers W (1932) Polymers and polyfunctionality. Trans Faraday Soc 32:39–49

Chaplin M (2014) Water structure and science. http://www1.lsbu.ac.uk/water/. Accessed 7 Nov 2016

Christiansen JA (1967) On the reaction between hydrogen and bromine. In: Back MH, Laidler K (eds) Selected readings in chemical kinetics. Pergamon, Oxford, pp 119–126

Clapeyron PBC (1834) Mémoire sur la puissance motrice de la chaleur. Journal de l'École polytechnique 23:153–190

Clausius R (1850) Über die bewegende Kraft der Wärme und die Gesetze, welche sich daraus für die Wärmelehre selbst ableiten lassen. Ann Phys 155:500–524

Donald WA, Leib RD, O'Brien JT et al (2008) Absolute standard hydrogen electrode potential measured by reduction of aqueous nanodrops in the gas phase. J Am Chem Soc 130:3371–3381. https://doi.org/10.1021/ja073946i

Ehl RG, Ihde A (1954) Faraday's electrochemical laws and the determination of equivalent weights. J Chem Educ 31:226–232

Eibenberger S, Gerlich S, Arndt M et al (2013) Matter-wave interference of particles selected from a molecular library with masses exceeding 10,000 amu. Phys Chem Chem Phys 15:14696–14700. https://doi.org/10.1039/c3cp51500a

Eigen M (1954) Über die Kinetik sehr schnell verlaufender Ionenreaktionen in wässriger Lösung. Z Phys Chem (N F) 1:176

Eigen M, Kurtze G, Tamm K (1953) Zum Reaktionsmechanismus der Ultraschallabsorption in wässrigen Elektrolytlösungen. Z Elektrochem 57:103

Einstein A (1905) Über einen die Erzeugung und Verwandlung des Lichtes betreffenden heuristischen Gesichtspunkt. Ann Phys 17:132–148

Elster J, Geitel HF (1882) Über die Electricität der Flamme. Ann Phys Chem 19:193–222

Emmrich M, Huber F, Pielmeier F et al (2015) Surface structure. Subatomic resolution force microscopy reveals internal structure and adsorption sites of small iron clusters. Science 348:308–311. https://doi.org/10.1126/science.aaa5329

© Springer International Publishing AG, part of Springer Nature 2018
A. Hofmann, *Physical Chemistry Essentials*,
https://doi.org/10.1007/978-3-319-74167-3

Evans MG, Polanyi M (1935) Some applications of the transition state method to the calculation of reaction velocities, especially in solution. Trans Faraday Soc 31:875–894

Eyring H (1935) The activated complex in chemical reactions. J Chem Phys 3:107–115

Franck J, Hertz G (1914) Über Zusammenstöße zwischen Elektronen und Molekülen des Quecksilberdampfes und die Ionisierungsspannung desselben. Verh Dtsch Phys Ges 16:457–467

Guldberg CM (1864) Concerning the laws of chemical affinity. Forhandlinger i Videnskabs-Selskabet i Christiania 111

Hertz H (1887) Über einen Einfluss des ultravioletten Lichtes auf die electrische Entladung. Ann Phys 267:983–1000

Hofmann A, Simon A, Grkovic T, Jones M (2014) Methods of molecular analysis in the life sciences. Cambridge University Press, Cambridge

Hubinger S, Nee JB (1995) Absorption spectra of Cl2, Br2 and BrCl between 190 and 600 nm. J Photochem Photobiol A: Chem 86:1–7

Huygens C (1690) Traite de la lumiere. Gauthier-Villars et Cie, Paris

International Union of Pure and Applied Chemistry (2013) A critical review of the proposed definitions of fundamental chemical quantities and their impact on chemical communities. https://iupac.org/projects/project-details/?project_nr=2013-048-1-100. Accessed 4 Nov 2016

Keesom WH (1915) The second viral coefficient for rigid spherical molecules whose mutual attraction is equivalent to that of a quadruplet placed at its center. Proc R Acad Sci 18:636–646

Khalifah RG (1971) The carbon dioxide hydration activity of carbonic anhydrase. I. Stop-flow kinetic studies on the native human isoenzymes B and C. J Biol Chem 246:2561–2573

Le Chatelier H, Boudouard O (1898) Limits of flammability of gaseous mixtures. Bulletin de la Société Chimique de France (Paris) 19:483–488

Lenard P (1903) Über die Absorption von Kathodenstrahlen verschiedener Geschwindigkeit. Ann Phys 317(12):490

Lewis GN (1916) The atom and the molecule. J Am Chem Soc 38:762–785

London F (1930) Zur Theorie und Systematik der Molekularkräfte. Z Phys 63:245–279

Maxwell JC (1860a) Illustrations of the dynamical theory of gases. Part I. On the motions and collisions of perfectly elastic spheres. Philos Mag 19:19–32

Maxwell JC (1860b) Illustrations of the dynamical theory of gases. Part II. On the process of diffusion of two or more kinds of moving particles among one another. Philos Mag 20:21–37

Meitner L (1922) Über die beta-Strahl-Spektra und ihren Zusammenhang mit der gamma-Strahlung. Z Phys A: Hadrons Nucl 11:35–54

Michaelis L, Menten ML (1913) Die Kinetik der Invertinwirkung. Biochem Z 49:333–369

Millikan RA (1913) On the elementary electric charge and the Avogadro constant. Phys Rev 2:109–143

Moseley HGJ (1913) The high frequency spectra of the elements. Philos Mag 26:1024–1034

Mößbauer RL (1958) Kernresonanzfluoreszenz von Gammastrahlung in Ir191. Z Phys 151:124–143. https://doi.org/10.1007/BF01344210

Mpemba EB, Osborne DG (1969) Cool? Phys Educ 4:172–175

Newton I (1704) Opticks: or, a treatise of the reflexions, refractions, inflexions and colours of light. Also two treatises of the species and magnitude of curvilinear figures. Re-published 1998. Octavo Corporation, Palo Alto, CA

Pauling L (1932) The nature of the chemical bond. IV. The energy of single bonds and the relative electronegativity of atoms. J Am Chem Soc 54:3570–3582

Peslherbe GH, Ladanyi BM, Hynes JT (2000) Structure of NaI ion pairs in water clusters. Chem Phys 258:201–224

Poole PH, Sciortino F, Essmann U, Stanley HE (1992) Phase behaviour of metastable water. Nature 360:324–328. https://doi.org/10.1038/360324a0

Rossini FD (1968) A report on the international practical temperature scale of 1968. In: International Union of Pure and Applied Chemistry Commission I.2: Thermodynamics and Thermochemistry. p 557–570

Simon A, Cohen-Bouhacina T, Porté MC, Aimé JP, Amédée J, Bareille R, Baquey C (2003) Characterization of dynamic cellular adhesion of osteoblasts using atomic force microscopy. Cytometry A 54:36–47

Slater JC (1930) Atomic shielding constants. Phys Rev 36:57–64

Smekal A (1923) Zur Quantentheorie der Dispersion. Naturwissenschaften 11:873–875

Stohner J, Quack M (2015) Fixierte Konstanten. Nachr Chem Tech Lab 63:515–521

Stranski IN (1928) Zur Theorie des Kristallwachstums. Z Phys Chem 136:259–278

Sugimoto Y, Pou P, Abe M et al (2007) Chemical identification of individual surface atoms by atomic force microscopy. Nature 446:64–67. https://doi.org/10.1038/nature05530

Thomson JJ (1897) Cathode rays. The Electrician 39:104

van't Hoff JH (1877) Die Grenzebene, ein Beitrag zur Kenntnis der Esterbildung. Berichte der Berliner Chemischen Gesellschaft 10:669–678

Waage P (1864) Experiments for determining the affinity law. Forhandlinger i Videnskabs-Selskabet i Christiania 92

Waage P, Guldberg CM (1864) Studies concerning affinity. Forhandlinger i Videnskabs-Selskabet i Christiania 35

Weeratunga S, Hu N-J, Simon A, Hofmann A (2012) SDAR: a practical tool for graphical analysis of two-dimensional data. BMC Bioinf 13:201. https://doi.org/10.1186/1471-2105-13-201

Weisskopf VF (1975) Of atoms, mountains, and stars: a study in qualitative physics. Science 187:605–612. https://doi.org/10.1126/science.187.4177.605

Wojdr M (2010) Fityk: a general-purpose peak fitting program. J Appl Cryst 43:1126–1128

Young T (1802) The Bakerian lecture: on the theory of light and colours. Phil Trans Royal Soc 92:12–48

Index

© Springer International Publishing AG, part of Springer Nature 2018
A. Hofmann, *Physical Chemistry Essentials*,
https://doi.org/10.1007/978-3-319-74167-3

Printed in the United States
By Bookmasters